Organic chemistry for students of biology and medicine

Organic Chemistry for Students of Biology and Medicine

Third Edition

G. A. Taylor

Copublished in the United States with
John Wiley & Sons, Inc., New York

Longman Scientific & Technical
Longman Group UK Limited,
Longman House, Burnt Mill, Harlow
Essex CM20 2JE, England
and Associated Companies throughout the world.

Copublished in the United States with
John Wiley & Sons Inc., 605 Third Avenue, New York, NY 10158

First published 1971
Second edition 1978
Third edition 1987
Reprinted 1989

British Library Cataloguing in Publication Data

Taylor, G. A.
Organic chemistry for students of biology
and medicine.—3rd ed.
1. Chemistry, Organic
I. Title
547′.0024574 QD251.2

ISBN 0-582-44708-9

Library of Congress Cataloguing in Publication Data Available
ISBN 0-470-20711-6 (USA only)

Produced by Longman Group (FE) Ltd
Printed in Hong Kong

Contents

Preface to the first edition

This book is the result of several years' experience of teaching organic chemistry to medical students, and is based upon the lecture course which evolved over this time. Its purpose is to give a sufficient knowledge of organic chemistry to enable students of biology or medicine to understand those chemical aspects of biochemistry which they encounter during their training. It is not intended for specialists in chemistry, for whom a more detailed and rigorous course is desirable. It is expected that students using this book will have some prior acquaintance with descriptive organic chemistry, usually up to A-level standard, but without any knowledge of reaction mechanisms.

With this purpose in mind, the book has been written in two sections. The first includes introductory chapters on atomic and molecular structure, and the principles behind the mechanistic approach to organic chemistry. This is followed by a combination of the descriptive chemistry of simple monofunctional compounds and the mechanisms of their reactions. Topics in this section are selected for their significance in biological reactions, thus the chemistry of petroleum, coal, etc., is neglected, Grignard reagents are not mentioned, and in general the emphasis is on aliphatic rather than aromatic compounds. Some subjects such as nitro-alkanes, aryl diazonium salts, and sulphate esters are included to achieve coherence but receive only brief mention. The second part of the book is concerned with the chemistry of selected groups of polyfunctional compounds, which culminate in subjects of obvious biochemical importance. A description of stereochemistry is introduced at two traditional points, which, it is thought, serves the purpose of the book better than the alternative of developing descriptive, mechanistic, and stereochemical treatments simultaneously from the start. This latter alternative, although admirable for the specialist course, can be very confusing to students whose previous acquaintance with organic chemistry is scanty. Adoption of this approach requires the treatment of the stereochemical aspects of reactions as a separate chapter, which has the disadvantage of allowing this important topic to be ignored. The book concludes with a brief description of prochirality.

The author thanks Dr. E. Haslam and Dr. R. Brettle for their comments on the manuscript.

Preface to the third edition

The passage of eight years since the production of the second edition of this book brings the necessity for a number of changes. The principal ones consist of bringing together the three chapters on stereochemical topics which, in the context of this book, I still prefer to introduce after much of the chemistry of simple functional groups. It is, however, easily possible for those who believe that stereochemistry should follow immediately after the introductory chapters on bonding to direct students to read the book in that sequence. The treatment of the conformations of open-chain and saturated cyclic compounds has been expanded, and the representation of saturated six-membered rings in the chair form is used where appropriate as well as the older flat form. A new chapter dealing with four groups of physiologically active compounds seeks to introduce and develop the relationship between structure and biological activity in areas of great pharmacological potential and also to indicate very briefly the biosynthetic origins of the naturally occurring compounds. It is hoped that in this way the book will continue to fulfil the original goal and give a better perspective of the chemical aspects of biology without trespassing too far into topics which are properly the province of the biochemist.

G. A. Taylor

1

Atomic and molecular structure

The **atom** (the smallest particle of an element retaining the chemical properties of the element) is the basic structural unit of chemical compounds. All matter is composed of atoms, approximately one hundred types being known, most of which occur naturally.

Atoms can combine to form molecules, and the methods and patterns of combination, and the circumstances in which combination occurs, are the concern of the chemist. A knowledge of the structure of atoms is necessary to understand the reasons why reactions occur, and the phenomenon of valency (the combining power of atoms).

Atomic structure

The structure of the atom can be divided into two distinct features. The **nucleus** is a small, dense, positively charged body at the centre of the atom, and almost all the mass of the atom is concentrated in the nucleus. A diffuse zone containing **electrons** (light particles, with unit negative charge) surrounds the nucleus, and it is these electrons which are responsible for the formation of chemical bonds.

The nucleus is itself composed of two types of particle:

(a) **Protons**, heavy particles with unit positive charge;
(b) **Neutrons**, heavy particles with no electrical charge.

Both the proton and the neutron have masses approximately two thousand times that of the electron, whose contribution to the total mass of the atom is extremely small.

The number of protons in the nucleus of an atom determines the total positive charge on the nucleus, and therefore the number of electrons required to produce a neutral atom. The number of protons, which is the same as the atomic number of the element concerned, thus determines the chemistry of the atom, and is the principal source of difference between the atoms of the elements in the periodic table. The number of neutrons in the nucleus is not so significant, as these have very little effect on the chemistry of the atom, and neutrons may be regarded as an optional extra in the construction of the nucleus. Thus the atoms of an element of atomic number N can be constructed in a variety of ways. A common unit of N protons in the nucleus and N electrons in the surrounding space can have m, $m + 1$, $m + 2$, etc., neutrons in the nucleus, giving rise to chemically indistinguishable species, differing only in the relative atomic mass‡. Such atoms, differing only in the number of nuclear neutrons, are known as **isotopes,** and many of the naturally occurring elements are composed of more than one isotope (see Table I), though some, like fluorine and sodium, occur naturally* only as one isotope. When the isotopes are described by symbols, as in Table I, the superscript prefix refers to the mass of the nucleus (i.e. the number of protons and neutrons).

The practical value of the relative atomic mass of an element is a mean value determined by the natural proportions of the isotopes (e.g. chlorine is approximately 75 per cent ^{35}Cl and 25 per cent ^{37}Cl and has a practical relative atomic mass of 35·5; similarly bromine, composed of nearly equal proportions of ^{79}Br and ^{81}Br has a relative atomic mass of 80).

The electrons are found outside the nucleus, and occupy a very much larger space. They are often visualised as minute spherical bodies moving in concentric circles, but this concept is quite wrong and very misleading. It is now known that it is quite impossible to treat electrons in a way which implies that their positions at any moment can be defined accurately. All that can be achieved is to define the probability of finding an electron in a region of space.

The electrons of a polyelectronic atom are not all equivalent. They may be divided firstly into groups known as **shells**, which differ greatly in energy, and which can accommodate differing maximum numbers of electrons. The shells are normally described by a quantum number 1, 2, 3,† etc., where shell no. 1 has the lowest energy, no. 2 has the next to the lowest energy, and so on. A polyelectronic atom can be built up by taking the nucleus and bringing electrons from an infinite distance (where their energy is arbitrarily defined as zero) to fill the vacant shells in much the same way that a set of drawers might be filled, starting from the bottom. The first two electrons go into the

* Artificially produced radioactive isotopes are known.

† Sometimes known as K, L, M, where $K \equiv 1$, $L \equiv 2$, etc.

‡ Previously known as the atomic weight.

Table I. *The composition of some natural isotopes*

NAME	SYMBOL	ATOMIC NUMBER	NUCLEUS		NO. OF ELECTRONS	APPROX. RELATIVE ATOMIC MASS	% NATURAL ABUNDANCE
			NO. OF PROTONS	NO. OF NEUTRONS			
Hydrogen*							
(Protium)	^{1}H	1	1	0	1	1	99·98
(Deuterium)	^{2}H (D)	1	1	1	1	2	0·02
(Tritium)	†^{3}H (T)	1	1	2	1	3	
Helium	^{3}He	2	2	1	2	3	$1·3 \times 10^{-4}$
	^{4}He	2	2	2	2	4	99·99
Lithium	^{6}Li	3	3	3	3	6	7·3
	^{7}Li	3	3	4	3	7	92·7
Sodium	^{23}Na	11	11	12	11	23	100
Potassium	^{39}K	19	19	20	19	39	93·3
	†^{40}K	19	19	21	19	40	0·01
	^{41}K	19	19	22	19	41	6·7
Fluorine	^{19}F	9	9	10	9	19	100
Chlorine	^{35}Cl	17	17	18	17	35	75·4
	^{37}Cl	17	17	20	17	37	24·6
Bromine	^{79}Br	35	35	44	35	79	50·5
	^{81}Br	35	35	46	35	81	49·5
Carbon	^{12}C	6	6	6	6	12	98·9
	^{13}C	6	6	7	6	13	1·1
	†^{14}C	6	6	8	6	14	2×10^{-10}
Nitrogen	^{14}N	7	7	7	7	14	99·6
	^{15}N	7	7	8	7	15	0·4
Oxygen	^{16}O	8	8	8	8	16	99·76
	^{17}O	8	8	9	8	17	0·04
	^{18}O	8	8	10	8	18	0·2

* For historical reasons the isotopes of hydrogen have acquired individual names.
† Radioactive isotope.

lowest energy shell (no. 1), which is then full. Subsequently the second shell (eight electrons) and third shell (eighteen electrons) will be filled, and so on until a sufficient number of electrons has been added to make an electrically neutral atom.

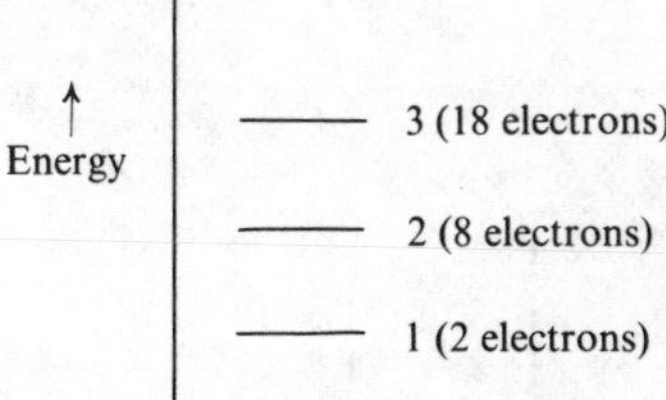

Diagrammatic representation of the relationship of the first three energy shells of an atom.

The first distinction between the electrons in an atom is thus seen to be the shells which they occupy. Even within these shells the electrons are not equivalent, as each shell is divided into **orbitals**. Each orbital can accommodate only two electrons, and is distinguished from other orbitals in the same shell by its geometry. An orbital can be regarded as a mathematically defined region of space in which there is a high probability of finding an electron.

We need be concerned here only with the orbitals associated with the first two shells (i.e. found in the first ten elements of the periodic table). These are of two types known as *s* and *p* orbitals. *s* orbitals are spherically symmetrical about the nucleus, but *p* orbitals are symmetrical about an axis and a perpendicular plane, both of which pass through the nucleus. The orbitals are generally described by the quantum number of the shell, and the letter designating the type of orbital, e.g. 2*s*, 2*p*.

The electrons in hydrogen and helium are found in the 1*s* orbital, this being

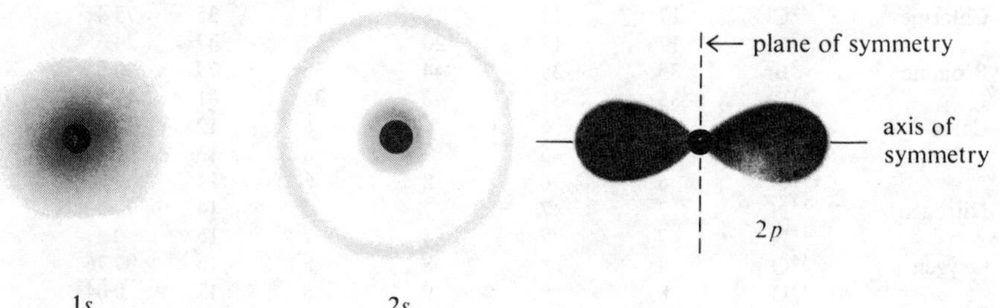

Sections through *s* and *p* orbitals. The probability of finding an electron in the orbital is greatest in the most densely shaded region, but there is a small probability of the electron being found outside these zones.

Table II. *The electronic structure of atoms of the first short period*

ATOM	SYMBOL	NO. OF ELECTRONS IN THE ORBITALS 1*s*	2*s*	$2p_x$	$2p_y$	$2p_z$
Hydrogen	H	1				
Helium	He	2				
Lithium	Li	2	1			
Beryllium	Be	2	2			
Boron	B	2	2	1		
Carbon	C	2	2	1	1	
Nitrogen	N	2	2	1	1	1
Oxygen	O	2	2	2	1	1
Fluorine	F	2	2	2	2	1
Neon	Ne	2	2	2	2	2

the only orbital in the first shell and the one responsible for the formation of chemical bonds to hydrogen atoms. When building the atoms of the elements from lithium to neon (atomic numbers 3 to 10) the first shell is filled initially and then electrons start to fill the vacant orbitals of the second shell. There are four orbitals available in the second shell, one *s* type, and three *p* type differing in the orientation of their axes of symmetry, which are mutually perpendicular (see diagram below). These 2*p* orbitals are distinguished by the suffixes *x*, *y*, and *z*.

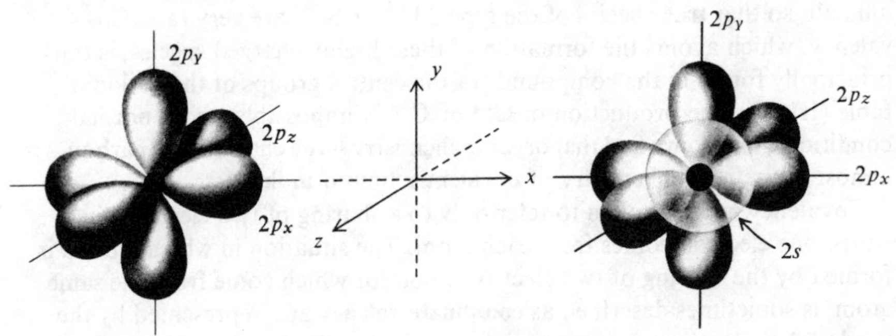

The three 2*p* orbitals.

The relationship of 2*s* and $2p_x$, $2p_y$, and $2p_z$ orbitals.

Because of a very slight energy difference between the 2*s* and 2*p* orbitals, the 2*s* orbital is filled first, and then the 2*p* orbitals are filled as shown in Table II. Neon with all the orbitals of the second shell filled terminates the elements of the first short period.

Molecular structure

The formation of compounds from elements is achieved by the combination of atoms into groups (**molecules**) whose structure is characteristic of the compound concerned. Some aspects of the formation of molecules and their structures will now be examined.

Valency is the combining power of an element, measured by the number of hydrogen atoms (or their equivalent) with which an atom of the element will combine to form a stable molecule. It is well known that the valency of an element is related to its position in the periodic table in a way which suggests that atoms with unfilled outer shells of electrons attempt to achieve a 'rare gas structure', i.e. a completed outer shell. There are two principal ways of achieving this stable state:

(a) **Electrovalency** caused by the loss or gain of electrons, to form charged species (**ions**) with complete outer shells.
(b) **Covalency** in which atoms achieve the equivalent of a rare gas structure by sharing electrons.

Electrovalency is mostly found in compounds involving the elements of groups I, II, VI, and VII, since here the gain or loss of no more than two electrons is required to obtain a complete outer shell. The successive removal of electrons from a cation, or addition to an anion becomes progressively more difficult, so that true species of the type Al^{+3} or N^{-3} are very rare. Covalency, which avoids the formation of these highly charged species, is thus principally found in the compounds of the central groups of the periodic table (III–V). The production of C^{+4} or C^{-4} is impossible under normal conditions, and so we find that organic chemistry—the chemistry of carbon—is almost entirely the chemistry of covalently bound molecules.

Covalency is often taken to refer only to a sharing of two electrons in which one electron comes from each atom. The situation in which a bond is formed by the sharing of two electrons, both of which come from the same atom, is sometimes described as **coordinate valency** and represented by the symbol $A \rightarrow B$ where A is supposed to donate both electrons. However, the bond formed in this way is identical with the normal covalent bond except that a separation of charges occurs. This can be seen to arise if the bond is formed in two steps:

(a) donation of one electron from A: to B giving $A^{\dot{+}}$ and $B^{\dot{-}}$
(b) sharing of one electron each from $A^{\dot{+}}$ and $B^{\dot{-}}$ to form a covalent bond.

The **semipolar bond** so formed is much better represented as $A^{+}\!\!-\!B^{-}$ which emphasises both the true covalent character of the link, and also the charge relationship between the two atoms.

The formation of bonds. We are now in a position to discuss the formation of covalent bonds, and understand some of the characteristic features of molecules constructed in this way. The formation of covalent bonds by the sharing of electrons results from the overlapping and interaction of partly filled atomic orbitals. The **molecular orbitals** (bonds) so formed are represented adequately by a simple sum of the geometrical properties of the individual atomic orbitals. This is indicated by the diagram below, illustrating the molecular orbitals (covalent bonds) which join hydrogen and fluorine atoms in the molecules H_2, F_2, and HF. Only the orbitals involved in bonding are represented, the complete picture should include the filled orbitals of the first and second shells of fluorine as well. A semipolar bond arises by an identical interaction between an empty atomic orbital and a full one.

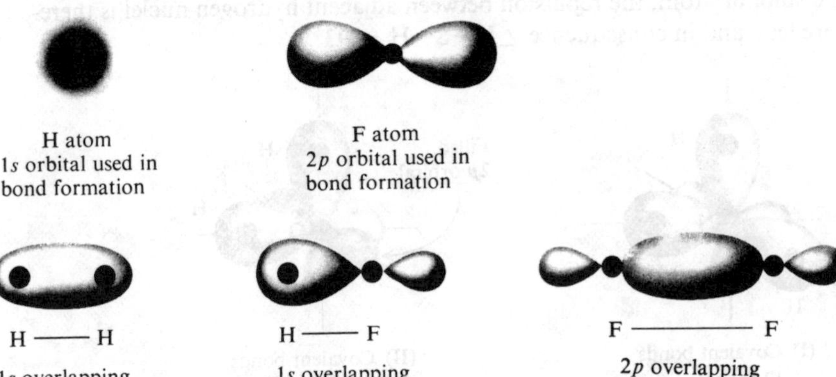

Bonds such as those illustrated above, in which the electrons are predominantly found in the region between the two nuclei are known as **σ bonds** (Greek 'sigma'). It should be noted that in these bonds the molecular orbital extends beyond the joined nuclei, which means that there is a finite probability that the electrons will be found outside the internuclear zone.

The three-dimensional structure of molecules. One of the characteristic features of polyatomic molecules is that they have a definite three-dimensional structure, in which the relative positions in space of the covalently linked atoms are fixed. Since bond formation is produced by the interaction of orbitals, which can be highly directional, the most effective bonds will be formed when the relative spatial positions of the atoms are such as to produce the best possible overlap of orbitals. Any distortion of the molecule which, by moving atoms from these most favourable sites, reduces the effectiveness of orbital overlap, will result in a weakening of the bond. Merely by considering the relative orientation of the 2*p* orbitals, we can make a rough prediction of the geometry of some simple molecules.

In molecules such as NH_3 (I) or H_2O (II) the N—H or O—H bonds are formed from the 1*s* orbital of hydrogen and the 2*p* orbitals of nitrogen or oxygen. We have seen earlier that the three 2*p* orbitals are mutually perpendicular (p. 5). We might therefore expect that the three N—H bonds or two O—H bonds would also be at right angles, and this is approximately correct, ∠H—N—H = 107°, ∠H—O—H = 104°. The deviation from 90° is probably caused by mutual repulsion of the positively charged hydrogen nuclei. In H_2S where the H—S bonds are formed from 3*p* orbitals (similar to 2*p* in geometry) the hydrogen atoms are further apart owing to the greater size of the sulphur atom, the repulsion between adjacent hydrogen nuclei is therefore less, and in consequence $\angle$H—S—H = 93°.

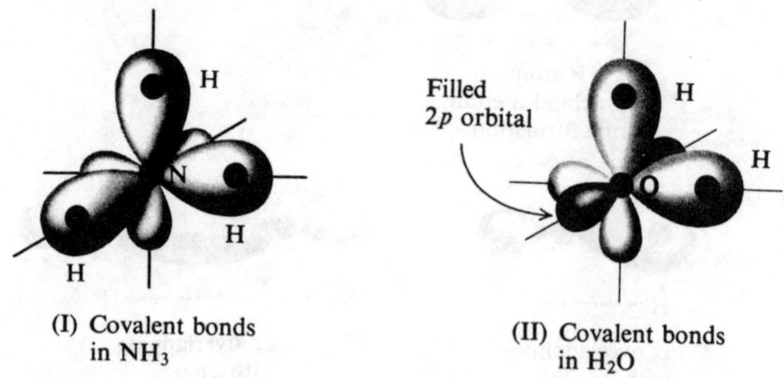

(I) Covalent bonds in NH_3

(II) Covalent bonds in H_2O

Hybridisation of orbitals. We have seen that in compounds of trivalent nitrogen, divalent oxygen, and monovalent fluorine *p* orbitals can be used for bond formation. The formation of compounds by tetravalent carbon or nitrogen is not adequately explained simply by use of the 2*s* orbital to form a fourth bond. The most effective orbitals for bond formation are those with the best geometry for overlap, and *p* orbitals have the desired directional shape, whereas *s* orbitals with spherical symmetry have the least suitable shape. A compromise solution is found to occur with tetravalent atoms. Interaction of the *s* and *p* orbitals produces sets of **hybrid orbitals** whose directional properties are somewhat less desirable than pure *p* orbitals, but very much better than *s* orbitals. This hybridisation can occur in three ways and the type adopted in any particular case is that which leads to the lowest overall energy (i.e. the strongest bonding).

(a) All four orbitals can interact to give four identical hybrid orbitals described as sp^3 orbitals (since they are produced by the mixing of one *s* and three *p* orbitals). The four sp^3 orbitals have the geometry shown (III), and are arranged about the nucleus with tetrahedral symmetry.

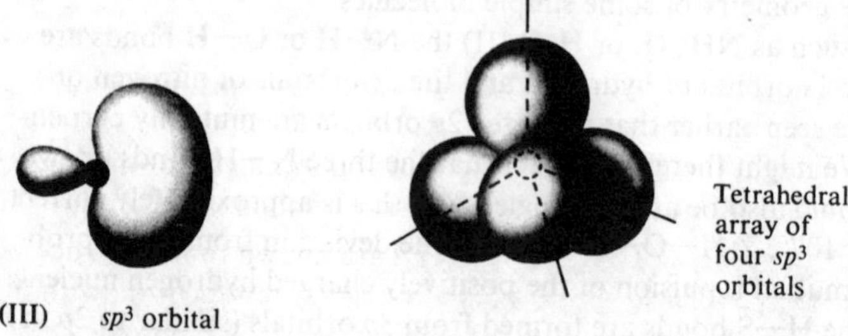

(III) sp^3 orbital

(b) The *s* orbital and two of the *p* orbitals can interact to form three sp^2 orbitals. These are arranged symmetrically in a plane, perpendicular to the unused *p* orbital (IV).

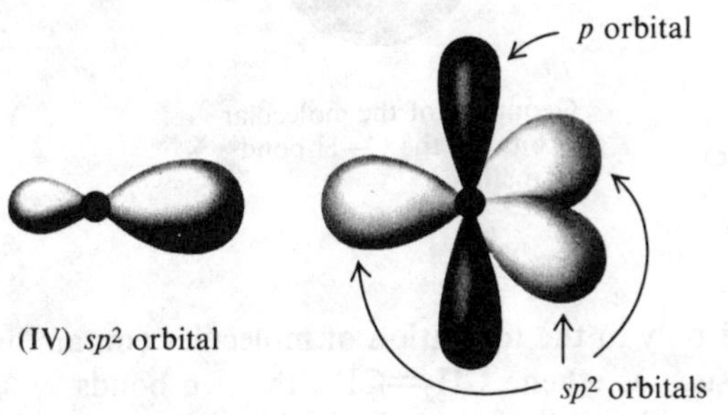

(IV) sp^2 orbital

The minor lobes of the hybrid orbitals are omitted from the right hand diagram.

(c) The *s* orbital and one of the *p* orbitals can interact to give two *sp* orbitals. These are arranged linearly, and are perpendicular to the remaining two *p* orbitals (V).

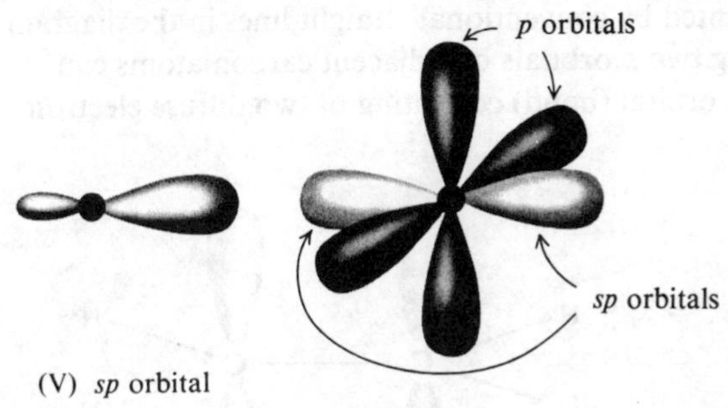

(V) *sp* orbital

The minor lobes of the hybrid orbitals are omitted from the right hand diagram

Occurrence of hybrid orbitals

sp³ *orbitals.* When an atom of the first short period forms a compound in which it is attached to four neighbouring atoms by single bonds, sp^3 hybrid orbitals are used to form these bonds. The tetrahedral symmetry of the sp^3 orbitals about the nucleus means a tetrahedral array of bonds, and all the species of type AX_4 (e.g. BF_4^-, CH_4, CCl_4, $\overset{+}{N}H_4$) are known to have structures in which the atoms X lie at the corners of a regular tetrahedron, with atom A at the centre, and all the angles $\angle X—A—X = 109{\cdot}5°$ approximately (VI).

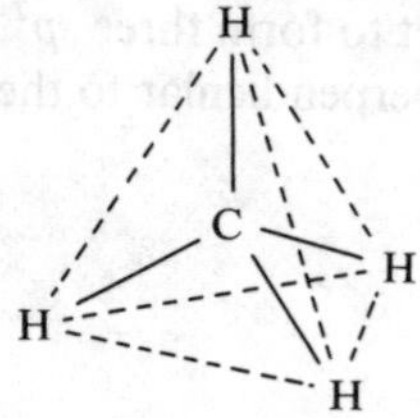

(VI) The structure of CH_4 (methane)
Dotted lines show the regular tetrahedron defined by the H atoms

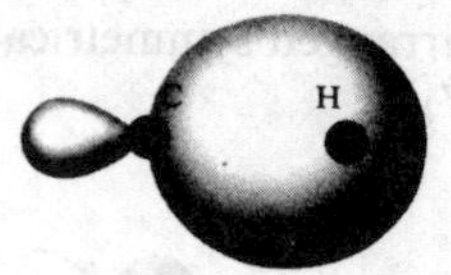

Geometry of the molecular orbital of the C—H bond in methane

sp[2] *orbitals*. These are utilised only in the formation of molecules containing double bonds. In a molecule such as ethene, $CH_2{=}CH_2$, the two bonds joining the carbon atoms are not identical, and are formed in different ways. The atomic orbitals of carbon are hybridised to form three sp^2 orbitals and one p orbital, and a framework of single bonds is built up by σ bonds (p. 7) resulting from the overlap of sp^2 orbitals of carbon and s orbitals of hydrogen. (The molecular orbitals of these σ bonds are illustrated separately, but for convenience are represented by conventional straight lines in the diagram (VII) below.) The remaining two p orbitals on adjacent carbon atoms can overlap to give a molecular orbital (bond) consisting of two diffuse electron

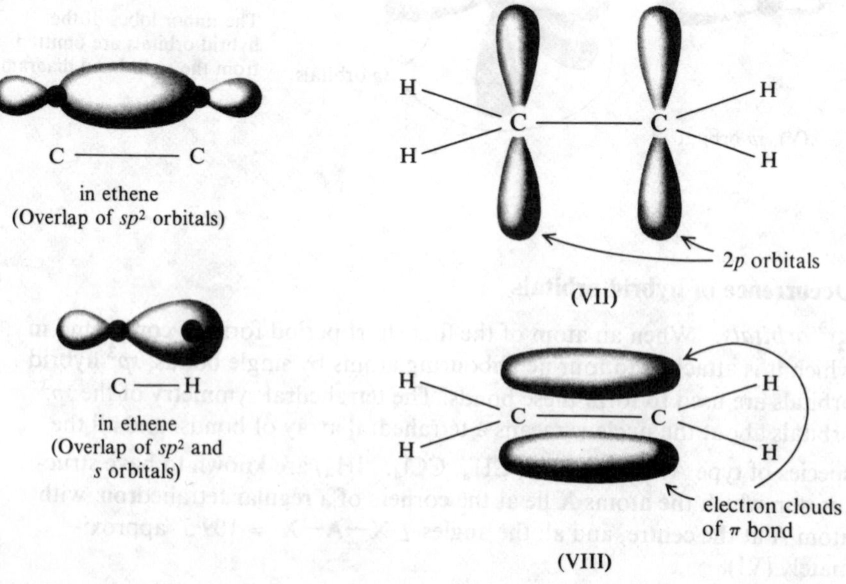

clouds lying on opposite sides of the C—C axis (VIII). The electrons in this type of bond are not found between the nuclei of the bonded atoms, and such a bond is known as a **π bond** (Greek 'pi'). All double bonds formed by the elements of the first short period (e.g. C=O, C=C, —N=N—) are constructed in this way.

The molecular geometry of compounds with double bonds is determined by two factors. In the first place the planar array of sp^2 orbitals about the nucleus (p. 9) means that the central atom and its three substituent atoms will be coplanar, with bond angles of 120°. In addition, for the *p* orbitals on adjacent atoms to overlap effectively they must be aligned parallel in space. This means that all the sp^2 orbitals of the two doubly bonded atoms, and all their substituent atoms must be coplanar. Physical measurements confirm this in many cases, e.g. in ethene, all six atoms lie in the same plane.

sp *orbitals.* These are used principally where atoms are joined by triple bonds. In ethyne, H—C≡C—H, the central σ bond skeleton is built of *sp* orbitals of carbon and *s* orbitals of hydrogen, which are similar to the corresponding bonds in ethene. The remaining pairs of *p* orbitals on the carbon

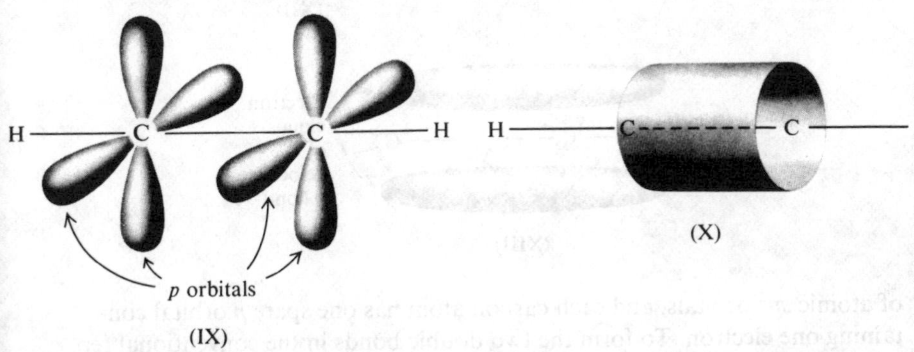

(IX)

(X)

atoms (IX) can overlap to form two π bonds, which together form a cylindrical electron cloud surrounding the C—C axis (X). Where atoms of the first short period are joined by triple bonds (e.g. C≡C, C≡N, N≡N) the mole cular orbitals are constructed in this way. The linear array of *sp* orbitals about the nucleus, means that the two triply bonded atoms, and their substituents are collinear, as is found to be the case in ethyne and HCN.

Conjugation. It has already been mentioned that the electrons of a σ bond are confined predominantly to a zone between the nuclei of the atoms joined. The electrons of a π bond are by comparison much less tightly held, and the electron clouds of the π bond are consequently much more diffuse and can sometimes interact with the electron clouds of adjacent π bonds. This interaction is particularly important in molecules which contain chains of atoms linked alternately by single and double bonds.

We will consider the molecule butadiene, $CH_2{=}CH{-}CH{=}CH_2$, whose carbon skeleton is joined by just such an alternate sequence of bonds. Neglecting the CH bonds, which are irrelevant for the present considerations, the basic skeleton (XI) of the molecule can be represented by (XII) in which each carbon atom is joined to its neighbours by σ bonds produced by the overlap

$$>C=\overset{|}{C}-\overset{|}{C}=C<$$

(XI)

(XII)

(XIII)

of atomic sp^2 orbitals, and each carbon atom has one spare p orbital containing one electron. To form the two double bonds in the conventional representation the p orbitals on atoms 1 and 2 must overlap, and likewise the p orbitals on atoms 3 and 4. However, the p orbital on atom 2 is equally close to those on atoms 1 and 3 and interaction could occur in either direction. The result of this proximity is that all four p orbitals combine to produce two molecular orbitals which stretch the whole length of the chain of four carbon atoms. Each of these molecular orbitals (π bonds) contains two electrons, and one is illustrated in diagram (XIII).

An alternate sequence of double and single bonds, such as we have just considered, is known as conjugated and will always produce extended ('delocalised') molecular orbitals in which electrons are free to move the whole length of the unsaturated system. In β-carotene, one of the colouring materials of carrots, these molecular orbitals stretch over twenty-two carbon nuclei and are responsible for the colour of the compound.

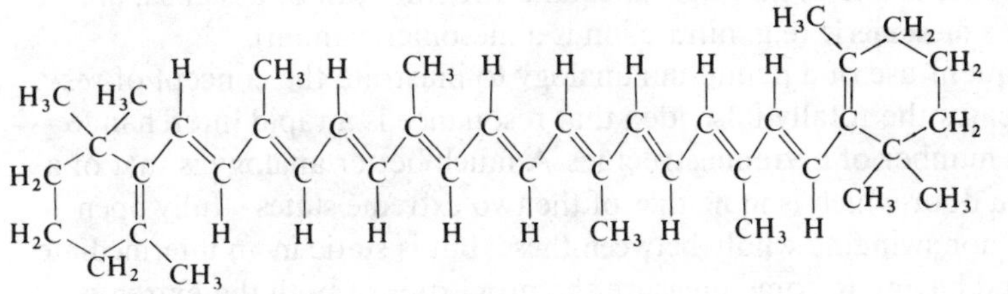

β-Carotene

Where double bonds are separated by two or more single bonds, e.g. $CH_2{=}CH{-}CH_2{-}CH{=}CH_2$, the π bonds are too far apart to interact with each other and no extended molecular orbital can be produced.

Resonance. It is not always possible to represent the structure of a compound adequately by a single, conventional, structural formula compatible with all the properties of the compound, and in these circumstances the correct description often appears to be an average of several conventional structures. A simple example of this is the nitrate ion, which can be represented in three ways:

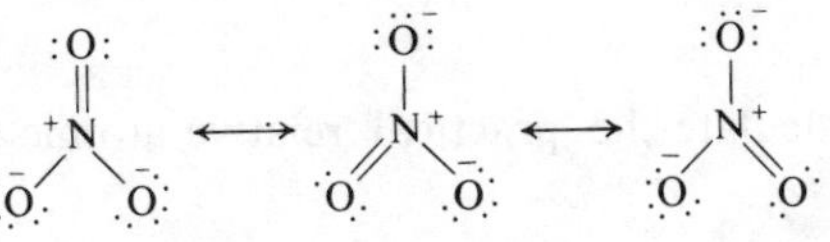

(XIV)

To obey the normal rules of valency one of the three oxygen atoms must be joined to the nitrogen atom by double bonds, one by a semipolar (coordinate) bond, and one—bearing a negative charge—by a single covalent bond. There are three different ways of fulfilling these requirements, distinguished by which of the three oxygen atoms is doubly bonded (the other two oxygen atoms turn out to be indistinguishable on account of the structure of the semipolar bond), but none of these is correct as it is known from X-ray examination of nitrates that all three N—O bonds are of identical length, and therefore electronically identical. In view of this we can say only that the true structure of the NO_3^- ion is an average of these three structures, which cannot be represented adequately by conventional symbols. The nitrate ion is described as a **resonance hybrid** (= average) of the three **canonical structures** (orthodox structures). Where it is required to indicate that resonance occurs between canonical structures (which can differ only in the distribution of electrons) the double-headed arrow is employed as shown in the diagram (XIV). Species, to which no single adequate structure can be assigned, are described as **mesomeric** (e.g. nitrate ion is a mesomeric anion).

The frequent use of a pendulum analogy to illustrate the concept of resonance creates the totally false idea that resonance is a rapid interchange between a number of extreme structues. A much better analogy is that of a partly-open door which is in neither of the two extreme states—fully open and shut—nor swinging wildly between these, but is static in an intermediate position and having in some measure the properties of both the extreme states.

The shapes of complex molecules. Earlier sections of this chapter describe how simple molecules have particular shapes arising from the geometry of the orbitals employed in bonding. In large and complex molecules the cumulative effects of the geometry of bonding at many sites lead to elaborate three-dimensional structures, the shapes of which can have profound effects upon the chemical reactions and biological properties of the compounds concerned. Detailed description of various aspects of molecular shape, 'stereochemistry', is deferred until later, but in the interval it should be remembered that there may be important three-dimensional aspects of even simple molecular structures and reactions.

Problems

1. From the figures given in Table I, calculate the 'practical' relative atomic mass of lithium.

2. In the second short period of the periodic table (Na—Ar), one 3*s* and three 3*p* orbitals are being filled. Write down the electronic configuration of Mg, Si, S, Cl, Ar.

3. Assign all the electrons in ethanal (CH_3—CH=O) to their atomic or molecular orbitals. By means of diagrams show the spatial distribution of electrons 'shared' by more than one nucleus.

4. The compound allene has a structure CH_2=C=CH_2, in which the three carbon atoms are collinear. What is the hybridisation of the central carbon atom?

   Bearing in mind that in ethene, CH_2=CH_2, all the nuclei are coplanar, what can you say about the geometrical relationship between the two CH_2 groups in allene?

5. Draw canonical structures for the carbonate ion (CO_3^{-2}), bicarbonate ion (HCO_3^-), and nitrite ion (NO_2^-).

2

Reactions and reagents

The structural relationship between the reagents and products of a chemical change has long been recognised by the organic chemist, but it is only comparatively recently that an understanding of the mechanistic basis of organic chemistry has developed. This knowledge of how reactions occur, and what factors determine their course constitutes a most important advance in chemistry, whose significance to the biologist lies in the fact that, although the chemistry of the living cell may sometimes appear to differ widely from normal 'laboratory reactions', we have no reason to suppose that the types of reaction or the factors influencing them are fundamentally different. Biochemists have already unravelled significant sections of the complex pattern of processes occurring in cells, and their discoveries can generally be correlated with the normal chemical reactivity of the functional groups involved.

Organic chemistry—the chemistry of carbon—is almost entirely concerned with covalent molecules (p. 6). It follows that reactions in organic chemistry will consist of the formation and breaking of covalent bonds, and the movements of the associated electrons. In order to understand why reactions follow certain courses we must examine factors which affect the distribution of electrons in covalent bonds and their availability for the formation of new bonds.

Electronegativity. The electronegativity of an element is a measure of the ability of an atom of the element to acquire an electron to form an anion, i.e. the ease with which the reaction $A + \epsilon \rightarrow A^{-}$ proceeds. It is generally found that electronegativity increases from left to right across the periodic table and decreases on going down any group. The most electronegative elements are therefore found in the top right-hand corner (O, F, Cl), and the least electronegative (i.e. the most electropositive) in the bottom left-hand corner (Ba, Cs, Rb). A qualitative list of relative electronegativity is given in Table III, but it should be realised that the molecular environment of an atom may affect its electronegativity to a limited extent.

Table III. *The relative electronegativities of some common elements*

F > O > Cl, N > Br > I, S, C > P, H > B > Mg > Li > Na.

The inductive effect. If we consider the distribution of electrons in a single covalent bond between two atoms, it can be seen that the sharing of electrons is not necessarily equal. In a symmetrical molecule A—A (e.g. H_2, Cl_2, HO—OH) the two nuclei, in whose neighbourhood the bonding electrons are found, are indistinguishable, and in the absence of any external effect the electron distribution will be symmetrical (i.e. the electrons will be 'equally shared'). If, however, we consider a molecule A—X where A and X are different (e.g. HF, ICl) then the nuclei are distinct and the atoms A and X may differ greatly in electronegativity. In these circumstances the distribution of electrons may be asymmetric (i.e. 'unequal sharing') and the electron density will be greater near the more electronegative element. This electron displacement carried to an extreme leads to the production of ions, but in many covalent bonds causes only slight **polarisation** of the bond, represented by A→—X (not to be confused with A→X, see p. 6) or $A^{\delta+}$—$X^{\delta-}$ where X is the more electronegative atom, and $\delta+$, $\delta-$ represent small electrical charges produced by the polarisation. Such a polarised bond may be regarded as a resonance hybrid of the purely covalent bond, in which there is a symmetrical electron distribution, and the purely ionic bond in A^+ X^- where both electrons have been transferred entirely to the more electronegative atom. A molecular orbital picture, which is complementary to the resonance description, could indicate the electron density in the bond by shading, in which case the bond in hydrogen fluoride would be represented as shown in (I).

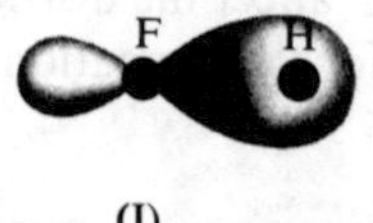

(I)

In a slightly more complex group of atoms such as A—A—X, the electron distribution in the bond A—A will not be completely symmetrical, as the polarisation of the A—X bond produces a partial positive charge on the central atom, i.e. A—$A^{\delta+}$——$^{\delta-}$X. This charge makes the central atom slightly more electronegative than its left-hand neighbour, resulting in another smaller electron displacement, i.e. $A^{\delta\delta+}$——$^{\delta\delta-}A^{\delta+}$——$^{\delta-}$X which might be represented by A→—A→→—X. This transmission of polarisation to adjacent bonds is known as the inductive effect, and because the electrons of σ bonds are localised (p. 7) the inductive effect in a chain of singly bonded atoms dies away very rapidly indeed.

The mesomeric or conjugative effect. The effect of relative electronegativity on the electron distribution in a double bond is very similar to that described above for the case of a single bond, except that the diffuse electron clouds of the π bond can be polarised to a very much greater extent. In the carbonyl group >C=O this can be regarded as resonance between the two structures shown in the diagram (II) below, or as the molecular orbital with asymmetric electron distribution as in diagram (III). The polarisation of the σ bond in these circumstances is insignificant and can be ignored.

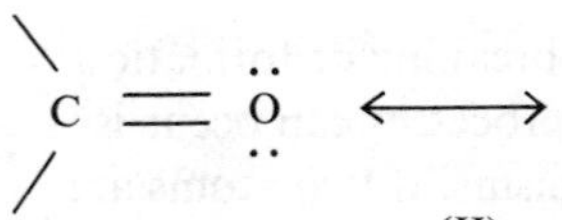

$$\text{>C=}\ddot{\text{O}}\text{:} \longleftrightarrow \text{>}\overset{+}{\text{C}}\text{—}\ddot{\underset{..}{\text{O}}}\text{:}^{-}$$

(II)

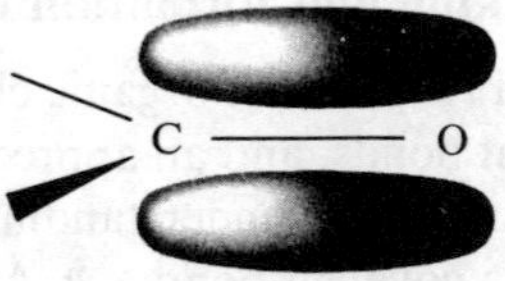

(III)

The polarisation of π bonds differs from that in σ bonds not only by its magnitude, but also by the extent to which it can be transmitted. The interaction of adjacent double bonds in a conjugated system has already been described, (p. 11) and if such a conjugated system contains an electronegative atom then very extensive electron displacements may occur along the conjugated chain of double bonds. In the case of propenal (CH_2=CH—CH=O) the molecular structure can be conventionally represented as (IVa), but on account of the electronegativity of the oxygen atom the correct description of

$$\underset{(3)}{CH_2}{=}\underset{(2)}{CH}{-}\underset{(1)}{CH}{=}\ddot{O} \longleftrightarrow CH_2{=}CH{-}\overset{+}{C}H{-}\ddot{O}{:}^{-} \longleftrightarrow \overset{+}{C}H_2{-}CH{=}CH{-}\ddot{O}{:}^{-}$$

(IVa) (IVb) (IVc)

the molecule is a resonance hybrid of the three canonical structures (IVa, b, c). The contribution from (IVc) must be considered, as the partial positive charge on carbon atom 1 will polarise the electron clouds of the adjacent π bond very considerably. The molecular orbital picture of this system is simply an asymmetric distribution of electrons in the extended molecular orbitals of the conjugated system, one of which is illustrated in diagram (V). The transmission of electron displacement in conjugated double bond systems is known as the **mesomeric** (or **conjugative**) **effect**, and differs from the inductive effect both in the greater polarisation possible with π bonds, and the much greater distance over which the polarisation can be transmitted.

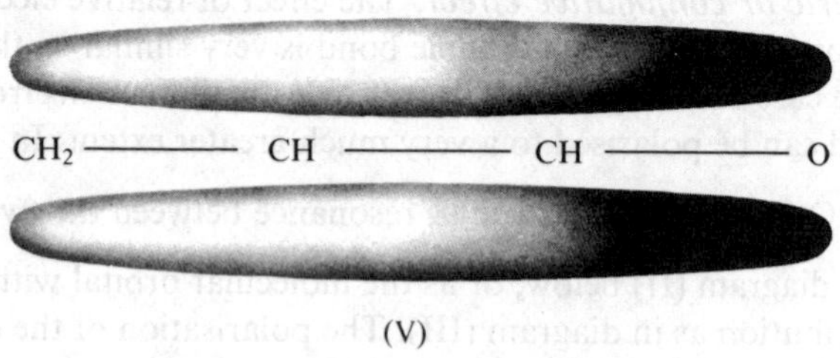

(V)

The breaking and formation of covalent bonds

Almost all reactions in organic chemistry involve the breaking or formation of covalent bonds, and an appreciation of how these processes can occur is fundamental to the understanding of reaction mechanisms. If two atoms are joined by a covalent bond, e.g. A—B, and during a reaction this bond is broken, then there are three ways in which this can occur, distinguished by the fate of the bonding electrons. **Homolytic fission** is the name given to the breaking of a covalent bond, in which each atom retains one of the shared

$$A{-}B \longrightarrow A\cdot + \cdot B \qquad \text{Homolytic fission}$$

$$\left.\begin{array}{l} A{-}B \longrightarrow A{:}^- + B^+ \\ A{-}B \longrightarrow A^+ + {:}B^- \end{array}\right\} \text{Heterolytic fission}$$

electrons. Each of the species produced has an odd or 'unpaired' electron, but bears no electrical charge, and is known as a **radical***. **Heterolytic fission** is the breaking of a covalent bond in which both of the bonding electrons remain with one of the atoms. This process, which produces electrically charged species, can occur in two ways, as illustrated above. Clearly if B is significantly more electronegative than A the latter mode of heterolytic fission is more probable.

The formation of covalent bonds can occur by exactly the reverse of the processes described above, and although organic chemistry contains reactions which involve more complex processes, a great many reactions can be explained by the use of these simple electron movements. It is convenient, when

* These species were once described as 'free radicals', but this usage is becoming obsolete. The term 'radical' is nowadays confined to species with unpaired electrons and is not employed with its older meaning of a discrete group of atoms in a molecule.

attempting to represent a reaction mechanism graphically, to use symbols indicating the direction of movement of electrons. The only two such symbols employed in this book are the curly arrows ⌒ and ⌒ which indicate the movement of one and two electrons respectively. When using these symbols, students should ensure that the source and destination of the electrons are

$$A{-}B \longrightarrow A\cdot + \cdot B$$
$$A{-}B \longrightarrow A^+ + :B^-$$

clearly indicated by the positions of the root and tip of the arrow, e.g. the reaction of ammonia with hydrogen chloride can be represented:

$$H_3N: \; H{-}Cl \longrightarrow H_3\overset{+}{N}{-}H + Cl^-$$

the lone pair of the nitrogen being donated to form a bond to the hydrogen atom, with simultaneous transfer of the electrons of the H—Cl bond to the chlorine atom.

Types of reagent. In later chapters we shall be considering the mechanism of some reactions in detail, and it is convenient to classify the reagents into three groups, with obvious relationships to the methods of bond formation and breakage described earlier.

Radicals are species with an odd number of electrons. They are normally very reactive, and attack molecules at positions of high electron density. The atoms of hydrogen and chlorine are simple examples of radicals.

Electrophilic reagents are species with a deficiency of electrons, e.g. a vacant atomic orbital, sometimes bearing a positive charge. These reagents attack positions of high electron density or negative charge. H^+, $AlCl_3$, BF_3, SO_3 are examples of electrophiles.

Nucleophilic reagents are species with a pair of electrons (e.g. a lone pair) available for bond formation, sometimes bearing a negative charge, which attack positions of low electron density or positive charge. OH^-, I^-, NH_3 are examples of nucleophiles.

The course of complex reactions

Chemistry contains many reactions which, when represented by balanced equations, appear to require the simultaneous interaction of several molecules. However, there are good reasons for believing that such reactions occur by a series of comparatively simple steps. The molecules of a gas or liquid are in continual random motion, and reactions between molecules can occur only during the brief period of a collision. The probability of three or more molecules colliding simultaneously is so small that a reaction requiring such a step would proceed only very slowly. We are therefore compelled to explain complex reactions, which proceed at observable rates, in terms of a sequence of simple reactions, each of which requires at the most two interacting species.

Intermediate species in reactions. As a consequence of the reasoning in the previous paragraph, it is necessary to propose that, during the course of some reactions, intermediate species of high reactivity and very short lifetime are formed. Some of the less familiar types of intermediate are listed below. It should be noted that although they may contain atoms with incomplete outer shells (e.g. carbonium ions), they never require an atom of the first short period to have more than eight electrons in the outer shell.

Types of reaction intermediate

<table>
<tr><th>STRUCTURE</th><th>NAME</th><th>NO. OF ELECTRONS IN THE OUTER SHELL</th></tr>
<tr><td>R
|
R—C$^+$
|
R</td><td>Carbonium ion</td><td>6</td></tr>
<tr><td>R
|
R—C:$^-$
|
R</td><td>Carbanion</td><td>8</td></tr>
<tr><td>R—$\ddot{\underset{\cdot\cdot}{\mathrm{N}}}^-$—R</td><td>Amide ion</td><td>8</td></tr>
<tr><td>R
|
R—N$^+$—R
|
R</td><td>Ammonium ion</td><td>8</td></tr>
<tr><td>R—$\ddot{\underset{\cdot\cdot}{\mathrm{O}}}$:$^-$</td><td>Oxide ion</td><td>8</td></tr>
<tr><td>R
|
R—O:$^+$
|
R</td><td>Oxonium ion</td><td>8</td></tr>
</table>

R = any monocovalent atom or group.

Problems

1. Predict the direction of polarisation of the following bonds, considering only the relative electronegativity of the atoms involved.

   C—H; B—H; N—Br; I—Cl

2. Do you expect BH_3 to be a nucleophile or an electrophile? Explain your answer.

3. Explain why:

   (a) $(CH_3)_4\overset{+}{N}$ is neither a nucleophile nor an electrophile;

   (b) $H_3\overset{+}{O}$ is not a nucleophile, even though there is a lone pair on oxygen.

4. All of the following species are nucleophiles. With what sites might an electrophile be expected to react in each case?
   H_2NOH (hydroxylamine); NO_2^- (nitrite ion); N_3^- (azide ion); NCS^- (thiocyanate ion).

3

Hydrocarbons

Organic chemistry is the chemistry of carbon compounds. The division of descriptive chemistry into inorganic and organic sections is based on the fact that of all the elements, only carbon has the ability to form stable compounds containing long chains of similar atoms. As a result, the number of known compounds of carbon exceeds by many thousands the total number of compounds formed by all the other elements.

Although the immense number of carbon compounds makes it impossible to study each one individually, fortunately the chemistry of most organic compounds is dominated by the reactions of one or more 'functional groups' (small reactive groups of atoms). It is frequently found that the chemistry of a functional group is not appreciably affected by changes in the rest of the molecule and so organic compounds can conveniently be classified and studied according to the functional groups present in the molecule.

Hydrocarbons are compounds composed solely of carbon and hydrogen atoms, and on account of the simplicity of their atomic composition they may be regarded as the parent compounds of organic chemistry from which all others may, in principle, be derived. We shall consider the chemistry of six classes of hydrocarbon.

Alkanes (paraffins)

Alkanes are hydrocarbons whose structures are based on carbon chains devoid of multiple bonds ('saturated') or rings. In consequence their general formula is C_nH_{2n+2}.

Isomerism (the phenomenon of two compounds of identical molecular formula having different molecular structures) is possible in alkanes containing more than three carbon atoms, when the possibility of straight or branched chain structures arises, e.g. C_4H_{10} can have two structures:

```
  H  H  H  H          H  H  H
  |  |  |  |          |  |  |
H—C——C——C——C—H  or  H—C——C——C—H
  |  |  |  |          |  |  |
  H  H  H  H          H  |  H
                       H—C—H
                         |
                         H
```

and C_5H_{12} has three:

```
                                 H                          H
                                 |                          |
  H  H  H  H  H          H  H—C—H  H  H           H  H—C—H  H
  |  |  |  |  |          |     |     |  |           |     |     |
H—C——C——C——C——C—H      H—C————C————C——C—H       H—C————C————C—H
  |  |  |  |  |          |     |     |  |           |     |     |
  H  H  H  H  H          H     H     H  H           H  H—C—H  H
                                                          |
                                                          H
```

With increasing values of n, the number of possible structures increases very rapidly, and $C_{20}H_{42}$ has approximately 4×10^5 isomers.

The diagrams above are known as the 'constitutional formulae' of the molecules represented. The **constitution** of a molecule describes the atomic sequence and nature of bonding between adjacent atoms in the molecule, but gives no information about the three-dimensional aspects of the molecular structure, such as shape or relative spatial arrangements of atoms or groups in the molecule.

Nomenclature. In order to simplify the naming of the vast number of organic compounds, a systematic form of nomenclature has been devised, in which the name of a compound is composed of syllables indicative of the functional groups present. All saturated hydrocarbons have the suffix 'ane' and in the unbranched ('normal') alkanes the previous syllable indicates the number of carbon atoms in the molecule:

FORMULA	NAME OF UNBRANCHED ALKANE	CARBON SKELETON OF THE ALKANE
CH_4	**meth**ane	C
C_2H_6	**eth**ane	C—C
C_3H_8	**prop**ane	C—C—C
C_4H_{10}	**but**ane	C—C—C—C
C_5H_{12}	**pent**ane	C—C—C—C—C
C_6H_{14}	**hex**ane	C—C—C—C—C—C
C_9H_{20}	**non**ane	C—C—C—C—C—C—C—C—C

The first four members of the series have old, non-systematic (trivial) names, and it will be found in other groups of compounds that the simpler species still retain their old trivial names, whilst the more complex molecules have been renamed systematically.

The branched chain alkanes can be regarded as derived from the unbranched alkanes by replacement of hydrogen atoms by 'substituent groups', each of which, consisting of an alkane minus a hydrogen atom, is known as an alkyl group, e.g. C_2H_5 — ethyl; C_9H_{19} — nonyl. If the carbon atoms of an unbranched alkane are numbered from one end, then the position as well as the structure of the substituent group can be described:

$$\underset{5}{CH_3}-\underset{4}{CH_2}-\underset{3}{\overset{\overset{\displaystyle CH_3}{|}}{CH}}-\underset{2}{CH_2}-\underset{1}{CH_3} \quad \text{3-Methylpentane}$$

For this purpose, the parent alkane is the one corresponding to the longest continuous chain of carbon atoms in the molecule, and the numbering of the carbon atoms of the parent alkane is started at the end which will give the

$$\underset{1}{CH_3}-\underset{2}{CH_2}-\underset{3}{\overset{\overset{\displaystyle CH_3}{|}}{CH}}-\underset{4}{CH_2}-\overset{5}{CH}(-{}_6CH_2-{}_7CH_2-{}_8CH_3)-CH_3 \quad \text{3,5-Dimethyloctane}$$

[NOT
4,6-Dimethyloctane
3-Methyl-5-propylhexane
2-Ethyl-4-propylpentane]

lowest aggregate of numbers in the systematic name. Where two substituents are attached to the same atom, both are numbered, e.g. 2,2-dimethylbutane.

$$CH_3-\underset{\underset{\displaystyle CH_3}{|}}{\overset{\overset{\displaystyle CH_3}{|}}{C}}-CH_2-CH_3$$

Preparation of alkanes

1. Hydrogenation of unsaturated hydrocarbons using a finely-divided metal catalyst and hydrogen:

$$R-CH{=}CH_2 \xrightarrow{Ni/H_2} R-CH_2-CH_3$$

$$R-C{\equiv}C-R' \xrightarrow{Ni/H_2} R-CH{=}CH-R' \xrightarrow{Ni/H_2} R-CH_2-CH_2-R'$$

Nickel, palladium, and platinum are the customary catalysts, and it is thought that hydrogen molecules, adsorbed on the catalyst surface, dissociate into atoms which can add to adsorbed unsaturated compounds.

2. Reduction of alkyl halides:

(i) by catalytic hydrogenation;

$$R{-}Br + Pd/H_2 \longrightarrow R{-}H + HBr$$

(ii) by dissolving metal (i.e. metal + acid) systems (p. 61);

$$R{-}I + Na/C_2H_5OH \text{ (or HI; Zn/HCl)} \longrightarrow RH$$

(iii) by complex metal hydrides;

$$R{-}Br + LiAlH_4 \longrightarrow R{-}H + LiBr + AlBr_3$$

Lithium aluminium hydride is an ether-soluble, ionic compound $Li^+ (AlH_4)^-$, which can be regarded as a source of hydride ions (H^-).

3. Reduction of carbonyl compounds (p. 114).

4. Wurtz reaction. Alkyl halides in ether solution react with sodium to give the sodium halide and an alkane produced by combination of the alkyl radicals.

$$2\,R{-}X + 2\,Na \longrightarrow 2\,NaX + R{-}R$$

$$\underset{\text{2-Iodopropane}}{(CH_3)_2CHI} + Na \longrightarrow NaI + \underset{\text{2,3-Dimethylbutane}}{(CH_3)_2CH{-}CH(CH_3)_2}$$

This reaction is useful only where symmetrical alkanes are required. If a mixture of different alkyl halides is used, differences in reactivity normally result in very poor yields of the mixed product R—R′.

$$R{-}X + R'{-}X + Na \longrightarrow NaX + R{-}R + R'{-}R' + R{-}R'$$

5. Electrolysis of solutions of the alkali salts of carboxylic acids results in the production of alkyl radicals at the anode, which combine to form the alkane. The steps of the process are thought to be:

$$R{-}C(=\ddot{O}:){-}\ddot{O}:^- \xrightarrow{-1\text{ electron}} R{-}C(=\ddot{O}:){-}\ddot{O}\cdot$$

$$R{-}C(=\ddot{O}:){-}\ddot{O}\cdot \longrightarrow R\cdot + CO_2$$

$$R\cdot + R\cdot \longrightarrow R{-}R$$

$$\text{e.g. } \underset{\text{Propionate anion}}{C_2H_5CO_2^-} \xrightarrow{\text{electrolysis}} \underset{\text{Butane}}{C_2H_5{-}C_2H_5} + CO_2$$

Properties of alkanes. The alkanes are neutral compounds, of density less than one, insoluble in water but soluble in organic solvents. The members of the series below pentane are gases, and those above pentadecane ($C_{15}H_{32}$) are waxy solids.

Reactions of alkanes. The alkanes are relatively inert substances with few simple reactions. The relatively small difference between the electronegativities of carbon and hydrogen means that heterolytic fission of C—H bonds is not a very likely process, and all the simpler reactions probably involve radicals.

1. Like many organic compounds the alkanes will burn in air or oxygen when heated to a sufficiently high temperature, forming water and carbon dioxide, or, in the absence of sufficient oxygen, carbon monoxide, or carbon.

2. *Halogenation.* If a mixture of an alkane and chlorine is exposed to diffuse light, hydrogen chloride is evolved and halogen-substituted compounds are obtained.

$$CH_4 + Cl_2 \xrightarrow{\text{light}} CH_3Cl, CH_2Cl_2, CHCl_3, CCl_4 + HCl$$

$$CH_3CH_2CH_3 + Cl_2 \xrightarrow{\text{light}} CH_3CH_2CH_2Cl, CH_3CHClCH_3, \text{etc.}, + HCl$$

This is a radical reaction, and is known to proceed by the following mechanism:

(a) Chlorine molecules absorb light and dissociate into atoms:

$$Cl{-}Cl \overset{\text{light}}{\rightleftharpoons} Cl\cdot + \cdot Cl$$

(b) A chlorine atom can abstract a hydrogen atom from an alkane molecule to form hydrogen chloride and an alkyl radical

$$Cl\cdot\ H{-}R \longrightarrow Cl{-}H + \cdot R$$

(c) The alkyl radical attacks a chlorine molecule producing a molecule of alkyl halide and a chlorine atom, which can start another stage of this **chain reaction**. The chain reaction can be terminated by the combination of any

$$R\cdot\ Cl{-}Cl \longrightarrow R{-}Cl + \cdot Cl$$

two of the radicals involved, e.g. $R\cdot + Cl\cdot \rightarrow R{-}Cl$, or $R\cdot + R\cdot \rightarrow R{-}R$.

It is not generally possible to obtain solely monosubstituted products from this reaction, as the alkyl halides are further attacked by chlorine atoms. In bright light the reaction may become very violent (i.e. explosive).

Similar reactions occur with bromine and alkanes. Fluorine reacts extremely violently. Iodination is possible only in the presence of an oxidising agent to remove the hydrogen iodide formed.

3. *Nitration* of alkanes is possible using nitric acid or nitrogen dioxide in the vapour phase at high temperatures. The nitroalkanes obtained are probably produced by a radical mechanism (cf. aromatic nitration p. 41).

$$R{-}H + HNO_3 \xrightarrow{400°C} R{-}NO_2 + H_2O$$

4. *Sulphonation* of alkanes by oleum gives sulphonic acids.

$$R{-}H + H_2SO_4/SO_3 \longrightarrow R{-}SO_3H$$

The petroleum industry is based upon an enormous range of alkane reactions, involving cracking and rearrangement processes. These are not relevant to the purposes of this book, and will be omitted.

Occurrence of alkanes. Many straight and branched chain alkanes occur in petroleum and natural gas, but have little biological significance. An exception is methane (marsh gas) which is produced by the anaerobic bacterial decomposition of cellulose. In view of its low chemical reactivity it is remarkable that certain micro-organisms can metabolise methane, and can grow in the absence of any alternative source of carbon. In recent years there has been much interest in the possibility of commercial farming of micro-organisms grown on oil based media as a source of food.

Cycloalkanes

Cycloalkanes are hydrocarbons in which the carbon skeleton is in the form of a ring rather than an open chain, but which are otherwise saturated. The systematic names indicate the ring size as shown, with the prefix 'cyclo-':

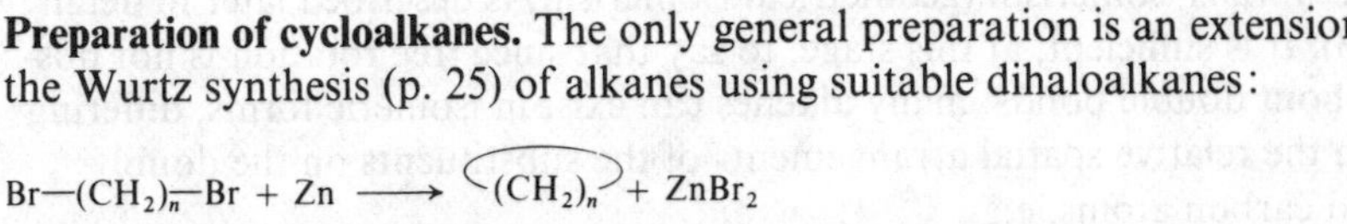

CH_2 / H_2C—CH_2 (three-membered ring)
Cyclopropane

H_2C—CH_2 / H_2C—CH_2 (four-membered ring)
Cyclobutane

CH_2 / H_2C CH_2 / H_2C—CH_2 (five-membered ring)
Cyclopentane

CH_2 / H_2C CH_2 / H_2C CH_2 / CH_2 (six-membered ring)
Cyclohexane

Preparation of cycloalkanes. The only general preparation is an extension of the Wurtz synthesis (p. 25) of alkanes using suitable dihaloalkanes:

$$Br{-}(CH_2)_n{-}Br + Zn \longrightarrow (CH_2)_n \text{ (ring)} + ZnBr_2$$

Yields vary greatly with n, and are satisfactory only when $n = 5$ or 6. In addition, there are many specific methods for the preparation of rings of particular sizes, which are outside the scope of this book, e.g. cyclohexane can be prepared by the catalytic hydrogenation of benzene.

(N.B. Throughout this book the symbol for benzene will be the Kekulé structure on the left, whilst the saturated six-membered ring of cyclohexane will be represented by the open hexagon on the right.)

$$\text{Benzene} \xrightarrow{Ni/H_2} \text{Cyclohexane}$$

Benzene Cyclohexane

Properties and reactions of cycloalkanes. The properties and reactions of cycloalkanes are very similar to those of the open-chain alkanes. The simple presence of a ring does not imply aromatic properties, e.g.

$$\text{Cyclohexane } (CH_2)_6 \xrightarrow[\text{light}]{+Cl_2} \text{Chlorocyclohexane } (CH_2)_5CHCl$$

Cyclohexane Chlorocyclohexane

Cyclopropane has been used as an anaesthetic, its inflammability being its principal disadvantage.

Alkenes (olefins)

Alkenes are hydrocarbons containing carbon atoms joined by double bonds, and systematic nomenclature indicates the presence of this type of linkage by the suffix 'ene', e.g. $CH_3{-}CH{=}CH_2$ propene. Isomerism may arise in olefins in a number of ways:

(a) Straight and branched chain isomers may occur as in the case of alkanes.

$CH_3{-}CH_2{-}CH_2{-}CH{=}CH_2$ Pent-1-ene

$CH_3{-}CH(CH_3){-}CH{=}CH_2$ 3-Methylbut-1-ene

$CH_3{-}CH_2{-}C(CH_3){=}CH_2$ 2-Methylbut-1-ene

(b) The position of the functional group may vary in any one carbon skeleton:

$CH_3{-}CH_2{-}CH{=}CH_2$ But-1-ene (or 1-Butene)

$CH_3{-}CH{=}CH{-}CH_3$ But-2-ene (or 2-Butene)

(c) *Cis-trans* isomerism (geometrical isomerism) is described later in detail (p. 193). It is sufficient, at this stage, to say that since free rotation is not possible about double bonds, many alkenes can exist in isomeric forms, differing only in the relative spatial arrangements of the substituents on the doubly bonded carbon atoms, e.g.

$(H_3C)(H)C{=}C(H_3C)(H)$ *cis*-But-2-ene

$(H_3C)(H)C{=}C(H)(CH_3)$ *trans*-But-2-ene

Cis-trans isomers have different chemical and physical properties and can be separated by normal physical methods, e.g. crystallisation, distillation, or chromatography.

Alkenes are isomeric with cycloalkanes, e.g. propene and cyclopropane, C_3H_6; hexene and cyclohexane, C_6H_{12}.

Preparation of alkenes

1. Catalytic hydrogenation of alkynes cannot normally be stopped at the alkene stage. However, with special catalysts (e.g. Lindlar's catalyst, $Pd/CaCO_3/Pb(OCOCH_3)_2$ high yields of *cis*-alkene can be obtained.

$$R{-}C{\equiv}C{-}R' \xrightarrow[\text{Lindlar's catalyst}]{H_2} R{-}CH{=}CH{-}R$$

Alkenes can also be produced by the following general methods:

2. Dehydration of alcohols by heating with sulphuric or phosphoric acids:

$$CH_3{-}CH(OH){-}CH_3 \xrightarrow{-H_2O} CH_3{-}CH{=}CH_2$$

3. Dehydrohalogenation of alkyl halides by strong bases (alkoxides, sodium amide):

$$R{-}\underset{H}{\overset{R}{C}}{-}\underset{Br}{\overset{R}{C}}{-}R \xrightarrow[KOH/C_2H_5OH]{-HBr} R_2C{=}CR_2$$

4. Pyrolysis of quaternary ammonium hydroxides:

$$CH_3{-}CH_2{-}CH_2{-}CH_2{-}\overset{+}{N}(CH_3)_3\ \overset{-}{O}H \xrightarrow{\text{heat}} CH_3{-}CH_2{-}CH{=}CH_2 + (CH_3)_3N + H_2O$$

5. Dehalogenation of vicinal* dihalides by metals (zinc or magnesium):

$$R{-}\underset{Br}{CH}{-}\underset{Br}{CH_2} + Zn \longrightarrow R{-}CH{=}CH_2 + ZnBr_2$$

Although preparations 2–4 appear quite diverse, they are in fact mechanistically similar. In general, the reactions can be represented by:

$$-\underset{H}{\overset{|}{C}}-\underset{X}{\overset{|}{C}}- \longrightarrow \ >C{=}C< \quad B{-}H\ :X^- \qquad \bar{B}: \qquad B\bar{:} = \text{a base}$$

The electronegativity of X produces a slight inductive displacement of electrons away from hydrogen in the nearby C—H bond. The elimination consists of a concerted donation of two electrons to the hydrogen atom by the base (B^-), transfer of the electrons of the C—H bond to the neighbouring

* See footnote, p. 98.

carbon atom, forming the double bond, and loss of the electronegative substituent X as an anion by heterolytic fission of the C—X bond. In preparations 3 and 4 the base is obviously hydroxide or ethoxide ion, whilst in preparation no. 2 the elimination is preceded by protonation of the hydroxyl group of the alcohol to form the alkyl oxonium ion, or esterification of the alcohol (via this oxonium ion) to the alkyl hydrogen sulphate (or phosphate), followed by attack of HSO_4^- acting as a base and loss of H_2O or HSO_4^-.

$$CH_3CH_2{-}O{-}H + H_2SO_4 \rightleftharpoons CH_3CH_2{-}\overset{H}{\overset{|+}{O}}{-}H + HSO_4^- \rightarrow CH_3CH_2{-}OSO_3H + H_2O$$

$$H{-}\underset{H}{\overset{H}{C}}{-}\underset{H}{\overset{H}{C}}{-}OSO_3H \;\; {}^{-}OSO_3H \longrightarrow \underset{H}{\overset{H}{>}}C{=}C\underset{H}{\overset{H}{<}} \;\; {}^{-}OSO_3H \;\; H{-}OSO_3H$$

The ease of this type of elimination is dependent on the electronegativity of X (the 'leaving group') and the strength of the base B^- (i.e. its ability to form a strong bond to a hydrogen atom). Protonation or esterification by mineral acid converts the hydroxyl group into a better leaving group.

Preparation no. 5 follows a similar mechanism to these others, but here the base is replaced by a metal atom, which can donate electrons to a covalently bound halogen atom.

$$Zn{:} \;\; Br{-}\overset{|}{\underset{|}{C}}{-}\overset{|}{\underset{|}{C}}{-}Br \longrightarrow \overset{+}{Zn}Br \;\; >C{=}C< \;\; {:}Br^-$$

Stereochemical aspects of alkene-forming eliminations are described later (p. 205).

Properties of alkenes. Alkenes are neutral compounds whose physical properties are similar to those of the corresponding alkanes.

Reactions of alkenes. The alkenes are much more reactive than the alkanes, on account of the π bond, which acts as a reservoir of electrons. Consequently, the chemistry of alkenes is dominated by reactions with electrophilic reagents.

A *Electrophilic additions to alkenes.* Many of the addition reactions of alkenes fit the general scheme:

$$>C{=}C< + X{-}Y \longrightarrow -\underset{X}{\overset{|}{\underset{|}{C}}}-\underset{Y}{\overset{|}{\underset{|}{C}}}-$$

This type of addition, whose mechanism is well established, is initiated by electrophilic attack.

If Y is more electronegative than X (so that XY is polarised $X \overset{\delta^+ \quad \delta^-}{\text{———}} Y$, then initially electrons are donated from the π bond to the electrophilic X atom (or group) to form an intermediate carbonium ion and an anion Y^-.

$$H_2C{=}CH_2 + X{-}Y \longrightarrow H{-}\underset{X}{\underset{|}{\overset{H}{\overset{|}{C}}}}{-}\overset{+}{C}H_2 \;+\; :Y^- \longrightarrow H{-}\underset{X}{\underset{|}{\overset{H}{\overset{|}{C}}}}{-}\underset{Y}{\underset{|}{\overset{H}{\overset{|}{C}}}}{-}H$$

The carbonium ion so formed immediately reacts with an anion to form the addition compound.

Addition reactions of this type are:

$H_2C{=}CH_2$ + $\overset{\delta+}{X}{-}\overset{\delta}{Y}$:

$$H{-}Cl \longrightarrow H{-}\underset{H}{\underset{|}{\overset{H}{\overset{|}{C}}}}{-}\underset{Cl}{\underset{|}{\overset{H}{\overset{|}{C}}}}{-}H$$

$$H{-}Br \longrightarrow H{-}\underset{H}{\underset{|}{\overset{H}{\overset{|}{C}}}}{-}\underset{Br}{\underset{|}{\overset{H}{\overset{|}{C}}}}{-}H$$

$$H{-}I \longrightarrow H{-}\underset{H}{\underset{|}{\overset{H}{\overset{|}{C}}}}{-}\underset{I}{\underset{|}{\overset{H}{\overset{|}{C}}}}{-}H$$

$$\underset{(Cl_2+H_2O)}{Cl{-}OH} \longrightarrow H{-}\underset{Cl}{\underset{|}{\overset{H}{\overset{|}{C}}}}{-}\underset{OH}{\underset{|}{\overset{H}{\overset{|}{C}}}}{-}H \text{ (a 'chlorohydrin')}$$

$$\underset{(Br_2+H_2O)}{Br{-}OH} \longrightarrow H{-}\underset{Br}{\underset{|}{\overset{H}{\overset{|}{C}}}}{-}\underset{OH}{\underset{|}{\overset{H}{\overset{|}{C}}}}{-}H \text{ (a 'bromohydrin')}$$

$$H{-}OSO_3H \longrightarrow H{-}\underset{H}{\underset{|}{\overset{H}{\overset{|}{C}}}}{-}\underset{OSO_3H}{\underset{|}{\overset{H}{\overset{|}{C}}}}{-}H \xrightarrow{+H_2O} H{-}\underset{H}{\underset{|}{\overset{H}{\overset{|}{C}}}}{-}\underset{OH}{\underset{|}{\overset{H}{\overset{|}{C}}}}{-}H + H_2SO_4$$

$$Cl{-}Cl \longrightarrow H{-}\underset{Cl}{\underset{|}{\overset{H}{\overset{|}{C}}}}{-}\underset{Cl}{\underset{|}{\overset{H}{\overset{|}{C}}}}{-}H$$

$$Br{-}Br \longrightarrow H{-}\underset{Br}{\underset{|}{\overset{H}{\overset{|}{C}}}}{-}\underset{Br}{\underset{|}{\overset{H}{\overset{|}{C}}}}{-}H$$

e.g. $CH_3{-}CH{=}CH{-}CH_3 + HI \longrightarrow CH_3{-}CH_2{-}CHI{-}CH_3$

$CH_3{-}CH_2{-}CH{=}CH_2 + Br_2 \longrightarrow CH_3{-}CH_2{-}CHBr{-}CH_2Br$

The addition of sulphuric acid to alkenes can be followed by hydrolysis of the sulphate ester, so that this reaction sequence can be used to add the elements of water across a double bond.

If an asymmetrical molecule adds to an asymmetrical alkene, two products are possible, but in practice only one is formed. The two possible reaction sequences are:

$$RHC{=}CH_2 + H{-}I \longrightarrow R\overset{+}{C}H{-}CH_3 \;(+\,{:}I^-) \longrightarrow R{-}CHI{-}CH_3$$

$$RHC{=}CH_2 + H{-}I \;\not\longrightarrow\; R{-}CH_2{-}\overset{+}{C}H_2 \;(+\,{:}I^-) \longrightarrow R{-}CH_2{-}CH_2I$$

The factor which determines the course of the reaction is the ease of formation of the alternative carbonium ion intermediates. Carbonium ions are stabilised by alkyl or aryl groups attached to the carbon atom bearing the positive charge, and the more stable the carbonium ion, the lower the energy and hence the greater the ease of formation. Polarisation of the adjacent C—C bonds stabilises the positive charge by effectively distributing the charge over a larger volume of space (cf. the electric potential of a spherical condenser bearing a fixed charge is inversely proportional to the radius). Therefore the relative stability (i.e. ease of formation) of carbonium ions is:

$$R_3C^+ > R_2HC^+ > RH_2C^+ > H_3C^+ \qquad R = \text{alkyl or aryl group.}$$

The product formed in the addition to an alkene is the one obtained via the more stable carbonium ion, i.e. R—CHI—CH_3 in the case above. This is summarised by the empirically discovered **Markownikoff rule**: 'In an addition reaction of an asymmetrical alkene the more positive section of the addendum adds to the carbon atom bearing the greater number of hydrogen atoms'; e.g.

$$CH_3{-}CH{=}CH_2 + \overset{\delta-}{HO}{-}\overset{\delta+}{Cl} \longrightarrow CH_3{-}CH(OH){-}CH_2Cl$$

B *Other reactions of alkenes, whose mechanisms will not be discussed.*

1. *Ozonolysis.* Alkenes react with ozone to form ozonides, which on reduction give carbonyl compounds.

$$R_2C{=}CR_2 + O_3 \longrightarrow \left[\text{R}_2\text{C}(\text{O–O–O})\text{CR}_2\right] \xrightarrow{\text{rearrangement}}$$

Unstable intermediate

$$\text{R}_2\text{C}(\text{O–O})(\text{O})\text{CR}_2 \xrightarrow{\text{reduction}} R_2C{=}O + O{=}CR_2$$

Ozonide

e.g.

$$(H_3C)_2C{=}C(C_2H_5)H \xrightarrow[\text{(ii) Pd/H}_2 \text{ or Zn/CH}_3\text{CO}_2\text{H}]{\text{(i) O}_3} (H_3C)_2C{=}O + O{=}C(C_2H_5)H$$

2. *Hydroxylation.* Dilute potassium permanganate converts alkenes into dihydroxy-compounds.

$$R{-}CH{=}CH_2 + KMnO_4 \rightarrow R{-}CH(OH){-}CH_2(OH)$$

This can also be achieved by reaction of the alkene with osmium tetroxide, and decomposition of the addition compound with sodium sulphite:

$$R{-}CH{=}CH{-}R' \xrightarrow{+OsO_4} R{-}CH(O)\!-\!CH(O){-}R'\ [\text{cyclic } O_2OsO_2 \text{ ester}] \xrightarrow[+H_2O]{Na_2SO_3} R{-}CH(OH){-}CH(OH){-}R'$$

3. Catalytic hydrogenation of alkenes produces alkanes (p. 24).

In all these addition reactions the conversion of sp^2 hybridised carbon atoms into sp^3 hybridisation has important three-dimensional consequences which are described in a later chapter (p. 207).

Occurrence. Few simple olefins are known to occur naturally, other than in natural gas and petroleum. It has recently been shown that ripening tomatoes and apples evolve ethene ($CH_2{=}CH_2$), and in the latter case traces of propene ($CH_3CH{=}CH_2$) and ethyne have also been detected. Some more complex polyenes (compounds containing several C=C groups) are found in nature, such as *β*-carotene (p. 13), one of the colouring matters of carrots, peaches, and green leaves, and lycopene, the colouring compound of tomatoes. Squalene, a polyene found extensively in nature in minute traces and in large quantities in the liver oil of the basking shark, is an intermediate in the biosynthesis of steroids (p. 326) from ethanoic acid (acetic acid).

$$\begin{array}{l} \quad CH_3 \qquad\quad CH_3 \qquad\quad CH_3 \qquad\quad CH_3 \\ \quad C \quad CH_2 \quad C \quad CH \quad C \quad CH \quad C \quad CH \quad CH \quad CH \quad CH \quad CH \quad CH \quad CH_2 \quad CH \quad CH \\ H_3C \quad CH \quad CH_2 \quad CH \quad CH \quad CH \quad CH \quad CH \quad CH \quad C \quad CH \quad C \quad CH \quad C \quad CH_2 \quad C \\ \qquad\qquad\qquad\qquad\qquad\qquad\qquad\qquad\qquad CH_3 \quad CH_3 \quad CH_3 \quad CH_3 \end{array}$$

Lycopene

$$\begin{array}{l} \quad CH_3 \qquad CH_3 \qquad CH_3 \\ \quad C \quad CH_2 \quad C \quad CH_2 \quad C \quad CH_2 \quad CH \quad CH_2 \quad CH \quad CH_2 \quad CH \quad CH_3 \\ H_3C \quad CH \quad CH_2 \quad CH \quad CH_2 \quad CH \quad CH_2 \quad C \quad CH_2 \quad C \quad CH_2 \quad C \\ \qquad\qquad\qquad\qquad\qquad\qquad CH_3 \quad CH_3 \quad CH_3 \end{array}$$

Squalene

Alkynes

Alkynes are hydrocarbons containing triple bonds, the systematic suffix indicating this group being '-yne', e.g.

$$CH_3{-}C{\equiv}C{-}CH_3 \qquad CH_3{-}\underset{}{\overset{CH_3}{\overset{|}{C}H}}{-}CH_2{-}C{\equiv}CH$$

But-2-yne   4-Methylpent-1-yne

The first member of the series, ethyne, C_2H_2, retains its old, trivial name acetylene, but all other alkynes are named systematically, e.g. propyne, butyne, pentyne, octyne, etc. Since the two triply bonded carbon atoms and the two neighbouring atoms are collinear (p. 11), no possibility of geometrical isomerism arises (p. 28).

Preparation of alkynes. The dehydrohalogenation of compounds containing suitable groups, e.g.

$$\left.\begin{array}{l} R{-}CH_2{-}CX_2{-}R' \\ R{-}CHX{-}CHX{-}R' \\ R{-}CH{=}CX{-}R' \end{array}\right\} \xrightarrow[\substack{\text{strong base} \\ \text{(e.g. } KNH_2)}]{-HX} R{-}C{\equiv}C{-}R' \qquad (X = \text{halogen})$$

is exactly analogous to the corresponding preparation of alkenes, as is also the dehalogenation of polyhalogen compounds by metals:

$$\left.\begin{array}{l} R{-}CX_2{-}CX_2{-}R \\ R{-}CX{=}CX{-}R' \end{array}\right\} \xrightarrow{+\,Mg \text{ or } Zn} R{-}C{\equiv}C{-}R' \qquad (X = \text{halogen})$$

Ethyne can be obtained by the action of water on calcium carbide, and

$$CaC_2 + H_2O \longrightarrow Ca(OH)_2 + H{-}C{\equiv}C{-}H$$

many alkynes can be prepared from ethyne via its alkali metal salts (see below).

Properties of alkynes. Ethyne is a gas, sweet smelling when pure, sparingly soluble in water and very soluble in propanone (acetone) under pressure (as in cylinders of acetylene). It burns with a very hot flame, hence the oxy-acetylene torch, and, although toxic, has been used in the past as a general anaesthetic (narcylene). The higher alkynes are gases, liquids, or solids, insoluble in water and neutral to indicators.

Reactions of alkynes

A *Addition reactions.* The triple bond of alkynes undergoes most of the electrophilic addition reactions of alkenes, with the direction of addition obeying Markownikoff's rule, e.g. the successive addition of two molecules of hydrogen chloride to propyne:

$$CH_3{-}C{\equiv}CH \xrightarrow{+HCl} CH_3{-}\underset{}{\overset{\displaystyle Cl}{\overset{|}{C}}}{=}CH_2 \xrightarrow{+HCl} CH_3{-}CCl_2{-}CH_3$$

One very important addition reaction is the hydration of the triple bond, catalysed by acid and mercury salts. Ethyne is hydrated to an aldehyde, ethanal, but all other alkynes form ketones.

$$H{-}C{\equiv}C{-}H \xrightarrow[+H^+/Hg^{2+}]{+H_2O} \underset{\text{Vinyl alcohol}}{[CH_2{=}CH{-}OH]} \xrightarrow{\text{rearranges}} \underset{\text{Ethanal}}{CH_3{-}CH{=}O}$$

$$R{-}C{\equiv}C{-}H \xrightarrow[+H^+/Hg^{2+}]{+H_2O} \left[\begin{matrix} R{-}C{=}CH_2 \\ | \\ OH \end{matrix}\right] \xrightarrow{\text{rearranges}} \begin{matrix} R{-}C{-}CH_3 \\ \| \\ O \end{matrix}$$

Alkynes are also susceptible to nucleophilic attack resulting in addition reactions, exemplified by the alkoxide catalysed addition of alcohols:

$$H{-}C{\equiv}C{-}H + C_2H_5OH \xrightarrow{+Na^+C_2H_5O^-} \underset{\text{Ethyl vinyl ether}}{C_2H_5O{-}CH{=}CH_2}$$

B *Reactions of terminal alkynes.* Compounds containing the ethynyl group ($-C{\equiv}CH$) undergo characteristic reactions caused by the acidity of hydrogen atoms in this environment. Complex, insoluble salts are produced by the reaction of terminal alkynes with mercury(II), copper(II), or silver salts, e.g.

$$R{-}C{\equiv}C{-}H + Cu^+ \longrightarrow R{-}C{\equiv}C{-}Cu$$

The precipitation of the dark-red cuprous salt is diagnostic of the ethynyl group.

Although the heavy metal salts are probably covalent, true ionic salts are formed by the action of very strong bases (e.g. $NaNH_2$, KNH_2) on terminal alkynes. The anions, so obtained, can be alkylated by alkyl halides, affording a general route for synthesis of alkynes. Liquid ammonia is frequently employed as a solvent for these reactions.

$$\text{H—C}\equiv\text{C—H} \xrightarrow[\text{in liquid NH}_3]{\text{Na}^+\text{NH}_2^-} \overset{+}{\text{Na}}\,\overset{-}{\text{C}}\equiv\text{CH} \xrightarrow{\text{R—Br}} \text{R—C}\equiv\text{CH} + \text{NaBr}$$

$$\text{R—C}\equiv\text{CH} \xrightarrow{\text{Na}^+\text{NH}_2^-/\text{NH}_3} \text{R—C}\equiv\text{C}^-\text{Na}^+ \xrightarrow{\text{R}'\text{—I}} \text{R—C}\equiv\text{C—R}' + \text{NaI}$$

This acidity of terminal alkynes is attributable to the orbitals involved in formation of the $\equiv$C—H bond, which utilises an *sp* hybrid orbital of the carbon atom (p. 9). C—H bonds in alkenes and alkanes formed from sp^2 and sp^3 orbitals are at least 10^{20} times less readily ionised.

Occurrence of alkynes. Simple alkynes are scarcely known in nature (but see p. 33). However, the seed oils and sap of some higher plants contain an extensive series of highly unsaturated compounds, and complex poly-ynes are also produced by certain fungi, e.g.

$$\text{CH}_3\text{—C}\equiv\text{C—C}\equiv\text{C—C}\equiv\text{C—CH}=\text{CH—CO}_2\text{CH}_3$$

'Matricaria ester' (from *Compositae* species)

$$\text{HC}\equiv\text{C—C}\equiv\text{C—CH}=\text{C}=\text{CH—CH}=\text{CH—CH}=\text{CH—CO}_2\text{H}$$

Mycomycin (a fungal metabolite)

Benzene

Benzene, C_6H_6, is the simplest of a large number of highly unsaturated, cyclic or polycyclic hydrocarbons, whose chemical behaviour is distinct from that of normal alkenes, and is known as 'aromatic'. The structure of benzene was a long-standing puzzle, which is now fully resolved, and physical studies (e.g. X-ray investigation of benzene crystals) have shown that benzene consists of a regular plane hexagon of carbon atoms, each one attached to a hydrogen atom. All C—C bond lengths are equal in this structure, whose symmetry is consistent with many studies showing that in benzene all carbon atoms are chemically equivalent, (i.e. no positional isomers are possible in monosubstituted derivatives of benzene).

The early proposals for the structure of benzene are not of relevance here, but the current molecular orbital picture of benzene will be described briefly.

The skeleton of benzene can be built up of six carbon and six hydrogen atoms arranged in a regular hexagon and joined by σ bonds, formed by the overlap of sp^2 orbitals of carbon and *s* orbitals of hydrogen. The geometry of sp^2 hybridisation (p. 9) requires all $\angle$ C—C—C $= 120°$, which coincides with the requirement for hexagonal symmetry. This leaves one electron in a *p* orbital on each carbon atom, as indicated by the diagram (I). These *p* orbitals

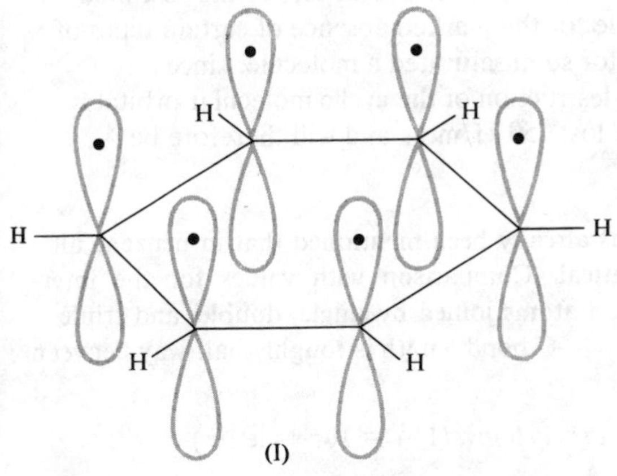

(I)

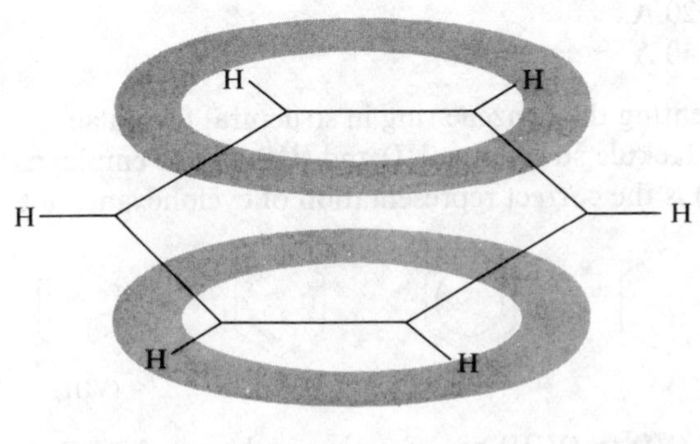

(II)

can interact to form extended π orbitals, which stretch around the ring (cf. conjugation, p. 11), and the six electrons will require three such orbitals of which the simplest consists of two circular electron clouds, one above and one below the ring of carbon atoms (II). The other two orbitals have somewhat different symmetry, but in all of them electrons can move freely around the six carbon atoms. The characteristic chemical properties of aromatic compounds are attributable to the presence of these cyclic orbitals.

Resonance energy. It is found that when benzene is oxidised to carbon dioxide and water, the energy released is very much less than that calculated for the burning of the hypothetical cyclohexa-1,3,5-triene (i.e. 'Kekulé benzene'). The energy deficit of 150 kJ/mole can be regarded as the energy released when three isolated double bonds interact to form the cyclic delocalised molecular orbitals described above. This loss of the so-called **resonance energy** is responsible for the marked absence of certain types of chemical behaviour expected for so unsaturated a molecule, since any reaction, which results in the destruction of the cyclic molecular orbitals, requires the restoration of the lost 150 kJ/mole and will therefore be energetically unfavourable.

Bond lengths in benzene. It has already been mentioned that in benzene all C—C bond lengths are identical. Comparison with values for the internuclear distance between carbon atoms joined by single, double, and triple bonds shows that the aromatic C—C bond length is roughly half way between those of ethane and ethene.

C—C *bond lengths in some hydrocarbons.* (1 Å = 10^{-10} metres)

$H_3C{-}CH_3$	1·54 Å
$H_2C{=}CH_2$	1·33 Å
$HC{\equiv}CH$	1·20 Å
C_6H_6 (Benzene)	1·40 Å

Several methods of representing the benzene ring in structural formulae are in use, of which only the Kekulé structures (III) and (IV) will be employed in this book. The symbol (V) is the correct representation of cyclohexane, not

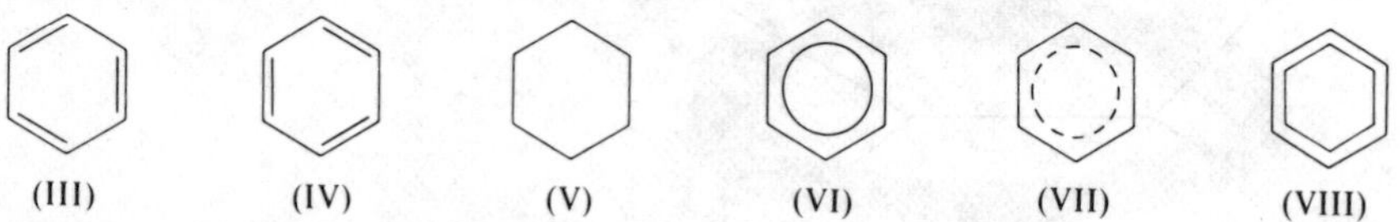
(III) (IV) (V) (VI) (VII) (VIII)

benzene. The three diagrams (VI) to (VIII) are sometimes advocated on the grounds of convenience and freedom from the implication of alkene-like double bonds suggested by (III). However, the extension of this representation to polycyclic aromatic hydrocarbons has serious disadvantages, since it is not easy to see how many electrons there are in the π orbitals of anthracene (IX) when represented by (X), and it is possible to write meaningless structures such as (XI), without realising it.

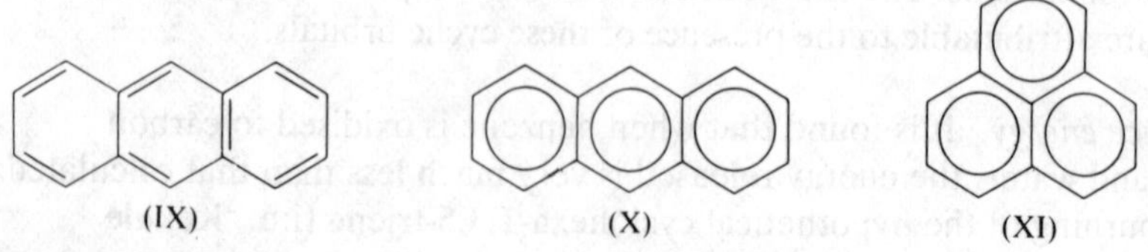
(IX) (X) (XI)

Preparation and occurrence of benzene. Benzene is obtained in large amounts directly from coal tar, and indirectly from petroleum via the cracking of larger and more saturated hydrocarbons. Benzene is highly toxic, prolonged exposure to low concentrations of the vapour leading to aplastic anaemia and subsequently leukaemia. Alkylbenzenes are much less poisonous since these are oxidised in the body to the relatively harmless carboxylic acids (e.g. toluene, $C_6H_5CH_3$, $\rightarrow$ benzoic acid, $C_6H_5CO_2H$). Benzene, however, is oxidised to the highly poisonous phenol, C_6H_5OH, which is responsible for damage to the bone-marrow.

Reactions of benzene

A *Alkene-like reactions.* Despite wide dissimilarities between the chemistry of benzene and alkenes, in a few reactions benzene does behave like an alkene.

1. Catalytic reduction gives cyclohexane (p. 28).
2. Addition of chlorine or bromine in sunlight leads to hexahalo-derivatives. This reaction is only superficially similar to that of alkenes. The need

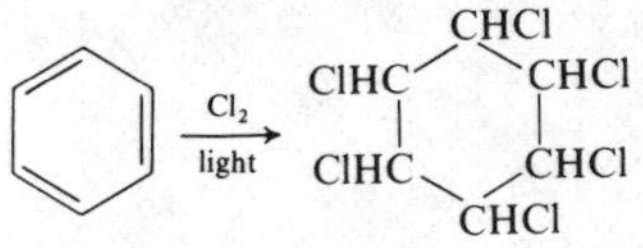

$$\text{C}_6\text{H}_6 \xrightarrow[\text{light}]{Cl_2} \text{(CHCl)}_6$$

for light indicates that the initial addition is probably a radical process (p. 26).

3. Ozone reacts with benzene to give a triozonide.

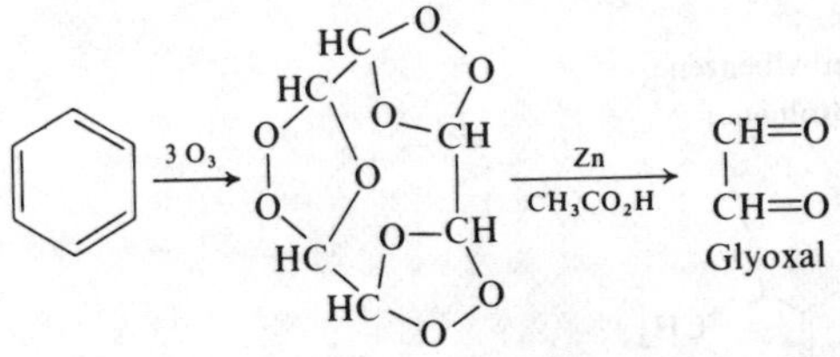

$$\text{C}_6\text{H}_6 \xrightarrow{3\,O_3} \text{triozonide} \xrightarrow[CH_3CO_2H]{Zn} \begin{matrix} CH{=}O \\ | \\ CH{=}O \end{matrix}$$

Glyoxal

4. Oxidation of benzene is extremely difficult in the absence of activating substituents. Benzene is only very slowly attacked by hot alkaline potassium permanganate (alkenes react instantly in the cold). Catalytic vapour-phase oxidation will, however, break open the ring.

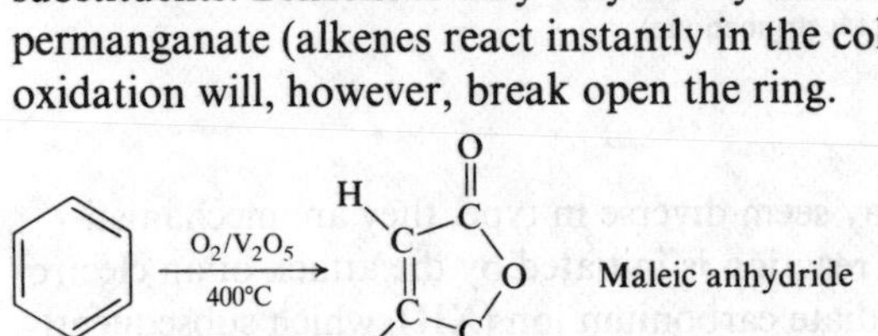

$$\text{C}_6\text{H}_6 \xrightarrow[400°C]{O_2/V_2O_5} \text{Maleic anhydride}$$

Maleic anhydride

B *Electrophilic substitution*. The reactions which distinguish benzene and other aromatic hydrocarbons from alkenes are substitution reactions, of which the most important are:

1. Nitration by a mixture of concentrated nitric and sulphuric acids.

2. Halogenation by chlorine or bromine in the presence of a catalyst e.g. $AlCl_3$, $FeCl_3$, I_2. Iodine does not react in this way, but iodination can be achieved by use of iodine monochloride, ICl.

3. The Friedel–Crafts reaction, in which alkyl or acyl halides, in the presence of aluminium halides, substitute the benzene ring with alkyl and acyl groups respectively.

4. Sulphonation by concentrated sulphuric acid or oleum.

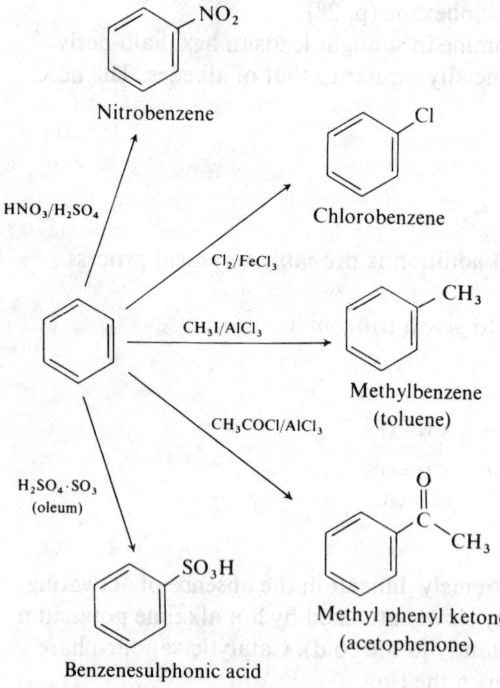

Nitrobenzene

HNO_3/H_2SO_4

Chlorobenzene

$Cl_2/FeCl_3$

$CH_3I/AlCl_3$

Methylbenzene (toluene)

$CH_3COCl/AlCl_3$

$H_2SO_4 \cdot SO_3$ (oleum)

Methyl phenyl ketone (acetophenone)

Benzenesulphonic acid

Although these reactions may seem diverse in type, they are mechanistically very similar. In all cases, reaction is initiated by the attack of an electrophile (X^+) to form an intermediate carbonium ion (XII), which subsequently loses a proton to form the substituted aromatic system (XIII).

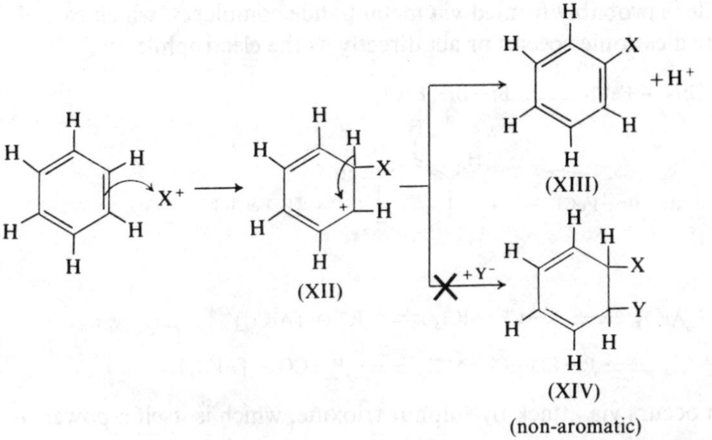

(XII)

(XIII)

(XIV)
(non-aromatic)

The difference between the reactions of alkenes and aromatic hydrocarbons lies in the second step. Whereas the carbonium ion formed in the electrophilic attack on an alkene promptly reacts with an anion (Y^-), if the corresponding reaction occurred with the carbonium ion (XII), the product (XIV) would no longer be aromatic. To convert an aromatic compound to a non-aromatic one requires the return of the 150 kJ/mole of resonance energy lost in forming the aromatic π orbitals (p. 37). Addition reactions of aromatic compounds akin to those of alkenes are, therefore, energetically very unfavourable, resulting in the easier alternative of substitution.

A great deal is known about the reactive species involved in the electrophilic substitutions listed above. Nitration has been studied extensively and is known to occur via attack by the nitronium ion $(NO_2)^+$. This is produced by the nitrating mixture as follows:

$$H{-}O{-}NO_2 + H_2SO_4 \rightleftharpoons H{-}\overset{H}{\underset{}{\overset{|+}{O}}}{-}NO_2 + HSO_4^-$$

$$H{-}\overset{H}{\overset{|+}{O}}{-}NO_2 \rightleftharpoons H_2O + NO_2^+$$

The role of sulphuric acid is not to remove the water formed during nitration, as nitration is rarely reversible. Any strong acid (e.g. $HClO_4$) which will protonate nitric acid, and thereby generate nitronium ions, will serve in its place. Since salts of the nitronium ion are known as crystalline solids (e.g. $NO_2^+ClO_4^-$; $NO_2^+BF_4^-$; $NO_2^+NO_3^-$ (solid N_2O_5)), there is no reason to doubt its existence.

The electrophilic species involved in halogenation and the Friedel–Crafts reaction are less well established than the nitronium ion. In these reactions the electrophile is probably formed via metal halide complexes, which can either generate a cationic species or act directly as the electrophile.

$$Br_2 + FeCl_3 \rightleftharpoons Br{-}\overset{+}{Br}{-}\overset{-}{Fe}Cl_3$$

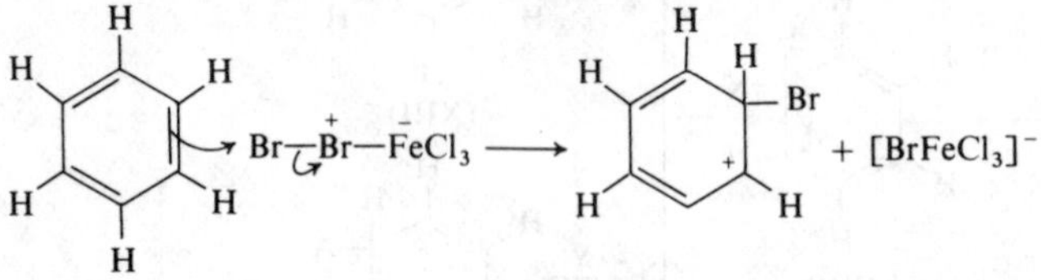

$$C_6H_6 + Br{-}\overset{+}{Br}{-}FeCl_3 \longrightarrow [C_6H_6Br]^+ + [BrFeCl_3]^-$$

$$R{-}Cl + AlCl_3 \rightleftharpoons R{-}\overset{+}{Cl}{-}\overset{-}{Al}Cl_3 \rightleftharpoons R^+ + [AlCl_4]^-$$

$$R{-}CO{-}Cl + AlCl_3 \rightleftharpoons R{-}CO{-}\overset{+}{Cl}{-}\overset{-}{Al}Cl_3 \rightleftharpoons R{-}\overset{+}{C}O + [AlCl_4]^-$$

Sulphonation occurs via attack by sulphur trioxide, which is itself a powerful electrophile.

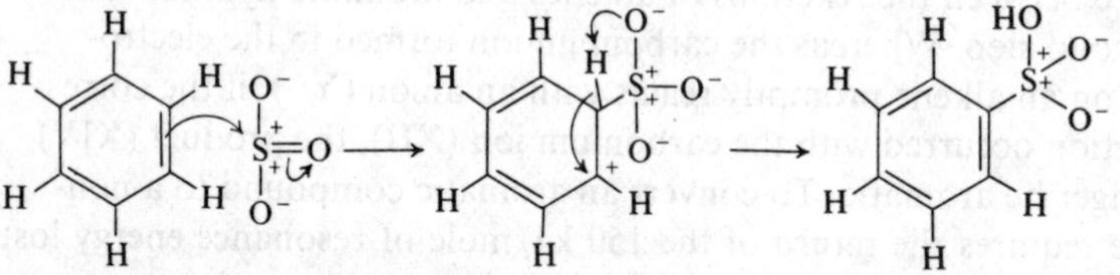

$$C_6H_6 + SO_3 \longrightarrow [C_6H_6SO_3] \longrightarrow C_6H_5\overset{+}{S}(O^-)_2OH$$

The orientation of substitution. It is convenient to consider, at this stage, the effect of substituents on the position of electrophilic attack. The well-established chemical equivalence of the six carbon atoms of benzene means that there is only one isomer of any monosubstituted benzene, but further substitution can lead to three isomeric disubstituted products:

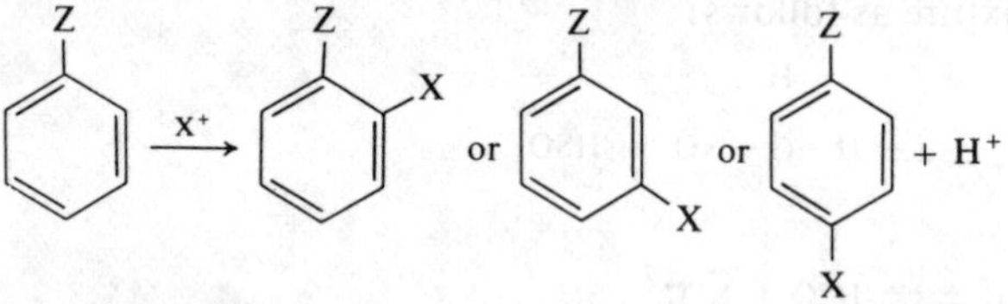

$$C_6H_5Z \xrightarrow{X^+} \text{ortho-} \text{ or } \text{meta-} \text{ or } \text{para-}C_6H_4ZX + H^+$$

(X^+ = any electrophile) *ortho*- *meta*- *para*-isomer

It is found that the position of attack of the electrophile X^+ is affected by the nature of Z in a way which can be explained by consideration of the structure of the intermediate carbonium ion. When an electrophile becomes attached to the aromatic system, the carbonium ion formed is a resonance hybrid of three canonical structures (p. 13) and is stabilised by the distribution of positive charge over several atoms. This is sometimes indicated by diagrams such as (XV). When considering the introduction of a second substituent, there will be three canonical structures for each of the intermediate carbonium ions involved in *ortho*-, *meta*-, and *para*-substitution, and these are shown in diagrams (XVI) to (XXIV). It should be noted that in the case of *ortho*- and *para*-substitution, canonical structures can be drawn for the intermediate, in

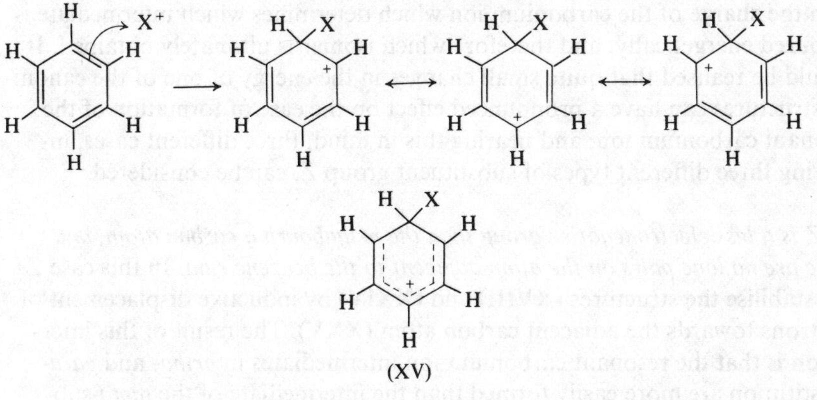

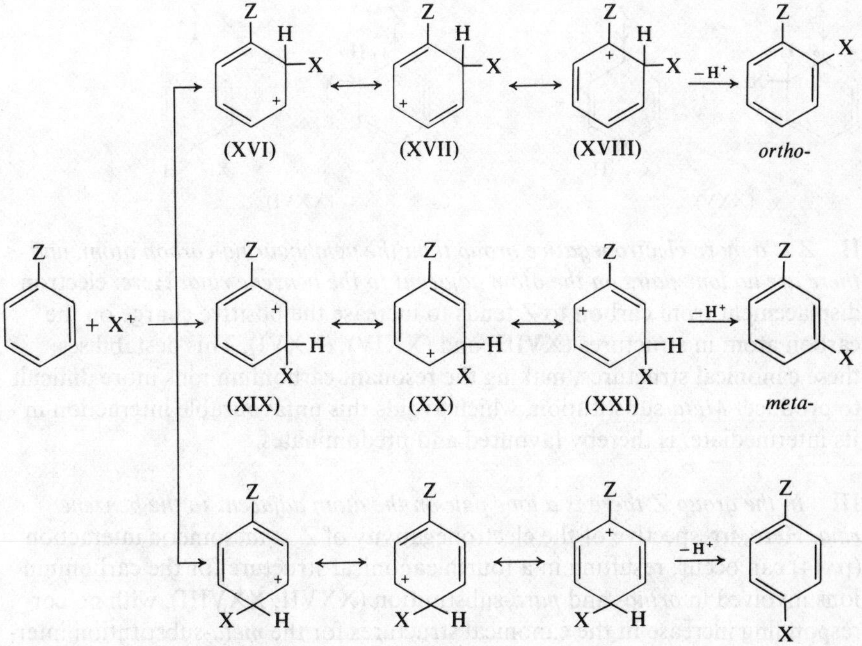

(XV)

(XVI) (XVII) (XVIII) *ortho*-

(XIX) (XX) (XXI) *meta*-

(XXII) (XXIII) (XXIV) *para*-

which the positive charge resides on the carbon atom bearing the substituent group Z (XVIII, XXIV), whilst no such structure is possible for the intermediate involved in *meta*-substitution. It is the interaction of the group Z with the charge of the carbonium ion which determines which intermediate is favoured energetically, and therefore which isomer is ultimately obtained. It should be realised that quite small changes in the energy of one of the canonical structures can have a pronounced effect on the ease of formation of the resonant carbonium ion, and bearing this in mind, three different cases, involving three different types of substituent group Z, can be considered.

I *Z is a less electronegative group than the neighbouring carbon atom, and there are no lone pairs on the atom adjacent to the benzene ring.* In this case Z can stabilise the structures (XVIII) and (XXIV) by inductive displacement of electrons towards the adjacent carbon atom (XXV). The result of this interaction is that the resonant carbonium ion intermediates in *ortho*- and *para*-substitution are more easily formed than the intermediate of the *meta*-substitution path, in which no such interaction is possible. This type of substituent group, therefore, leads to predominant *ortho*- and *para*-substitution by electrophiles.

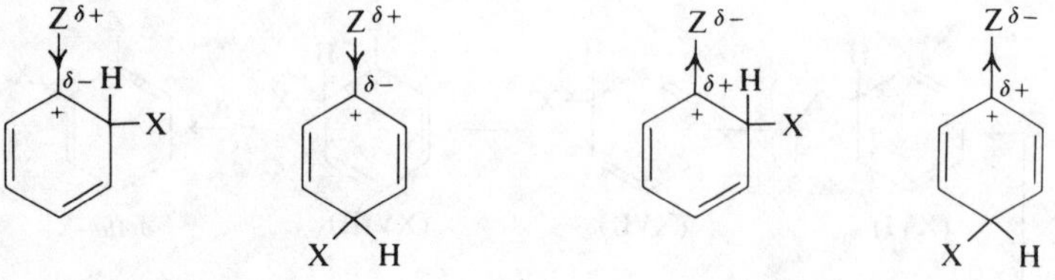

(XXV)

(XXVI)

II *Z is a more electronegative group than the neighbouring carbon atom, and there are no lone pairs on the atom adjacent to the benzene ring.* Here, electron displacement from carbon to Z tends to increase the positive charge on the carbon atom in structures (XVIII) and (XXIV), (XXVI). This destabilises these canonical structures, making the resonant carbonium ions more difficult to produce. *Meta*-substitution, which avoids this unfavourable interaction in its intermediate, is thereby favoured and predominates.

III *In the group Z there is a lone pair on the atom adjacent to the benzene ring.* Here, irrespective of the electronegativity of Z, a mesomeric interaction (p. 14) can occur, resulting in a fourth canonical structure for the carbonium ions involved in *ortho*- and *para*-substitution (XXVII, XXVIII), with no corresponding increase in the canonical structures for the *meta*-substitution intermediate. An increase in the number of canonical structures means a marked decrease in the energy of formation (i.e. increase in the ease of formation) of the intermediate carbonium ion, resulting in overwhelming predominance of *ortho-* and *para-*substitution by electrophiles. The double bond of an alkenyl group (e.g. $—CH{=}CH_2$) can serve in place of a lone pair in a similar way.

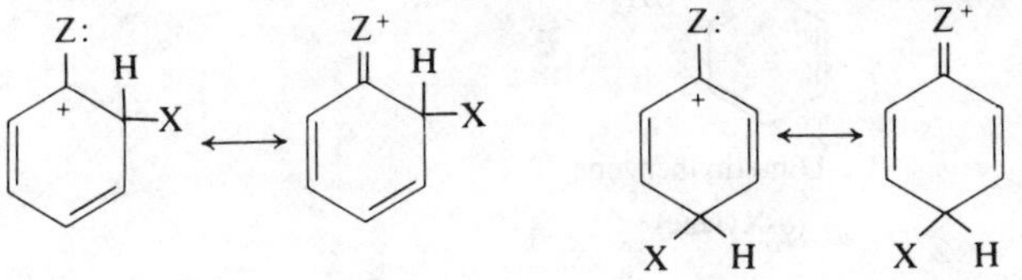

(XVIII) ⟷ (XXVII)   (XXIV) ⟷ (XXVIII)

The table below lists most of the common substituent groups, classified into the three types described above.

SUBSTITUENT GROUP Z	TYPE	POSITION OF ELECTROPHILIC ATTACK ON $C_6H_5—Z$
—Alkyl, e.g. $—CH_3$	I	*o, p*
$—CH_2Cl$	I	*o, p*
$—CHCl_2$*	I and II	*o, m, p*
$—CCl_3$	II	*m*
$—CH{=}CH_2$	III	*o, p*
—CHO	II	*m*
—CO.R	II	*m*
—COOH	II	*m*
—COOR	II	*m*
$—CONH_2$	II	*m*
$—NH_2$	III	*o, p*
—NHR	III	*o, p*
$—NR_2$	III	*o, p*
$—N^+R_3$	II	*m*
$—NO_2$	II	*m*
—OH	III	*o, p*
—OR	III	*o, p*
—SH	III	*o, p*
—SR	III	*o, p*
$—SO_3H$	II	*m*

* The electronegativity of $—CHCl_2$ is such that it is intermediate between types I and II, leading to approximately equal ease of attack at all positions.

Alkylbenzenes

Alkylbenzenes, which may be prepared by the Friedel–Crafts reaction (p. 40), are liquids or solids, obtained industrially either from coal-tar fractions or by a variety of processes from petroleum, whose physical properties resemble those of benzene. Some of the simpler compounds are known by trivial names: e.g.

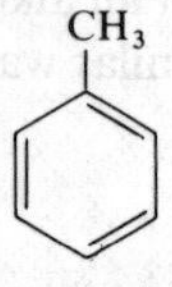

Methylbenzene (Toluene)

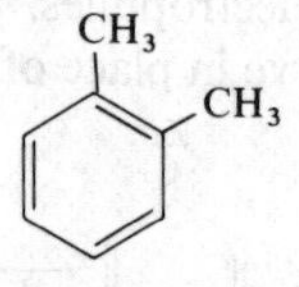

1,2-Dimethylbenzene (*o*-Xylene)

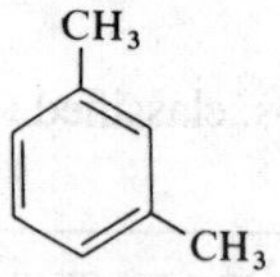

1,3-Dimethylbenzene (*m*-Xylene)

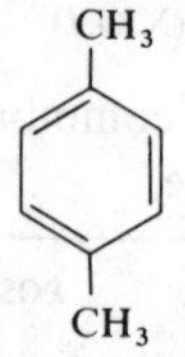

1,4-Dimethylbenzene (*p*-Xylene)

Reactions of alkylbenzenes

1. Oxidation of alkylbenzenes usually results in attack on the side chain. Powerful oxidising agents will convert them into the corresponding aromatic carboxylic acids.

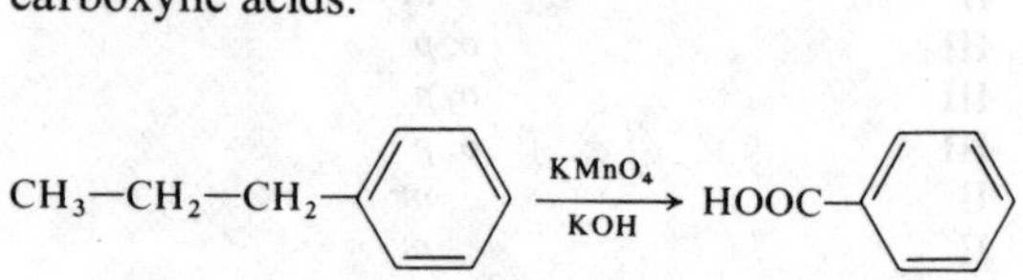

$$CH_3-CH_2-CH_2-C_6H_5 \xrightarrow[KOH]{KMnO_4} HOOC-C_6H_5$$

2. Catalytic reduction affords the corresponding cyclohexane derivative.

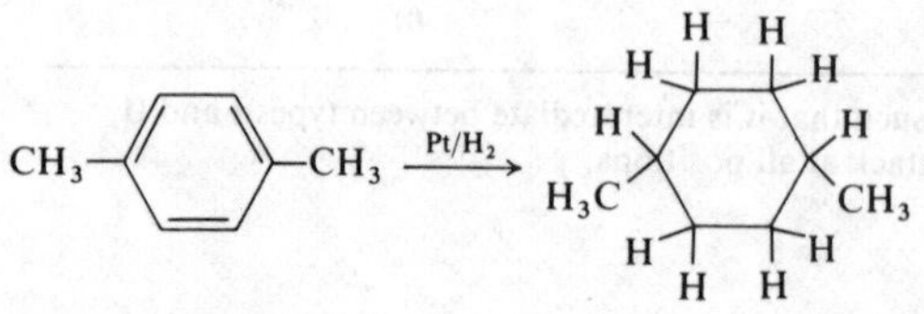

$$CH_3-C_6H_4-CH_3 \xrightarrow{Pt/H_2} H_3C-C_6H_{10}-CH_3$$

3. Halogenation can occur either in the side chain or on the benzene ring. In the presence of light a radical halogenation of the side chain occurs

(p. 26), whilst with catalysts electrophilic substitution of the ring predominates.

CH_3 — Cl_2 + light → CH_2Cl Benzyl chloride → $CHCl_2$ Benzal chloride → CCl_3 Benzotrichloride

$Cl_2 + FeCl_3$ → CH_3, Cl *o*-Chlorotoluene + CH_3, Cl *p*-Chlorotoluene

4. Ring substitution by electrophiles (p. 40) gives predominantly *ortho*- and *para*-substituted products.

CH_3 — NO_2^+ → CH_3, NO_2 + CH_3, NO_2 ultimately → CH_3, O_2N, NO_2, NO_2

SO_3 → CH_3, SO_3H + CH_3, SO_3H

Problems

1. Draw the constitutional formulae of all the isomeric compounds of formula C_7H_{16}, giving the systematic name in each case.
2. Give the constitutional formulae for all the monochloro derivatives of 2-methylbutane. By what mechanism are these formed from chlorine and 2-methylbutane in sunlight?
3. Draw the constitutional formulae of all the alkenes C_7H_{14}, which would give 3-methylhexane on catalytic hydrogenation. Which of these would react with ozone to give methanal (formaldehyde), HCHO?

4. How could you show, by chemical means, that two alkenes were related as *cis-trans* isomers?

5. What product would you expect to be formed from but-1-ene and iodine monochloride (ICl)?

6. Rationalise the formation of $HOCH_2CH_2Cl$ by the reaction of ethene with chlorine and water by a mechanism which does *not* involve the initial formation of hypochlorous acid (HOCl).

7. It is found that, if bromine is slowly added to a suspension of hex-1-ene in aqueous sodium chloride solution, the following products can be isolated: $CH_3(CH_2)_3CHBrCH_2Br$, $CH_3(CH_2)_3CHClCH_2Br$, and $CH_3(CH_2)_3CHOHCH_2Br$.

   Rationalise this in mechanistic terms. Why are $CH_3(CH_2)_3CHOHCH_2Cl$ and $CH_3(CH_2)_3CHBrCH_2Cl$ not formed?

8. Outline a reaction scheme for the synthesis of butan-2-one starting from ethyne (acetylene).

9. How many mononitro derivatives of 1-hydroxy-3-methylbenzene are there? Which of these is expected to be formed in lowest yield during nitration of 1-hydroxy-3-methylbenzene?

10. Propene and benzene react together in the presence of phosphoric acid to give 2-phenylpropane. What is the mechanism of this reaction and why is 1-phenylpropane not formed?

4

Simple organic halogen compounds

Alkyl halides

The alkyl halides are derived from alkanes by replacement of one of the hydrogen atoms by halogen. These compounds can be named in two ways, either by a combination of the name of the alkyl radical with the name of the halogen anion (even though the compounds are undoubtedly covalent), e.g. ethyl bromide, or by the more systematic method, indicating halogen substitution of the parent alkane, e.g. 2-iodobutane. The latter method is the only feasible one for halogen derivatives of higher alkanes, where many isomeric compounds must be distinguished.

$CH_3{-}CH_2{-}CH_2{-}CH_2{-}CH_2Cl$
1-Chloropentane
(1-Pentyl chloride)

$CH_3{-}CH_2{-}CH_2{-}CHCl{-}CH_3$
2-Chloropentane
(2-Pentyl chloride)

$$CH_3{-}CH_2{-}CH_2{-}CHBr{-}\underset{}{\overset{CH_3}{\overset{|}{C}}}H{-}CH_3$$ 3-Bromo-2-methylhexane

Preparation of alkyl halides

1. Direct halogenation of alkanes (p. 26) is not normally a suitable method for preparation of alkyl halides as complex mixtures of products are usually obtained.

2. The addition of hydrogen halides to olefins leads to alkyl halides (p. 31).

$$CH_3CH_2CH{=}CH_2 + HI \longrightarrow CH_3CH_2CHI{-}CH_3$$

3. Alcohols react with hydrogen halides to give alkyl halides:

$$R{-}OH + HBr \longrightarrow R{-}Br + H_2O$$

This reaction proceeds in two steps. Initial protonation of the hydroxyl group by the hydrogen halide forms an 'oxonium ion', which undergoes nucleophilic attack by the halide anion (see below) resulting in the elimination of water and formation of the alkyl halide.

$$R-\ddot{O}-H + H^+ \rightleftharpoons R-\overset{+}{\underset{|}{O}}(H)-H$$

$$:\ddot{Br}:^- \curvearrowright R-\overset{+}{O}(H)-H \longrightarrow :\ddot{Br}-R + :\ddot{O}(H)-H$$

e.g.

$$CH_3\,CH_2-\underset{OH}{CH}-CH_3 + HI \longrightarrow CH_3\,CH_2-\underset{\overset{+}{O}H_2}{CH}-CH_3 \xrightarrow{:\ddot{I}:^-} CH_3\,CH_2-\underset{I}{CH}-CH_3 + H_2O$$

Hydrogen chloride normally reacts very slowly, but hydrogen bromide and iodide will readily convert alcohols to the corresponding halides. Hydrogen fluoride does not react with alcohols in this way.

4. Alcohols react with a number of non-metal halides (principally of phosphorus) to give alkyl halides. Side reactions produce minor amounts of the esters of inorganic acids.

$$R-OH + PCl_3 \longrightarrow R-Cl + H_3PO_3$$
$$R-OH + PCl_5 \longrightarrow R-Cl + HCl + POCl_3$$
$$R-OH + POCl_3 \longrightarrow R-Cl + H_3PO_4$$
$$R-OH + SOCl_2 \longrightarrow R-Cl + SO_2 + HCl$$

Similarly with PBr_3 and PI_3

(but $R-OH + PCl_3 \longrightarrow (RO)_3P + HCl$
$R-OH + POCl_3 \longrightarrow (RO)_3PO + HCl$ may also occur simultaneously)

Properties of alkyl halides. The lower alkyl halides are neutral gases and liquids. They are insoluble in water, but readily soluble in organic solvents. Ethyl chloride, b.p. 12·5°C is used as a local anaesthetic, since the volatile liquid, when sprayed onto the skin, evaporates and anaesthetises by cooling the tissue.

Reactions of alkyl halides

A *Nucleophilic substitution reactions.* The most important reactions of alkyl halides are those in which the halogen atom is replaced by another atom or group. Many such reactions are known which fit the pattern of either of the following processes:

$$^-Y: + R-X \longrightarrow Y-R + :X^-$$

$$Z: + R-X \longrightarrow \overset{+}{Z}-R + :X^-$$

where R = alkyl group, X = halogen, and Y^- and Z are nucleophiles (p. 19). The reactions listed below are all of this type, and it should be noted that the cation, which accompanies Y^-, has no appreciable effect on the course of reaction, e.g. lithium, sodium, potassium, magnesium, calcium, or barium hydroxides are all equally effective in the conversion of an alkyl halide into the corresponding alcohol.

$$
R{-}X + \left\{\begin{array}{l}
\bar{O}H \\
\bar{O}R' \\
\bar{S}H \\
\bar{S}R' \\
\bar{S}^{=} \\
CN^- \\
NO_2^- \\
Cl^- \\
Br^- \\
I^- \\
H_2O \\
NH_3 \\
NH_2R' \\
NHR'_2 \\
NR'_3
\end{array}\right. \rightarrow X^- + \left\{\begin{array}{l}
R{-}OH \\
R{-}OR' \\
R{-}SH \\
R{-}SR' \\
R{-}S{-}R \text{ (in two steps)} \\
R{-}C{\equiv}N + R{-}\overset{+}{N}{\equiv}\overset{-}{C} \\
R{-}NO_2 + R{-}ONO \\
R{-}Cl \\
R{-}Br \\
R{-}I \\
R{-}\overset{+}{O}H_2 \rightleftharpoons ROH + H^+ \\
R{-}\overset{+}{N}H_3 \rightleftharpoons RNH_2 + H^+ \\
R{-}\overset{+}{N}H_2R' \rightleftharpoons RNHR' + H^+ \\
R{-}\overset{+}{N}HR'_2 \rightleftharpoons RNR'_2 + H^+ \\
R{-}\overset{+}{N}R'_3
\end{array}\right.
$$

R = alkyl group; X = halogen.

All the reactions listed above occur by one of two mechanisms. The greater electronegativity of the halogen atom (X) polarises the C—X bond in the alkyl halide, leaving a partial positive charge on the carbon atom. The nucleophile, which must possess a lone pair of electrons, can donate electrons to the positively charged carbon atom. However, as the carbon atom is limited to a maximum covalency of four, any donation of electrons by the nucleophile must result in a corresponding regression of electrons in the C—X

$$
{}^-Y\!:\curvearrowright \overset{\delta+}{C} \overset{\delta-}{\frown X} \longrightarrow Y{-}C + :X^-
$$

bond towards the halogen atom. Thus there is a simultaneous attack of the nucleophile on the carbon atom to form a new covalent bond, and loss of the halogen atom as an anion. Since the interaction of the nucleophile (Y^-) and the carbon atom is most favourable when Y^-, C, and X are collinear, the nucleophile must approach the alkyl halide from the side opposite to that from which the anion X^- ultimately departs. This mechanism is known as 'bimolecular nucleophilic substitution', often represented by S_N2, and an important aspect of this mechanism, which is described in detail later (p. 203), is that in the course of the reaction the pyramidal array of the other three groups attached to the carbon atom is turned inside out as shown in the diagram above.

An alternative mechanism operates where the carbon atom attached to the halogen also bears three alkyl groups, e.g. $(CH_3)_3C{-}Cl$. Here, the bulk of the alkyl groups prevents the close approach of a nucleophile to the electron-depleted carbon atom, and in such cases the reaction proceeds by two steps.

The alkyl halide ionises slowly to give a carbonium ion and a halide ion, and a subsequent very rapid reaction of the carbonium ion with the nucleophile gives the products of the reaction. It should be noted that the highly

$$R_3\overset{\delta+}{C}{-}\overset{\delta-}{X} \longrightarrow R_3\overset{+}{C} + :X^-$$

$$R_3\overset{+}{C} \leftarrow :Y^- \longrightarrow R_3C{-}Y$$

substituted carbonium ions involved in this mechanism are just those ones which are most easily obtained (p. 32). This mechanism is known as 'unimolecular nucleophilic substitution', S_N1 for short. The three-dimensional implications of this mechanism, which are different from those of the S_N2, are described later (p. 203).

The 'leaving group' in nucleophilic substitution of this type is not necessarily a halide anion. Any electronegative group, which on heterolytic fission (p. 18) of the bond to carbon can form a stable molecule or ion, will serve in place of a halogen atom. An example is quoted above in the preparation of alkyl halides where the oxonium ion ($R{-}\overset{+}{O}H_2$) undergoes typical nucleophilic attack by a halide ion, resulting in the loss of water and formation of the alkyl halide.

In general, the good leaving groups are those which depart as the conjugate bases* of strong acids. Thus $-Cl$ and $-Br$, which are good leaving groups, depart as the anions Cl^- and Br^- which are the conjugate bases of the strong acids HCl and HBr, whilst $-OH$, $-OCH_3$, and $-NH_2$ are very

* Where two species are related by loss or gain of a proton, e.g. $X + H^+ \rightleftharpoons HX^+$, then X is known as the 'conjugate base' of HX^+, and HX^+ is known as the 'conjugate acid' of X.

poor leaving groups since OH^-, OCH_3^-, and NH_2^- are the conjugate bases of the very weak acids H_2O, CH_3OH, and NH_3. Protonation of a group can, however, transform it into a good leaving group, as in the case of $-\overset{+}{O}H_2$, since H_2O is the conjugate base of the strongly acidic oxonium ion H_3O^+.

Some nucleophiles, which possess several lone pairs of electrons, can attack in more than one way, leading to two or more products. Examples of this behaviour are found in the cases of cyanide and nitrite anions, whose structures are illustrated below. Nucleophilic attack by these species leads to

$:\bar{C}{\equiv}N:$ $\qquad$ $\ddot{O}{=}\ddot{N}{-}\ddot{O}:^-$

mixtures of products, as listed above.

B Other reactions of alkyl halides described elsewhere are reduction to the alkane by catalytic hydrogenation (p. 25) or lithium aluminium hydride (p. 25); the Wurtz reaction (p. 25); the Friedel–Crafts reaction (p. 40); and dehydrohalogenation to alkenes by strong bases (p. 29).

Haloalkenes

Halogen derivatives of alkenes can be of two types, those in which the halogen atom is attached to one of the doubly bonded carbon atoms, and those in which the halogen atom is attached to a saturated (sp^3 hybridised) carbon atom, some distance from the double bond. Vinyl chloride and 5-bromohex-1-ene are typical examples:

$CH_2{=}CH{-}Cl$ — Vinyl chloride

$CH_3{-}CHBr{-}CH_2{-}CH_2{-}CH{=}CH_2$ — 5-Bromohex-1-ene

The reactivities of these two types of compound differ significantly. Vinyl halides and allied compounds react with nucleophiles very much more slowly than the corresponding saturated alkyl halides. The structure of vinyl chloride is more correctly described as a resonance hybrid of the two canonical structures:

$$CH_2{=}CH{-}\ddot{\underset{..}{Cl}}: \longleftrightarrow \bar{C}H_2{-}CH{=}\ddot{\underset{..}{Cl}}^+$$

and nucleophilic substitution, which involves the extrusion of a chloride anion, is made difficult by the partial positive charge on chlorine. This does not affect the reactions of the double bond with electrophiles, and vinyl halides show the normal alkene reactions (p. 30).

The haloalkenes, in which substitution occurs on carbon atoms remote from the double bond, behave like normal alkyl halides. However, where the double bond is adjacent to the carbon atom bearing the halogen substituent, e.g. allyl bromide $CH_2{=}CH{-}CH_2{-}Br$, reactivity towards nucleophiles is very much greater than in alkyl halides. Allyl halides usually undergo nucleophilic substitution by an S_N1 mechanism involving a resonance stabilised carbonium ion.

$$CH_2{=}CH{-}CH_2{-}Br \longrightarrow CH_2{=}CH{-}\overset{+}{C}H_2 \longleftrightarrow \overset{+}{C}H_2{-}CH{=}CH_2 + Br^-$$

Distribution of the charge over several atoms reduces the potential (i.e. energy) of this cation. Heterolytic fission of an allyl-halogen bond is therefore easier than for an alkyl-halogen bond since less energy is required to form the carbonium ion. As a result allyl halides and allied compounds tend to be vesicants and lachrymators, since the carbonium ions produced react readily with weakly nucleophilic centres in the proteins of living tissue.

Some aliphatic polyhalogen compounds

Chloroform (trichloromethane), $CHCl_3$, is a colourless heavy liquid b.p. 61°C, immiscible with water, and having a sickly sweet smell. It has been used extensively as an anaesthetic, but has been superseded on account of its toxicity.

Chloroform is prepared by the action of hypochlorites on ethanal, methylketones, or the corresponding alcohols, which are initially oxidised to the carbonyl compounds.

$$CH_3CH_2{-}CHOH{-}CH_3 \xrightarrow{NaOCl} CH_3CH_2{-}CO{-}CH_3 \xrightarrow{NaOCl} CH_3CH_2{-}CO{-}CCl_3$$

$$CH_3CH_2{-}CO{-}CCl_3 \xrightarrow{OH^-/H_2O} CH_3CH_2CO_2^- + CHCl_3$$

Normally bleaching powder, CaCl(OCl), and propanone (acetone) ($CH_3{-}CO{-}CH_3$), ethanal ($CH_3{-}CHO$), or ethanol (CH_3CH_2OH), are used for commercial preparation of chloroform.

Reactions of chloroform

1. Hydrolysis of chloroform with aqueous alkali slowly gives methanoates (formates):

$$CHCl_3 + H_2O/OH^- \longrightarrow H{-}C{\begin{smallmatrix}\nearrow O\\ \searrow O^-\end{smallmatrix}} + Cl^-$$

but reaction with alkali alkoxides gives esters of the hypothetical orthoformic acid:

$$CHCl_3 + CH_3O^- \longrightarrow \underset{\text{Trimethyl orthoformate}}{HC(OCH_3)_3}$$

These orthoesters are stable in alkaline conditions, but react rapidly with water in the presence of acid, giving the normal esters of methanoic (formic) acid.

$$HC(OCH_3)_3 + H_2O \xrightarrow{H^+} HCO_2CH_3 + 2CH_3OH$$

2. Oxidation. Chloroform is not flammable, but combines slowly with

oxygen in the presence of light, or at high temperatures, to give phosgene:

$$CHCl_3 + O_2 \xrightarrow{\text{light}} COCl_2 + HCl$$

Commercial chloroform is stored in dark bottles, and contains traces of ethanol to inhibit this reaction.

Iodoform. CHI_3, is a pale yellow solid, m.p. 120°C, with a characteristic smell. It is insoluble in water, and has been used as an antiseptic. Its preparation is identical with that of chloroform, alkali hypoiodites being used in place of hypochlorites. The formation of iodoform has been utilised as a test for the presence of the groups $CH_3—CO—R$ or $CH_3—CHOH—R$ (where R = hydrogen, alkyl, or aryl groups), but this test is now superseded by more reliable physical methods.

Iodoform is slowly oxidised by air in the presence of light, giving iodine, carbon monoxide, and carbon dioxide.

Carbon tetrachloride. CCl_4, is a colourless heavy liquid, b.p. 76°C, prepared by the chlorination of methane (p. 26) or carbon disulphide.

$$CS_2 + Cl_2 \longrightarrow CCl_4 + S_2Cl_2$$

It is immiscible with water, but is a good solvent for many organic compounds, hence its use in dry cleaning. It has an unpleasant smell and is toxic, prolonged exposure to the vapour resulting in severe hepatic and renal damage. It is also readily absorbed through the skin.

Carbon tetrachloride is remarkably inert to most reagents. It is completely non-flammable, hence its use in the past in fire extinguishers, but does react with oxygen at high temperatures to form phosgene, $COCl_2$. It can be hydrolysed with difficulty, reacting with superheated water at 250° to give carbon dioxide and hydrochloric acid.

A number of other polychloro derivatives of simple alkanes and alkenes are widely used as solvents, e.g. CH_2Cl_2, CH_2ClCH_2Cl, $CHCl{=}CCl_2$, $CCl_2{=}CCl_2$. These polychloro-compounds tend to be less reactive chemically than simple alkyl halides but, like chloroform and carbon tetrachloride, all are toxic and give phosgene, $COCl_2$, on high temperature oxidation (e.g. inhalation of vapour through a cigarette end) and may react with explosive violence with alkali metals or powdered aluminium or magnesium. Trichloroethene, $CHCl{=}CCl_2$, is used as an inhalation anaesthetic. A range of fluorinated alkane derivatives is also widely employed as refrigerants and aerosol propellants ($CFCl_3$, CF_2Cl_2, $CClF_2CClF_2$) and in fire extinguishers ($CBrClF_2$). These compounds are usually chemically inert and of low toxicity. One such compound, 'Halothane', $CF_3CHBrCl$, is widely used as an inhalation anaesthetic.

Aromatic halogen compounds

1. Aryl halides. Aryl halides are aromatic compounds having a halogen substituent on the aromatic ring, e.g.

Chlorobenzene

4-Bromoethylbenzene

They may be prepared by halogenation of the parent hydrocarbon in the presence of a suitable catalyst (p. 40), or from the corresponding primary amine via the diazonium salt (p. 93). The reaction of the corresponding hydroxy-compounds (phenols) with phosphorus halides gives very little aryl halide, the phosphite or phosphate esters being the principal products.

$$C_6H_5OH + PCl_3 \xrightarrow{\text{mostly}} (C_6H_5O)_3P + HCl$$

$$C_6H_5OH + PCl_3 \xrightarrow{\text{very little}} C_6H_5Cl + H_3PO_3$$

Properties of aryl halides. The aryl halides are neutral liquids or solids of density greater than one. They are insoluble in water, but readily soluble in most organic solvents.

Reactions of aryl halides. The aryl halides, like vinyl halides, are characterised by their inertness towards nucleophiles. Nucleophilic substitution reactions are not normally possible except under very vigorous conditions. Thus chlorobenzene cannot be hydrolysed with aqueous alkali at 100°C and reacts only above 300°C under pressure. Likewise cyanides will not readily replace the halogen atom by a cyano-group, bromobenzene (but not chlorobenzene) reacting with cuprous cyanide only at 200°C.

$$C_6H_5Cl \xrightarrow[300°C]{H_2O/OH^-} C_6H_5OH$$

$$C_6H_5Br \xrightarrow[200°C]{CuCN} C_6H_5CN$$

N.B. Neither of these reactions proceeds by the mechanisms described above for nucleophilic substitution of alkyl halides.

This stability is probably attributable to resonance with canonical structures of the type:

:C̈l: ⟷ :C̈l:⁺ ⟷ :C̈l:⁺ ⟷ :C̈l:⁺

the resultant partial positive charge on the halogen atom inhibiting nucleophilic displacement (cf. vinyl halides, p. 53)

Reactivity towards nucleophiles is greatly increased by electronegative substituents on the benzene ring in positions *ortho-* and *para-* to the halogen atom, e.g.

$C_6H_5Cl + H_2N{-}NH_2 \longrightarrow$ no reaction.

$$\text{Cl-2,4-(NO}_2)_2C_6H_3 + H_2N{-}NH_2 \xrightarrow{20^\circ C} \text{NH-NH}_2\text{-2,4-(NO}_2)_2C_6H_3$$

Chloro-2,4-dinitrobenzene   2,4-Dinitrophenylhydrazine

Electrophilic substitution of aryl halides occurs in the *ortho-* and *para-* positions relative to the halogen substituent (p. 42).

2. Side-chain substituted compounds. Halogen derivatives of alkylbenzenes can have substituents on either the aromatic nucleus or the side chain. The former are typical aryl halides, whilst the latter behave like alkyl halides. Typical examples of side-chain substituted compounds are:

$C_6H_5CH_2{-}Cl$ $C_6H_5CH_2{-}CH_2{-}Br$

Benzyl chloride   1-Bromo-2-phenylethane

The side-chain halogen derivatives may be prepared by methods similar to those employed for alkyl halides (p. 49). They react readily with nucleophiles (e.g. OH^-, CN^-, NH_3) like alkyl halides, and also undergo electrophilic substitution in the aromatic nucleus (p. 40).

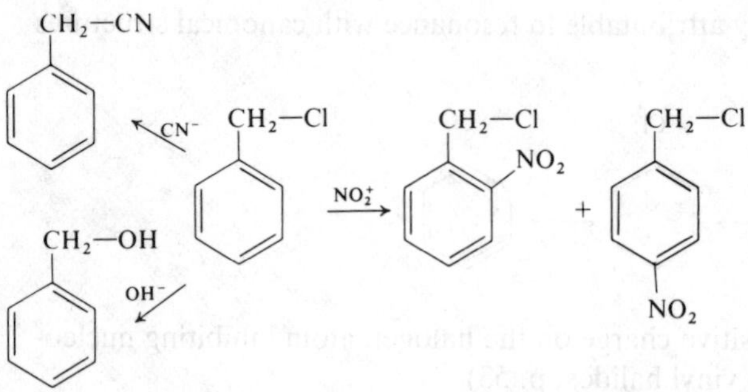

CH_2—CN

CN^- CH_2—Cl $\xrightarrow{NO_2^+}$ CH_2—Cl, NO_2 + CH_2—Cl, NO_2

CH_2—OH

OH^-

The compounds in which the halogen atom is attached to the carbon atom adjacent to the aromatic ring (e.g. benzyl bromide, C_6H_5—CH_2—Br; 1-chloro-1-phenylethane, C_6H_5—CHCl—CH_3) show a marked increase in activity compared with the compounds in which the site of halogen substitution is farther from the benzene ring (cf. allyl halides, p. 53). In consequence, benzyl halides and allied compounds tend to be lachrymators.

Problems

1. What would you expect to be the principal products of reaction between the following pairs of reagents? In each case the second reagent is considered to be present in large excess.

$BrCH_2CH_2CH_2I$ + KSH
NaOH + $ClCH_2CH$=CHCl

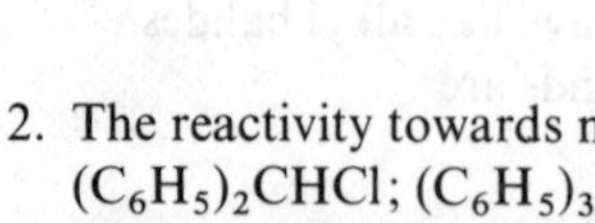

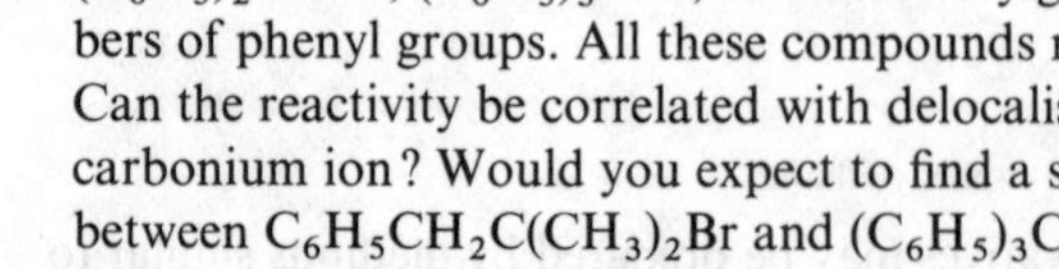

$(CH_3)_3N$ + ICH_2—C_6H_4—I

Cl—C_6H_4—$CH_2CH_2CH_2$—$C_6H_3(NO_2)$—Cl + NH_2NH_2

2. The reactivity towards nucleophiles of the series $C_6H_5CH_2Cl$; $(C_6H_5)_2CHCl$; $(C_6H_5)_3CCl$; increases very greatly with increasing numbers of phenyl groups. All these compounds react by the S_N1 mechanism. Can the reactivity be correlated with delocalisation of the charge on the carbonium ion? Would you expect to find a similar difference in reactivity between $C_6H_5CH_2C(CH_3)_2Br$ and $(C_6H_5)_3CC(CH_3)_2Br$?
3. What products might arise from reaction of methyl iodide with the following compounds? (see p. 21).
KNO$_2$; LiN$_3$; NaNCS.

5

Simple organic oxygen and sulphur compounds

Alcohols

The monohydroxy-derivatives of alkanes and cycloalkanes are known as alcohols, and the presence of a hydroxyl group in the molecule is indicated by the prefix 'hydroxy' or the suffix 'ol' in the systematic name:

$$CH_3{-}CH_2{-}CH_2{-}\underset{\displaystyle OH}{\underset{|}{CH}}{-}CH_3 \qquad CH_3{-}\overset{\displaystyle CH_3}{\overset{|}{CH}}{-}CH_2{-}\overset{\displaystyle OH}{\overset{|}{CH}}{-}CH_2{-}CH_3$$

2-Hydroxypentane
Pentan-2-ol

4-Hydroxy-2-methylhexane
2-Methylhexan-4-ol

In addition, some of the lower alcohols retain their old trivial names:

	SYSTEMATIC NAME	TRIVIAL NAME
CH_3OH	Methanol	Methyl alcohol
CH_3CH_2OH	Ethanol	Ethyl alcohol
$CH_3CH_2CH_2OH$	Propan-1-ol	*n*-Propyl alcohol
$CH_3CH(OH)CH_3$	Propan-2-ol	Isopropyl alcohol
$CH_3(CH_2)_2CH_2OH$	Butan-1-ol	*n*-Butyl alcohol
$CH_3CH_2CH(OH)CH_3$	Butan-2-ol	*sec*-Butyl alcohol
$(CH_3)_2CHCH_2OH$	2-Methylpropan-1-ol	Isobutyl alcohol
$(CH_3)_3COH$	2-Methylpropan-2-ol	*t*-Butyl alcohol

(*n*-, *sec*-, *t*- mean 'normal', 'secondary', and 'tertiary' respectively.)

Yet another system of nomenclature has been proposed, which names alcohols as derivatives of methanol, known in this system as 'carbinol'. The replacement of hydrogen atoms of the methyl group is indicated as follows:

$$\underset{\text{Methyl ethyl carbinol}}{C_2H_5-\overset{\overset{\displaystyle H}{|}}{\underset{\underset{\displaystyle OH}{|}}{C}}-CH_3} \qquad \underset{\text{Dimethyl phenyl carbinol}}{C_6H_5-\overset{\overset{\displaystyle CH_3}{|}}{\underset{\underset{\displaystyle OH}{|}}{C}}-CH_3}$$

Although this nomenclature persists in some cases, it is not suitable for naming complex compounds, and will not be employed in this book.

Because of some differences in the chemical behaviour, it is customary to divide alcohols into three classes—primary, secondary, and tertiary—distinguished by the number of hydrogen atoms attached to the carbon atom bearing the hydroxyl group. Primary alcohols contain the group $-CH_2OH$, secondary alcohols contain the group $>CHOH$, and tertiary alcohols, which have no hydrogen atoms on the carbon atom concerned, contain the group $-\!\!\!>COH$. Examples of these are found in the butyl alcohols, whose formulae are given above. *n*-Butyl alcohol and isobutyl alcohol are primary alcohols, whilst the names secondary and tertiary butyl alcohol are self-explanatory. Although the classes differ in some aspects of their chemistry, most of the reactions of alcohols are common to all three types.

Preparation of alcohols. It is not yet possible to introduce hydroxyl groups directly into alkanes, and all syntheses of alcohols start from compounds containing reactive functional groups. The following are amongst the commonest methods of preparation.

1. Hydration of olefins in the presence of acid (p. 31).

$$R-CH=CH-R + H_2SO_4/H_2O \longrightarrow R-CH_2-\underset{\underset{\displaystyle OH}{|}}{CH}-R$$

$$\text{e.g.} \qquad CH_2=CH_2 \xrightarrow{H^+/H_2O} CH_3-CH_2OH$$

2. Hydrolysis of alkyl halides by water or alkali (p. 51).

$$CH_3-CH_2-CH_2-Br \xrightarrow{H_2O/\bar{O}H} CH_3-CH_2-CH_2-OH + Br^-$$

3. Hydrolysis of ethers under strongly acidic conditions (p. 73).

$$CH_3CH_2-O-CH_3 \xrightarrow{H_2SO_4/H_2O} CH_3CH_2OH + CH_3OH$$

4. Hydrolysis of esters:

(a) acid catalysed hydrolysis;

$$CH_3{-}C(=O){-}O{-}C_2H_5 + H_2O \xrightarrow{H^+} CH_3{-}C(=O){-}OH + C_2H_5OH$$

Ethyl ethanoate / Ethyl acetate → Ethanoic acid / Acetic acid + Ethanol

(b) alkaline hydrolysis ('saponification');

$$CH_3{-}C(=O){-}O{-}C_2H_5 + H_2O/OH^- \longrightarrow CH_3{-}C(=O){-}O^- + C_2H_5OH$$

(for the mechanism of ester hydrolysis see p. 139).

5. Reduction of more highly oxidised compounds:

(a) $R{-}C(H){=}O \longrightarrow R{-}CH_2OH$

aldehyde → primary alcohol

(b) $R_2C{=}O \longrightarrow R_2CHOH$

ketone → secondary alcohol

(c) $R{-}C(=O){-}OH \longrightarrow R{-}CH_2OH$

carboxylic acid

(d) $R{-}C(=O){-}O{-}R' \longrightarrow R{-}CH_2OH + R'OH$

ester

Reductions (a) and (b) can be achieved by many reducing agents. Zinc and hydrochloric or acetic acid, sodium amalgam and water, and sodium and ethanol are typical 'dissolving metal'* reducing agents. Catalytic reduction with hydrogen and finely divided nickel or platinum catalysts will also convert aldehydes and ketones into the corresponding alcohols, and lithium

* Dissolving metal reductions employ a reactive metal and an acid. The reduction process consists of electron transfer from the metal to the substrate, followed or preceded by proton capture from the acid present.

aluminium hydride (p. 25) is frequently used. Of these reagents, only lithium aluminium hydride will reduce carboxylic acids and esters.

6. Primary amines react with nitrous acid to give alcohols (p. 86).

$$R{-}NH_2 + HONO \longrightarrow R{-}OH + N_2 + H_2O$$

Properties of alcohols. Alcohols are neutral liquids or solids, whose boiling points are very much higher than those of the parent alkanes (p. 156). The lower ones are miscible with water or very soluble, but solubility decreases with increasing size of the alkyl group. All the lower alcohols are more or less poisonous, ethanol being exceptional in its low toxicity.

Reactions of alcohols

1. Although alcohols are neutral to indicators and are virtually undissociated in aqueous solution, the hydrogen of the hydroxyl group can be replaced by the direct reaction of an alcohol with the electropositive metals of groups I and II of the periodic table:

$$C_2H_5OH + Na \longrightarrow \underset{\text{Sodium ethoxide}}{C_2H_5O^-Na^+} + H_2$$

$$CH_3OH + Mg \longrightarrow \underset{\text{Magnesium methoxide}}{(CH_3O^-)_2Mg^{++}} + H_2$$

Since alcohols are extremely weak acids, their salts—the alkoxides—are very strong bases. The basicity of alkoxides depends on the class of the parent alcohol, the tertiary alkoxides (e.g. $(CH_3)_3CO^-$) being the strongest bases, and methoxide the weakest, i.e. basicity $R_3CO^- > R_2CHO^- > RCH_2O^- > CH_3O^-$ (R = alkyl group). The alkoxides are also good nucleophiles, reacting with alkyl halides to form ethers (pp. 51, 73).

2. Alcohols react with carboxylic acids in the presence of mineral acid catalysts, to give esters (p. 137):

$$R{-}C(=O){-}OH + CH_3OH \underset{}{\overset{H_2SO_4}{\rightleftharpoons}} R{-}C(=O){-}O{-}CH_3 + H_2O$$

Some derivatives of carboxylic acids also react with alcohols to give esters, e.g. acyl chlorides and acid anhydrides:

$$\underset{\text{Butanoyl chloride}}{CH_3CH_2CH_2C(=O)Cl} + CH_3OH \longrightarrow \underset{\text{Methyl butanoate}}{CH_3CH_2CH_2C(=O)OCH_3} + HCl$$

$$CH_3-C(=O)-O-C(=O)-CH_3 + CH_3CH(OH)CH_3 \longrightarrow CH_3-CH(O-C(=O)-CH_3)-CH_3 + CH_3CO_2H$$

Ethanoic anhydride
Acetic anhydride

2-Propyl ethanoate
2-Propyl acetate

3. Oxidation of alcohols gives products which vary according to the class of alcohol. Primary alcohols can be oxidised to aldehydes, which can be further oxidised to carboxylic acids. Use of chromic acid ($Na_2Cr_2O_7 + H_2SO_4$; or CrO_3) as the oxidant permits the isolation of some

$$R-CH_2OH \xrightarrow{CrO_3} \underset{\text{aldehyde}}{R-C(=O)H} \xrightarrow{CrO_3} \underset{\text{carboxylic acid}}{R-C(=O)OH}$$

aldehyde, but further oxidation to carboxylic acid also occurs. Powerful oxidising agents such as permanganates or concentrated nitric acid give only carboxylic acids.

Secondary alcohols are oxidised readily to ketones, which are much more resistant to oxidation than aldehydes (p. 113). Propanone (acetone),

$$R-CHOH-R' \xrightarrow{CrO_3} \underset{\text{ketone}}{R-C(=O)-R'}$$

$CH_3-CO-CH_3$, can actually be used as a solvent for potassium permanganate or chromium trioxide in some oxidations.

Tertiary alcohols are resistant to oxidation under mild conditions, and a solution of chromium trioxide in tertiary butyl alcohol is employed as an oxidising reagent. However, under vigorous conditions, oxidative degradation of tertiary alcohols occurs, with cleavage of C—C bonds, producing a mixture of fragments which may be ketones or carboxylic acids:

$$CH_3-C(C_2H_5)_2-OH \xrightarrow[\text{Oxidation}]{\text{Vigorous}} \begin{matrix} C_2H_5-CO_2H \\ CH_3-CO_2H \\ CH_3-CO-C_2H_5 \\ C_2H_5-CO-C_2H_5 \end{matrix} + CO_2 + H_2O$$

4. Alcohols react with sulphuric acid in three ways. Under mild conditions alkyl hydrogen sulphates are formed:

$$R-OH + H_2SO_4 \xrightarrow{80^\circ C} R-O-SO_3H + H_2O$$

whilst under more vigorous conditions, dehydration to ethers (p. 72) or alkenes (p. 29) occurs. Tertiary alcohols are particularly easily dehydrated to alkenes.

Other reactions of alcohols described elsewhere are the conversion into alkyl halides by reaction with phosphorus halides or hydrogen halides (p. 50), and the formation of esters of mineral acids (p. 159).

Methanol (*methyl alcohol*). CH_3OH, is prepared commercially by the catalytic reduction of carbon monoxide:

$$CO + H_2 \xrightarrow[\text{catalyst}]{300\text{–}400°C} CH_3OH$$

It is a poisonous liquid, b.p. 65°C, miscible with water, occurring in minute traces in some potable spirits (arising from the degradation of pectin, p. 260). It is used to denature commercial ethanol, hence 'methylated spirit'.

It has most of the normal reactions of a primary alcohol, but cannot be dehydrated to an alkene. Oxidation gives successively methanal (formaldehyde),

$$CH_3OH \xrightarrow{[O]} \underset{\text{Methanal}}{H_2C{=}O} \xrightarrow{[O]} \underset{\text{Methanoic acid}}{H{-}C({=}O)OH} \xrightarrow{[O]} H_2O + CO_2$$

methanoic acid (formic acid), and carbon dioxide. It is methanal, produced in the body by enzymic oxidation, which is responsible for the high toxicity of methanol.

Ethanol (ethyl alcohol). CH_3CH_2OH, is obtained by the fermentation of glucose, though this has been superseded as a commercial source of ethanol by the hydration of ethene. It is a colourless liquid, b.p. 78°C, miscible with water, and of moderate toxicity when pure (commercial absolute ethanol contains traces of benzene and is much more poisonous). It is a typical primary alcohol.

During the alcoholic fermentation of grain, etc., proteins present in the organic material are converted, via amino acids, into a number of simple alcohols, which are found in 'fusel oil', a distillation fraction of higher boiling point than ethanol. Amongst the constituents of fusel oil are: propan-1-ol (*n*-propyl alcohol), butan-1-ol (*n*-butyl alcohol), 2-methylpropan-1-ol (isobutyl alcohol), pentan-1-ol, 2-methylbutan-1-ol, and 3-methylbutan-1-ol.

Cyclohexanol. $(CH_2)_5CHOH$, can be prepared by the hydrolysis of chlorocyclohexane, or by catalytic hydrogenation of phenol:

$$\text{C}_6\text{H}_{11}\text{Cl} \xrightarrow{^-OH/H_2O} \text{C}_6\text{H}_{11}\text{OH} \xleftarrow{Ni/H_2} \text{C}_6\text{H}_5\text{OH}$$

It is a typical secondary alcohol, e.g. oxidation gives a ketone cyclohexanone.

$$\text{C}_6\text{H}_{11}\text{OH (cyclohexanol)} \xrightarrow{CrO_3} \text{C}_6\text{H}_{10}\text{O (cyclohexanone)}$$

Polyhydric alcohols (polyols)

Polyhydric alcohols, or polyols, are aliphatic compounds containing two or more hydroxyl groups. Most of the biologically important compounds of this type are related to carbohydrates and will be described later. It is, however, convenient to consider the chemistry of some simple polyhydroxy compounds at this stage.

Ethane-1,2-*diol* (*ethylene glycol*). $(CH_2OH)_2$, is the simplest stable dihydroxy compound (diol). It is a colourless, poisonous liquid, b.p. 197°C, miscible with water (used as an 'antifreeze') but only sparingly soluble in ether. It can be prepared from ethene by direct hydroxylation (p. 33) or by the following routes:

$$CH_2{=}CH_2 \xrightarrow{Br_2} CH_2Br{-}CH_2Br \xrightarrow{^-OH/H_2O} CH_2OH{-}CH_2OH$$

$$CH_2{=}CH_2 \xrightarrow[(HOCl)]{Cl_2/H_2O} CH_2Cl{-}CH_2OH \xrightarrow{^-OH/H_2O} CH_2OH{-}CH_2OH$$

which represent general methods of converting alkenes into compounds with hydroxyl groups on adjacent carbon atoms (*vic*-diols, or 1,2-diols).

Ethane-1,2-diol reacts in most cases as a typical alcohol, in which the two similar functional groups can behave independently. This leads to a more complex pattern of derivatives than is found in simple alcohols, and two series of esters, ethers, halides, etc., can be formed from the diol.

$$\begin{array}{l}CH_2{-}OH\\ |\\ CH_2{-}OH\end{array} \xrightarrow{CH_3CO_2H/H^+} \begin{array}{l}CH_2{-}O{-}COCH_3\\ |\\ CH_2{-}OH\end{array}$$

$$\begin{array}{l}CH_2{-}O{-}COCH_3\\ |\\ CH_2{-}OH\end{array} \xrightarrow{CH_3CO_2H/H^+} \begin{array}{l}CH_2{-}O{-}COCH_3\\ |\\ CH_2{-}O{-}COCH_3\end{array}$$

$$\begin{array}{l}CH_2{-}O{-}COCH_3\\ |\\ CH_2{-}OH\end{array} \xrightarrow{C_6H_5CO_2H/H^+} \begin{array}{l}CH_2{-}O{-}COCH_3\\ |\\ CH_2{-}O{-}COC_6H_5\end{array}$$

[In the general case of a diol in which the two hydroxyl groups are not identical, e.g. $CH_3—CHOH—CH_2OH$, two different monoderivatives can be formed.]

Ethane-1,2-diol reacts with sodium to give a monosodium salt, $HOCH_2—CH_2O^-Na^+$. Further reaction of the second hydroxyl group is difficult, since this would produce a second negative charge on the anion, close to that already present.

Oxidation of ethane-1,2-diol (e.g. by HNO_3 or CrO_3) can lead to five possible products, depending upon the sequence and extent of oxidation of the two primary alcohol groups present. Only the final product, oxalic acid

$CH_2OH—CH_2OH \xrightarrow{[O]} CH{=}O—CH_2—OH$ (Glycolaldehyde)

$CH{=}O—CH_2—OH \xrightarrow{[O]} CH{=}O—CH{=}O$ (Glyoxal) $\xrightarrow{[O]}$ HO(O=)C—CH=O (Glyoxylic acid)

$CH{=}O—CH_2—OH \xrightarrow{[O]}$ HO(O=)C—CH_2OH (Glycollic acid) $\xrightarrow{[O]}$ HO(O=)C—CH=O (Glyoxylic acid)

HO(O=)C—CH=O (Glyoxylic acid) $\xrightarrow{[O]}$ HO(O=)C—C(=O)OH (Oxalic acid, Ethanedioic acid)

(ethanedioic acid) is readily obtained by this method, partial oxidation leading to mixtures of intermediates.

Diols, which have hydroxyl groups on adjacent carbon atoms (vicinal or *vic*-diols), are oxidised in a characteristic way by periodic acid, HIO_4, or lead tetra-acetate, $Pb(OCOCH_3)_4$. Either of these reagents cleaves the bond between the hydroxyl-substituted carbon atoms, leading to the production of carbonyl compounds:

$$\begin{array}{l} CH_2OH \\ | \\ CH_2OH \end{array} \xrightarrow[Pb(OCOCH_3)_4]{HIO_4 \text{ or}} \begin{array}{l} CH_2{=}O \\ \\ CH_2{=}O \end{array}$$

Periodic acid is reduced to iodic acid, HIO_3, in the process and lead tetra-acetate to lead (II) acetate and acetic acid. The reactions are quantitative and of great use in the study of the structure of polyols containing a sequence of *vic*-diol groups. Oxidation with either reagent leads to quantitative degradation to characteristic fragments:

$$\begin{array}{l} CH_2—OH \\ | \\ CH—OH \\ | \\ R—C—OH \\ | \\ CH—OH \\ | \\ R' \end{array} \xrightarrow{+3HIO_4} \begin{array}{l} CH_2{=}O \\ H—CO_2H \\ R—CO_2H \\ R'—CH{=}O \end{array} + 3HIO_3$$

Propane-1,3-diol. $HO—CH_2—CH_2—CH_2—OH$, is another diol with all the reactions expected of a primary alcohol. However, as the hydroxyl groups are not on adjacent carbon atoms, this substance is unaffected by lead tetra-acetate or periodic acid.

Glycerol (propane-1,2,3-triol). $HOCH_2CH(OH)CH_2OH$, is widely distributed in all living tissue. It is obtained by the hydrolysis of animal fats or plant oils, which are naturally occurring esters of glycerol and long chain carboxylic acids.

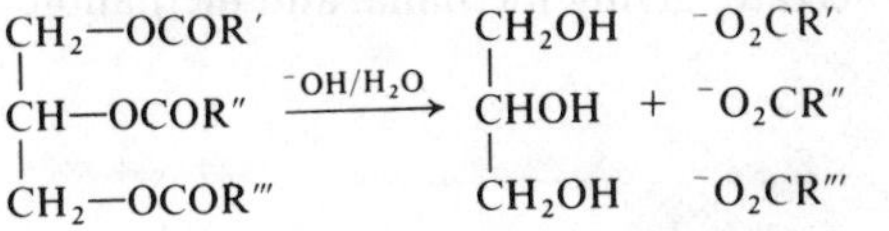

$$\begin{array}{l} CH_2—OCOR' \\ | \\ CH—OCOR'' \\ | \\ CH_2—OCOR''' \end{array} \xrightarrow{^-OH/H_2O} \begin{array}{l} CH_2OH \\ | \\ CHOH \\ | \\ CH_2OH \end{array} + \begin{array}{l} ^-O_2CR' \\ \\ ^-O_2CR'' \\ \\ ^-O_2CR''' \end{array}$$

Glycerol is also produced by the fermentation of glucose under special conditions. It is a colourless, viscous liquid of sweet taste, b.p. 290°C (decomp.), miscible with water, but insoluble in ether.

The reactions of glycerol are those expected of a compound, which is both a primary and a secondary alcohol. Several series of derivatives can be formed, e.g. there are five series of esters of a single carboxylic acid:

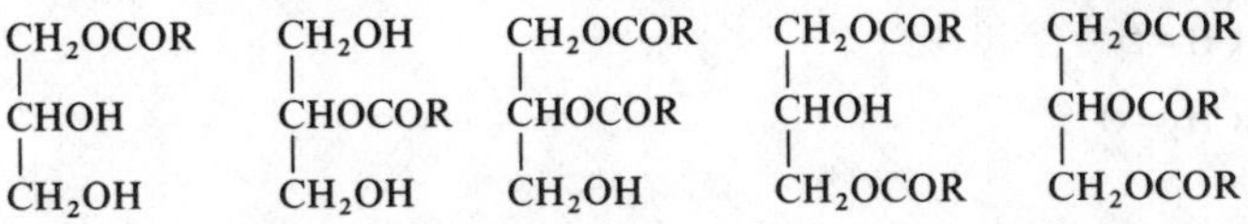

$$\begin{array}{l} CH_2OCOR \\ | \\ CHOH \\ | \\ CH_2OH \end{array} \quad \begin{array}{l} CH_2OH \\ | \\ CHOCOR \\ | \\ CH_2OH \end{array} \quad \begin{array}{l} CH_2OCOR \\ | \\ CHOCOR \\ | \\ CH_2OH \end{array} \quad \begin{array}{l} CH_2OCOR \\ | \\ CHOH \\ | \\ CH_2OCOR \end{array} \quad \begin{array}{l} CH_2OCOR \\ | \\ CHOCOR \\ | \\ CH_2OCOR \end{array}$$

and a more complex series of derivatives results where the diesters and triesters are derived from more than one carboxylic acid.

Glycerol reacts with sodium to form a monosodium salt, in which two alkoxide anions are in equilibrium, leading to mixtures of products from reactions of the sodium salts

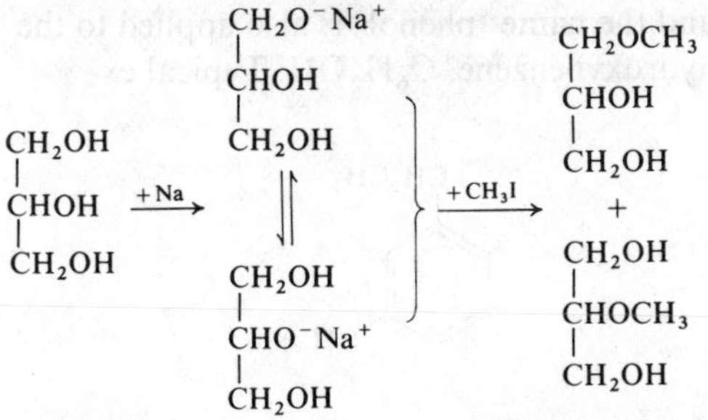

$$\begin{array}{l} CH_2OH \\ | \\ CHOH \\ | \\ CH_2OH \end{array} \xrightarrow{+Na} \left.\begin{array}{l} CH_2O^-Na^+ \\ | \\ CHOH \\ | \\ CH_2OH \\ \quad \Updownarrow \\ CH_2OH \\ | \\ CHO^-Na^+ \\ | \\ CH_2OH \end{array}\right\} \xrightarrow{+CH_3I} \begin{array}{l} CH_2OCH_3 \\ | \\ CHOH \\ | \\ CH_2OH \\ \quad + \\ CH_2OH \\ | \\ CHOCH_3 \\ | \\ CH_2OH \end{array}$$

Oxidation of glycerol with powerful oxidants, such as chromic acid, leads to complete degradation of the molecule. Milder oxidising agents (sodium hypobromite or dilute nitric acid) give an equilibrium mixture of the expected aldehyde and ketone, and further oxidation with nitric acid gives glyceric acid.

$$\begin{array}{l} CH_2OH \\ | \\ CHOH \\ | \\ CH_2OH \end{array} \xrightarrow{NaOBr} \begin{array}{l} CH_2OH \\ | \\ C{=}O \\ | \\ CH_2OH \end{array} \rightleftharpoons \begin{array}{l} CH{=}O \\ | \\ CHOH \\ | \\ CH_2OH \end{array} \xrightarrow{+HNO_3} \begin{array}{l} CO_2H \\ | \\ CHOH \\ | \\ CH_2OH \end{array}$$

Dihydroxy propanone — Glyceraldehyde — Glyceric acid

Since glycerol contains hydroxyl groups on adjacent carbon atoms, it is oxidised by periodic acid or lead tetra-acetate, giving methanal and methanoic acid:

$$\begin{array}{l} CH_2OH \\ | \\ CHOH \\ | \\ CH_2OH \end{array} \xrightarrow{2HIO_4} \begin{array}{l} CH_2{=}O \\ H{-}C(=O){-}OH \\ CH_2{=}O \end{array}$$

Use of these reagents will distinguish between the isomeric monosubstituted derivatives, as the derivative which lacks the central hydroxyl group will no longer have the required *vic*-diol structure.

$$\begin{array}{l} CH_2{-}OCH_3 \\ | \\ CH{-}OH \\ | \\ CH_2{-}OH \end{array} \xrightarrow{HIO_4} \begin{array}{l} CH_2{-}OCH_3 \\ | \\ CH{=}O \\ \\ CH_2{=}O \end{array}$$

$$\begin{array}{l} CH_2{-}OH \\ | \\ CH{-}OCH_3 \\ | \\ CH_2{-}OH \end{array} \xrightarrow{HIO_4} \text{no reaction}$$

Phenols

Aromatic hydroxy-compounds in which the hydroxyl group is attached to an aromatic ring are called phenols, and the name 'phenol' is also applied to the simplest compound of the series, hydroxybenzene, C_6H_5OH. Typical examples of phenols are:

CH_3 / OH — 4-Hydroxymethylbenzene (*p*-Cresol)

OH / OH — 1,2-Dihydroxybenzene (Catechol)

CH_2CH_3 / Cl / OH — 2-Chloro-5-hydroxyethylbenzene

Preparation of phenols. The only general methods of preparation of phenols are the reaction of aromatic primary amines with nitrous acid above 5°C

$$\text{4-}CH_3C_6H_4NH_2 \xrightarrow[+\,H^+]{+\,HNO_2} \text{4-}CH_3C_6H_4N_2^+ \xrightarrow[T\,>\,5°C]{+\,H_2O} \text{4-}CH_3C_6H_4OH$$

(p. 91), and the reaction of aromatic sulphonic acids with fused alkali hydroxides:

$$C_6H_5SO_3H \xrightarrow[200\text{–}250°C]{KOH} C_6H_5O^-\,K^+ \xrightarrow{+\,H^+} C_6H_5OH$$

The vigorous conditions of the latter reaction are similar to those required for the hydrolysis of aryl halides (p. 56), a reaction which is not a general preparation of phenols, though used industrially for the preparation of phenol itself.

Properties of phenols. Simple phenols are liquids or low-melting solids, frequently with a very characteristic odour, moderately soluble in water and very soluble in most organic solvents. Many phenols have been used as disinfectants, but simple phenols are generally very toxic, some, like phenol itself, being absorbed through the skin with the production of severe burns.

Reactions of phenols

A *Reactions of the —OH group.* The most characteristic property of phenols as a class is the feeble acidity of the hydroxyl group, and phenols will dissolve readily in dilute aqueous sodium hydroxide, producing the phenoxide anion:

$$\text{3-}CH_3C_6H_4OH + OH^- \rightleftharpoons \text{3-}CH_3C_6H_4O^- + H_2O$$

Phenol itself has a $pK_a = 9.8$ (p. 133) and simple alkyl substituted phenols have comparable acidities. Carbonic acid ($pK_1 = 6{\cdot}56$) is a thousand times more strongly acidic, so most simple phenols do not react with sodium bicarbonate solution, and can be precipitated from solutions of the phenoxide by saturation with carbon dioxide.

$$Ar\text{—}O^- + CO_2 + H_2O \longrightarrow Ar\text{—}OH + HCO_3^-$$
(Ar = aryl group)

The acidity of phenols is attributable to resonance stabilisation of the anion, for which four canonical structures (p. 13) are possible:

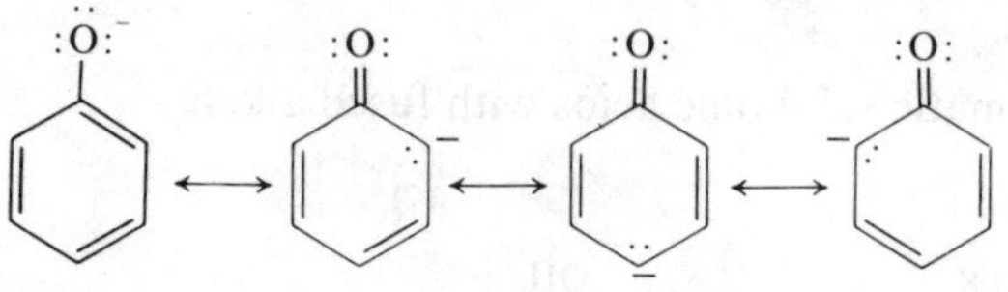

The effect of this resonance is to distribute the negative charge of the anion all over the molecule, rather than leaving it concentrated on one particular atom, as in the case of the alkoxide anions. The consequent decrease in the energy required to form a phenoxide anion shows itself in the ease of ionisation of the phenol, i.e. its greater acidity, compared with alcohols (cf. the ease of formation of carbonium ions, p. 32).

The acidity of a phenol is greatly increased by the presence of powerful electronegative substituents on the aromatic ring in positions *ortho-* or *para-* to the hydroxyl group. This is particularly the case with nitro-groups, as further canonical structures exist for the resonant anion, e.g. 4-nitrophenoxide ion is a resonance hybrid of the five canonical structures:

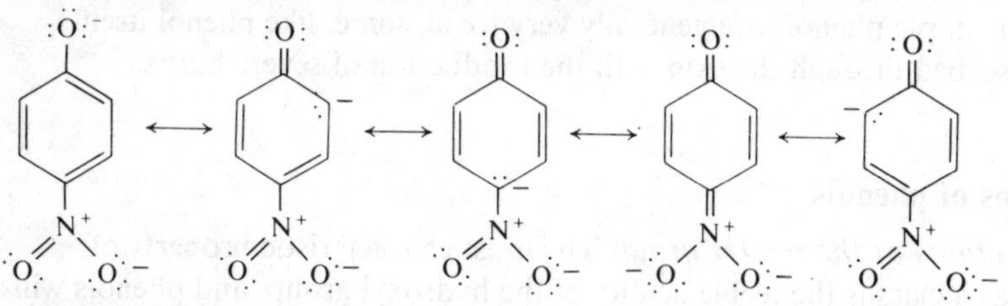

The effect of nitro-groups on the acidity of the phenol is shown by the table of *pK* values.

COMPOUND	pK_a(p. 133)
Phenol, C_6H_5OH	9·89
2-Nitrophenol, $C_6H_4(NO_2)OH$	7·12
4-Nitrophenol, $C_6H_4(NO_2)OH$	7·19
2,4-Dinitrophenol $C_6H_3(NO_2)_2OH$	4·00
2,6-Dinitrophenol $C_6H_3(NO_2)_2OH$	3·77
2,4,6-Trinitrophenol (Picric acid) $C_6H_2(NO_2)_3OH$	0·80

Few of the reactions of alcoholic hydroxyl groups occur with phenols. Esters are readily formed by reaction with acid anhydrides or chlorides, and direct esterification with carboxylic acids is also possible.

$$\text{o-}CH_3C_6H_4OH + (CH_3CO)_2O \xrightarrow{H^+} \text{o-}CH_3C_6H_4O{-}COCH_3 + CH_3CO_2H$$

The direct formation of ethers by the action of sulphuric acid is not possible, but preparation of alkyl aryl ethers via the phenoxide anion occurs normally:

$$C_6H_5O^-\,Na^+ + CH_3I \longrightarrow C_6H_5O{-}CH_3 + NaI$$

Oxidation of phenols occurs readily, but leads to complex products, often with simultaneous formation of much tarry material. In suitable circumstances, quinones (p. 120) are formed.

Phenols cannot normally be converted directly into the corresponding aryl halides. Hydrogen halides have no effect, and reaction with phosphorus halides leads predominantly to the aryl esters of phosphorus oxyacids (p. 56).

Phenols, like many other compounds containing the 'enol' group, $>C{=}C{-}OH$, form intensely coloured complexes with ferric ion in neutral solution. The blue, purple, or green colours produced are often used as a qualitative test for phenols, but are also given by certain aliphatic compounds, which exist in solution partly as the enol (p. 105).

B *Reactions of the aromatic ring*. Phenols are very readily attacked by electrophiles, substitution occurring in the *ortho*- and *para*- positions. It is often difficult to prevent substitution occurring more than once. Dilute nitric acid rapidly converts phenol into its *ortho*- and *para*-nitro derivatives, whilst under the conditions in which benzene is converted into nitrobenzene, phenol is nitrated three times, forming picric acid (2,4,6-trinitrophenol). In aqueous solution, chlorine or bromine water will give the corresponding trisubstituted derivatives. Even the feebly electrophilic nitrosonium ion, NO^+ produced in acidified nitrous acid solution, will convert phenol into its *p*-nitroso-derivative.

OH OH OH
NO$_2$
+
NaNO$_2$
+dil. HCl
NO OH NO$_2$
dil. HNO$_3$ HNO$_3$/H$_2$SO$_4$
OH
cold conc.
H$_2$SO$_4$ HNO$_3$/H$_2$SO$_4$ O$_2$N NO$_2$
Br$_2$/H$_2$O
OH OH
SO$_3$H OH
Br Br
NO$_2$
SO$_3$H
Br

Aromatic alcohols

Aromatic compounds, which contain a hydroxyl group on a side chain, behave like typical alcohols, save for the possibility of electrophilic substitution of the aromatic ring. Typical examples of these compounds are:

CH_2OH	$CH_2—CH_2—OH$	OH | $CH—CH_3$
Benzyl alcohol	2-Phenylethanol	1-Phenylethanol

Dialkyl ethers

Dialkyl ethers may be regarded either as the anhydrides of alcohols, or better, as the dialkyl derivatives of water, e.g.

H—O—H	R—O—H	R—O—R
water	alcohol	ether

Although ether linkages (i.e. C—O—C) are frequently found in naturally occurring compounds, simple ethers are of little biological significance.

Preparation of ethers

1. Direct dehydration of alcohols by sulphuric acid is possible with primary alcohols, but secondary and tertiary alcohols are too readily converted

into alkenes (p. 30). Initial conversion of part of the primary alcohol into the alkyl hydrogen sulphate is followed by alkylation of the residual alcohol by this ester of sulphuric acid (p. 159).

$$C_2H_5OH \xrightarrow[80°C]{H_2SO_4} C_2H_5{-}OSO_3H \xrightarrow[150°C]{C_2H_5OH} (C_2H_5)_2O + H_2SO_4$$

2. A more general preparation is the reaction of nucleophilic alkoxide anions with alkyl halides (Williamson's synthesis):

$$R'{-}\ddot{O}:^{-} \curvearrowright R{-}Br: \longrightarrow R'{-}\ddot{O}{-}R + :Br:^{-}$$

Since the alkoxide and the alkyl halide need not have the same alkyl group, this method is suitable for the production of mixed ethers,

$$\underset{\text{methoxide anion}}{CH_3{-}O^{-}} + \underset{\text{1-bromobutane}}{CH_3CH_2CH_2CH_2{-}Br} \longrightarrow \underset{\text{butyl methyl ether}}{CH_3CH_2CH_2CH_2{-}O{-}CH_3} + Br^{-}$$

Properties of ethers. The lower ethers are neutral gases, or volatile liquids of much lower boiling point than the isomeric alcohols. They are very sparingly soluble in water, but good solvents for many organic compounds.

Reactions of ethers. Ethers are inert to most of the reagents which attack alcohols, as they lack the chemically reactive hydroxyl group. They are completely inert to alkalis or alkali metals, but under strongly acidic conditions they are converted into oxonium cations by protonation of the oxygen atom, and these cations may react with nucleophiles. Thus hydrogen bromide or iodide cleaves the ether link in the following way, (cf. the mechanism of conversion of alcohols into alkyl halides, p. 49):

$$R{-}\ddot{O}{-}R + H^{+} \rightleftharpoons R{-}\overset{H}{\overset{|}{O}}{}^{+}{-}R$$

$$:\ddot{I}:^{-} \curvearrowright R{-}\overset{H}{\overset{|}{O}}{}^{+}{-}R \longrightarrow :\ddot{I}{-}R + H{-}\ddot{O}{-}R$$

$$[ROH + HI \longrightarrow R{-}I + H_2O]$$

Oxidising agents have no effect upon ethers, except under very vigorous conditions, when the oxidation products of the corresponding alcohols are formed. Radical halogenation (p. 26) occurs preferentially on the carbon atom adjacent to oxygen:

$$CH_3{-}CH_2{-}O{-}CH_2{-}CH_3 \xrightarrow[\text{light}]{Cl_2} CH_3{-}\underset{\underset{Cl}{|}}{CH}{-}O{-}CH_2{-}CH_3$$

In the presence of air and light, explosive peroxides may be formed:

$$(CH_3)_2CH{-}O{-}CH(CH_3)_2 \xrightarrow[\text{light}]{O_2} (CH_3)_2\underset{\displaystyle OOH}{\underset{|}{C}}{-}O{-}CH(CH_3)_2$$

Diethyl ether (ethoxyethane). $(C_2H_5)_2O$, is a colourless, inflammable liquid, b.p. 35°C, extensively used as a solvent and an anaesthetic.

Aromatic ethers

Aromatic ethers may be divided into two classes, the aryl alkyl ethers, which are formally derived from a phenol and an alcohol, and diaryl ethers, derived from two molecules of a phenol, e.g.

$C_6H_5{-}O{-}CH_3$ Methyl phenyl ether (Anisole)	$C_6H_5{-}O{-}C_6H_5$ Diphenyl ether

The former compounds may be obtained by the reaction of the phenoxide anion with alkyl halides, but not by the reaction of alkoxides and aryl halides (p. 56). Their reactions are similar to those of dialkyl ethers, and electrophilic substitution of the aromatic ring occurs predominantly *ortho-* and *para-* with respect to the alkoxy group.

Apart from electrophilic substitution of the aromatic rings, diaryl ethers are inert compounds, cleavage of the ether link being very difficult. Thyroxine—the thyroid hormone—is a naturally occurring diaryl ether.

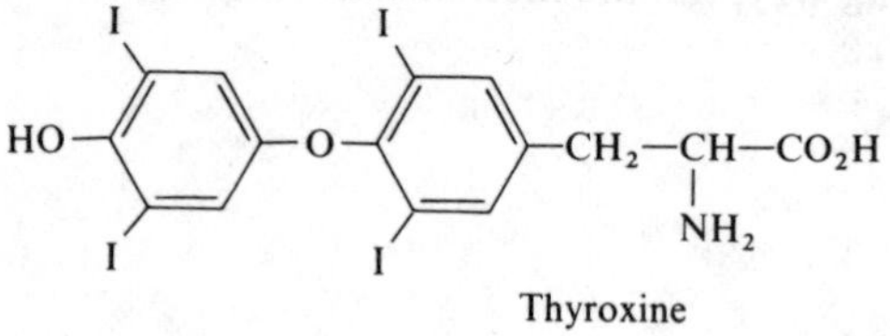

Thyroxine

Simple sulphur compounds

The replacement of the hydrogen atoms of water by alkyl or aryl groups leads to alcohols or phenols, and ethers. In precisely the same way the hydrides of sulphur, H_2S and H_2S_2, are the parents of three types of simple aliphatic sulphur compounds.

R—S—H Thiols (Mercaptans)	R—S—R Thioethers	R—S—S—R Disulphides

Thiols may be prepared by the reaction of alkali hydrosulphides with alkyl halides (p. 51), or by the action of phosphorus sulphides on alcohols

$$H{-}\ddot{\underset{..}{S}}{:}^{-} \curvearrowright R{-}\ddot{\underset{..}{I}}{:} \longrightarrow H{-}\ddot{\underset{..}{S}}{-}R + {:}\ddot{\underset{..}{I}}{:}^{-}$$

$$R{-}OH + P_2S_5 \rightarrow R{-}SH$$

Thiols are evil-smelling liquids, which are much more volatile than the corresponding alcohols. Although sulphur is less electronegative than oxygen, thiols are much more strongly acidic than alcohols (CH_3CH_2SH, $pK = 11$) and can form salts, 'thiolates' or 'mercaptides', corresponding to the alkoxides, even in aqueous alkali:

$$R{-}SH + {}^{-}OH \rightleftharpoons R{-}S^{-} + H_2O$$

Covalent heavy metal derivatives, e.g. $Hg(SC_2H_5)_2$ can also be precipitated from aqueous solutions.

Thiols, like alcohols, can be esterified by carboxylic acids, acyl chlorides, or acid anhydrides:

$$R{-}SH + R'{-}CO_2H \xrightleftharpoons{H^+} R{-}S{-}CO{-}R' + H_2O$$
$$R{-}SH + R'{-}COCl \longrightarrow R{-}S{-}CO{-}R' + HCl$$

Oxidation of thiols gives products quite different from those of alcohol oxidation. Mild oxidants produce disulphides, e.g.

$$CH_3{-}SH + I_2 \longrightarrow CH_3{-}S{-}S{-}CH_3 + HI$$

and thiols can be quantitatively titrated with iodine. More vigorous oxidation converts the disulphides into sulphonic acids:

$$C_2H_5SH \xrightarrow{HNO_3} [C_2H_5{-}S{-}S{-}C_2H_5] \xrightarrow{HNO_3} C_2H_5SO_3H$$

Thioethers may be obtained by the action of alkali sulphides on alkyl halides (p. 51), or by the alkylation of thiolate anions by alkyl halides (cf. Williamson's ether synthesis, p. 73):

$$R{-}Br + S^{--} \longrightarrow R{-}S{-}R \text{ (in two steps)}$$
$$R{-}S^{-} + R'{-}I \longrightarrow R{-}S{-}R' + I^{-}$$

Thioethers are liquids with disagreeable odours, which, unlike ethers, can be alkylated easily to form 'sulphonium' salts:

$$C_2H_5{-}\ddot{\underset{..}{S}}{-}C_2H_5 + C_2H_5{-}I \longrightarrow (C_2H_5)_3\overset{+}{\ddot{S}}\ \ I^{-}$$

Triethylsulphonium iodide

(cf. quaternary ammonium salts, p. 82).

Oxidation of thioethers, with powerful oxidising agents, gives successively sulphoxides and sulphones:

$$\underset{\text{Diethyl sulphide}}{(C_2H_5)_2\ddot{S}:} \xrightarrow{H_2O_2} \underset{\text{Diethyl sulphoxide}}{(C_2H_5)_2\ddot{S}^{+}\!-\!\ddot{O}:^{-}} \xrightarrow[\text{or } HNO_3]{H_2O_2} \underset{\text{Diethyl sulphone}}{(C_2H_5)_2S^{2+}(\!-\!\ddot{O}:^{-})_2} \longleftrightarrow (C_2H_5)_2S(=O)_2$$

Dialkyl disulphides are produced by the oxidation of the corresponding thiols (see above), into which they may be converted by reducing agents:

$$2R{-}SH \underset{\substack{\text{Reduction} \\ \text{e.g. Zn/HCl}}}{\overset{\substack{\text{Oxidation} \\ \text{e.g. } I_2}}{\rightleftharpoons}} R{-}S{-}S{-}R$$

Occurrence of organic sulphur compounds. Aliphatic sulphur compounds are widely distributed in nature, and are of great biological importance. The protein chains of enzymes frequently contain thiol groups, which are vital for the catalytic activity of the enzyme, and the poisonous properties of some heavy metals, e.g. arsenic, lead, and mercury are due to their ability to combine with these thiol groups, thereby interfering with cell reactions. These enzymes are also inhibited by treatment with iodoethanoic acid, ICH_2CO_2H, a powerful alkylating agent which converts the —SH group into —SCH_2CO_2H. Many carboxylic acids are utilised by cells in the form of their esters of coenzyme A, a complex nucleotide thiol (p. 300). Sulphonium salts are known to be intermediates in some biological alkylations. Disulphide groups are important structural features in many proteins and polypeptide hormones, e.g. insulin, and certain reduction-oxidation reactions of cells utilise thiol-disulphide interconversion as a redox system (see below). The evil smelling secretion of skunks contains much 3-methylbutane-1-thiol, $(CH_3)_2CHCH_2CH_2SH$, with *trans*-but-2-ene-1-thiol, $CH_3CH{=}CHCH_2SH$, and the disulphide derivative $CH_3CH{=}CHCH_2SSCH_3$. Sulphur compounds are also responsible for the characteristic odours of *Allium* species (onions and garlic); garlic oil contains much diallyl disulphide, $CH_2{=}CHCH_2({-}S{-})_nCH_2CH{=}CH_2$ ($n = 2$), with the corresponding tri- and tetrasulphides ($n = 3, 4$), and also a recently identified potent antithrombotic compound 'ajoene'. The lachrymatory component of onion juice has been identified as the *S*-oxide of thiopropanal.

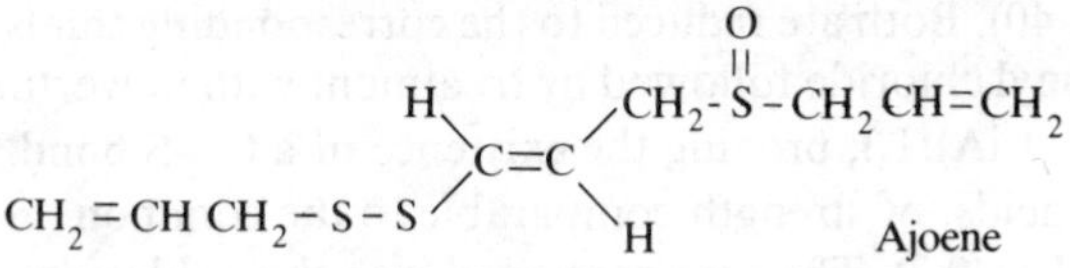

$$CH_2{=}CH\,CH_2{-}S{-}S{-}CH{=}C(H){-}CH_2{-}\overset{O}{\overset{\|}{S}}{-}CH_2CH{=}CH_2$$

Ajoene

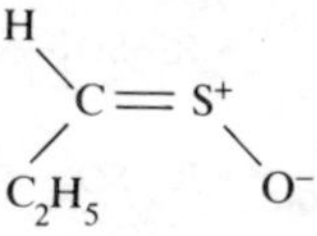

$$\underset{C_2H_5}{\overset{H}{>}}C{=}S^{+}{-}O^{-}$$

Thiopropanal –*S*–Oxide

Thioctic acid (*α-lipoic acid*) is a naturally occurring disulphide, which is a cofactor required for the enzymic oxidation of pyruvic acid to acetic acid in bacteria. The disulphide is the oxidising agent, being reduced to the corresponding thiol, which can subsequently be reoxidised.

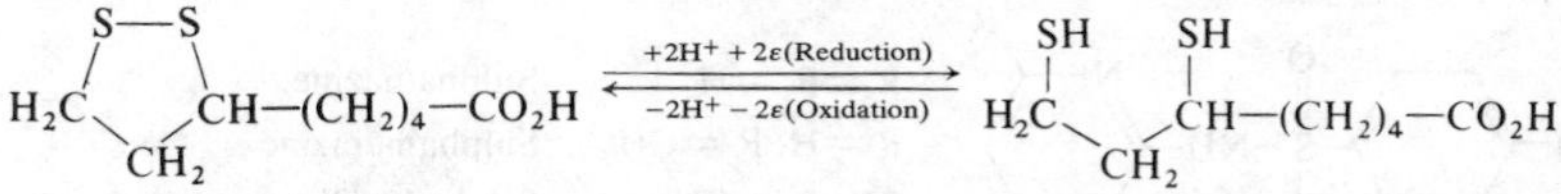

$$\text{H}_2\text{C(S–S)CH}_2\text{CH–(CH}_2)_4\text{–CO}_2\text{H} \underset{-2H^+ - 2\varepsilon(\text{Oxidation})}{\overset{+2H^+ + 2\varepsilon(\text{Reduction})}{\rightleftharpoons}} \text{H}_2\text{C(SH)–CH}_2\text{–CH(SH)–(CH}_2)_4\text{–CO}_2\text{H}$$

Thioctic acid   6.8-Dimercapto-octanoic acid

The overall reaction for the conversion of pyruvic acid into the ester of acetic acid and coenzyme A (acetyl CoA) is:

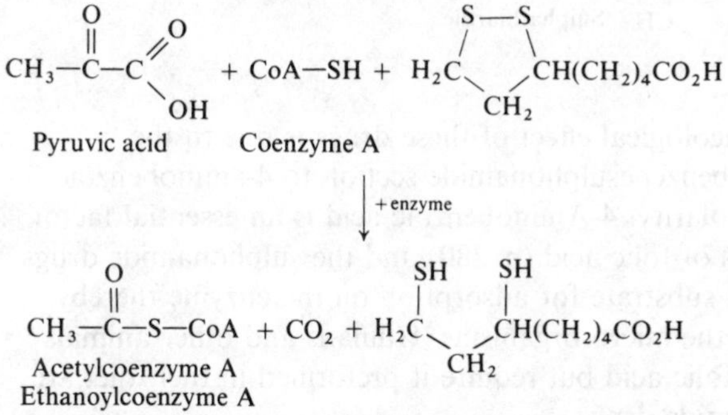

$$CH_3{-}\overset{O}{\overset{\|}{C}}{-}C(=O)OH + CoA{-}SH + H_2C(S{-}S)CH_2CH(CH_2)_4CO_2H$$

Pyruvic acid   Coenzyme A

$$\downarrow +\text{enzyme}$$

$$CH_3{-}\overset{O}{\overset{\|}{C}}{-}S{-}CoA + CO_2 + H_2C(SH){-}CH_2{-}CH(SH)(CH_2)_4CO_2H$$

Acetylcoenzyme A
Ethanoylcoenzyme A

(For the mechanism of this process see p. 283).

Sulphonic acids

Aliphatic sulphonic acids (e.g. CH_3SO_3H, methanesulphonic acid) can be obtained by the oxidation of thiols. Aromatic sulphonic acids (e.g. $C_6H_5SO_3H$, benzenesulphonic acid) can be prepared by direct sulphonation of aromatic hydrocarbons (p. 40). Both are reduced to the corresponding thiols by conversion into the sulphonyl chloride followed by treatment with powerful reducing agents (e.g. Zn/HCl, $LiAlH_4$), proving the existence of a C—S bond.

Sulphonic acids are strong acids, of strength comparable to the common mineral acids (pK $C_6H_5SO_3H$ = 0·7). They are converted into their chlorides by phosphorus pentachloride, and thence into esters and amides. The esters,

$$R-\overset{\overset{O}{\|}}{\underset{\underset{O}{\|}}{S}}-OH \xrightarrow{PCl_5} R-\overset{\overset{O}{\|}}{\underset{\underset{O}{\|}}{S}}-Cl \begin{cases} \xrightarrow{+R'NH_2} R-SO_2-NHR' \\ \xrightarrow[+R'OH]{} R-SO_2-OR' \end{cases}$$

which cannot be produced by direct esterification of the acids, are alkylating agents like the sulphate esters (p. 159).

$$RSO_3CH_3 + NH_3 \longrightarrow RSO_3^- + CH_3\overset{+}{N}H_3$$

The sulphonamide drugs are a series of compounds derived from the amide of 4-aminobenzenesulphonic acid, $H_2NC_6H_4SO_2NH_2$.

$H_2N-C_6H_4-SO_2-NH-$(pyrimidin-2-yl bearing R′ and R″)

R′=R″=H ....Sulphadiazine
R′=H, R″=CH_3...Sulphamerazine
R′=R″=CH_3 .....Sulphadimidine

$H_2N-C_6H_4-SO_2-NH-$(thiazol-2-yl) Sulphathiazole

The important pharmacological effect of these drugs is due to the similarity of the 4-aminobenzenesulphonamide section to 4-aminobenzoic acid in both shape and polarity. 4-Aminobenzoic acid is an essential factor in the microbial synthesis of folic acid (p. 280) and the sulphonamide drugs compete with the natural substrate for adsorption on the enzyme thereby restricting the growth of the micro-organisms. Humans and other animals which do not synthesise folic acid but require it preformed in their diet are not affected by sulphonamide drugs.

Problems

1. What reagents will convert butan-1-ol into the following compounds?
$CH_3CH_2CH_2CHO$, $CH_3CH_2CH{=}CH_2$, $CH_3CH_2CH_2CH_2I$, $CH_3CH_2CH_2CH_2OCH_3$, $CH_3CH_2CH_2CH_2OCOCH_3$.

2. What products do you expect to be formed by the reaction of methanol with concentrated sulphuric acid? Give the mechanisms for the reactions involved.

3. Suggest a method for preparation of $C_2H_5OCH_2CHO$ starting from glycerol and any other readily available reagents.

4. Treatment of $ClCH_2CH_2OH$ with solid potassium hydroxide gives 'ethylene oxide'

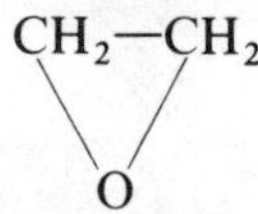

What is the mechanism of this reaction? What would be the product of reacting this compound with hydriodic acid?

5. Do you expect 2,4,6-tricyanophenol to be more acidic than phenol? Explain your answer.

6

Simple organic nitrogen compounds

Nitro-compounds

Nitroalkanes, as the name implies, are compounds in which a hydrogen atom of an alkane has been replaced by a nitro-group (—NO_2), the nitrogen being attached directly to a carbon atom, as opposed to the C—O link in the isomeric nitrite esters, e.g.

$$CH_3-\overset{+}{N}\begin{matrix}{\diagup\!\!\!\!\!\diagup} O \\ \diagdown O^- \end{matrix} \qquad CH_3-O-N{=}O$$

Nitromethane — Methyl nitrite

Nitroalkanes are obtained either by direct nitration of alkanes (p. 27), or by the reaction of silver nitrite with alkyl halides (p. 51):

$$R-I + AgNO_2 \longrightarrow R-NO_2 + AgI$$

The lower nitroalkanes are liquids of comparatively high boiling point (CH_3NO_2, b.p. 101°C) sparingly soluble in water, and generally stable to heat or mechanical shock. They are of little biological significance.

Reduction by catalytic hydrogenation or dissolving metal systems (p. 61) gives the corresponding primary amine:

$$R-NO_2 \xrightarrow{Ni/H_2} R-NH_2$$

Nitroalkanes containing the group $\gt CH-NO_2$ dissolve in aqueous alkali, forming the salts of the tautomeric 'aci' -compounds (p. 107), e.g.

2-Nitropropane $\xrightarrow{-H^+}$ Mesomeric anion

Aromatic nitro-compounds are readily obtained by direct nitration of aromatic compounds (p. 40). The simple nitro-compounds, e.g. nitrobenzene, $C_6H_5NO_2$, are liquids or low melting point solids, of high boiling point, frequently pale yellow in colour.

Reduction of aromatic nitro-compounds by catalytic hydrogenation or acid/metal systems (e.g. Sn + HCl) gives the corresponding primary amine.

4-Nitrotoluene $\xrightarrow[\text{or Ni/H}_2]{\text{Sn/HCl}}$ 4-Aminotoluene (*p*-Toluidine)

Under alkaline conditions (Na + C_2H_5OH; Zn + NaOH) a range of intermediate reduction products can be obtained, e.g.

Nitrobenzene $\xrightarrow{\text{Zn/NaOH}}$ Nitrosobenzene $\longrightarrow$ Phenylhydroxylamine

Azoxybenzene $\longrightarrow$ Azobenzene $\longrightarrow$ Hydrazobenzene

Electrophilic substitution of the aromatic ring of nitrobenzene is very much more difficult than in the case of benzene, and substitution occurs *meta* with respect to the nitro-group (p. 42).

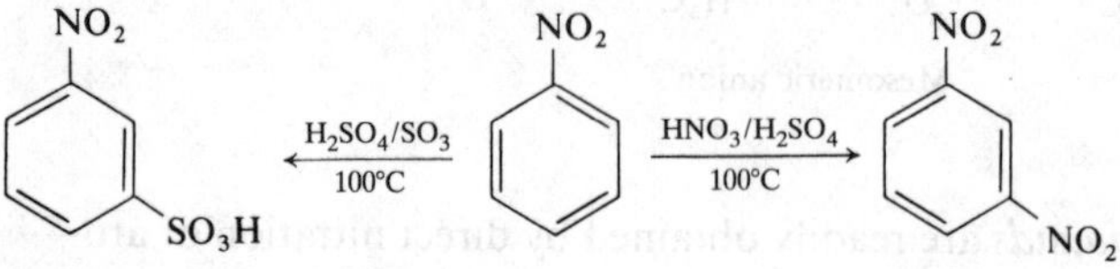

$$\text{m-}C_6H_4(NO_2)(SO_3H) \xleftarrow[100°C]{H_2SO_4/SO_3} C_6H_5NO_2 \xrightarrow[100°C]{HNO_3/H_2SO_4} \text{m-}C_6H_4(NO_2)_2$$

Although of much greater chemical importance than their aliphatic analogues, aromatic nitro-compounds have little biological significance. Some polynitrocompounds are utilised as high explosives.

e.g. 2,4,6-Trinitrotoluene (T.N.T) — CH_3 at C-1, NO_2 at C-2, C-4 and C-6 of the benzene ring

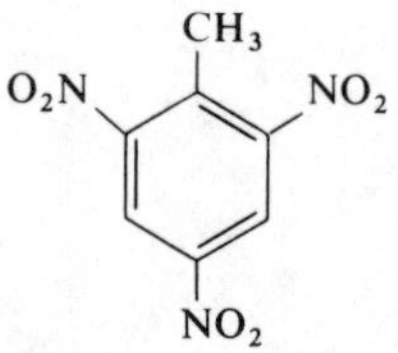

Aliphatic amines

Amines are derived from ammonia by replacement of the hydrogen atoms of NH_3 by organic groups, and three types of amine can be obtained, known as primary, secondary, and tertiary amines, distinguished by the number of substituent groups attached to nitrogen. In addition, an allied fourth class of compounds, quaternary ammonium salts, is obtained by replacement of all four hydrogen atoms of the ammonium cation by alkyl or aryl groups, e.g.

$CH_3CH_2CH_2{-}\ddot{N}H_2$
1-Propylamine
(Primary)

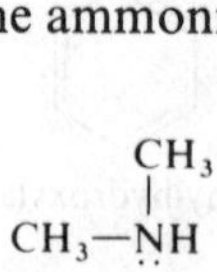

$(CH_3)_2\ddot{N}H$ — $CH_3{-}\ddot{N}H{-}CH_3$
Dimethylamine
(Secondary)

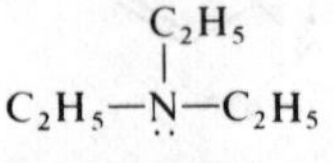

$(C_2H_5)_3\ddot{N}$ — $C_2H_5{-}\ddot{N}(C_2H_5){-}C_2H_5$
Triethylamine
(Tertiary)

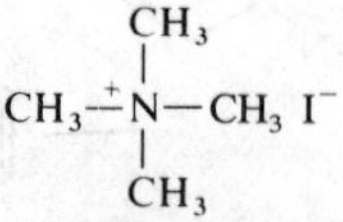

$(CH_3)_4\overset{+}{N}\ I^-$
Tetramethylammonium iodide
(Quaternary ammonium salt)

General preparations of aliphatic amines

1. The reaction of ammonia with alkyl halides gives amines and quaternary ammonium salts by a sequence of nucleophilic substitutions (p. 51). Normally a mixture of all possible products is formed:

$$
\begin{aligned}
NH_3 + RBr &\longrightarrow R\overset{+}{N}H_3 + Br^- \\
R\overset{+}{N}H_3 + NH_3 &\rightleftharpoons RNH_2 + \overset{+}{N}H_4 \\
RNH_2 + RBr &\longrightarrow R_2\overset{+}{N}H_2 + Br^- \\
R_2\overset{+}{N}H_2 + NH_3 &\rightleftharpoons R_2NH + \overset{+}{N}H_4 \\
R_2NH + RBr &\longrightarrow R_3\overset{+}{N}H + Br^- \\
R_3\overset{+}{N}H + NH_3 &\rightleftharpoons R_3N + \overset{+}{N}H_4 \\
R_3N + RBr &\longrightarrow R_4\overset{+}{N} + Br^-
\end{aligned}
$$

A large excess of ammonia favours the formation of primary and secondary amines, whilst an excess of alkyl halide favours the production of tertiary amine and quaternary ammonium salt.

2. The reduction of amides by lithium aluminium hydride can give primary, secondary, or tertiary amines, depending on the amide employed.

$$
\left.\begin{array}{l}
R{-}\overset{O}{\overset{\|}{C}}{-}NH_2 \\
R{-}\overset{O}{\overset{\|}{C}}{-}NHR' \\
R{-}\overset{O}{\overset{\|}{C}}{-}NR'_2
\end{array}\right\}
\xrightarrow{LiAlH_4}
\left\{\begin{array}{l}
R{-}CH_2{-}NH_2 \\
R{-}CH_2{-}NHR' \\
R{-}CH_2{-}NR'_2
\end{array}\right.
$$

3. Decarboxylation of amino acids is of little chemical importance, but occurs widely in living systems. The enzyme-catalysed decarboxylation utilises pyridoxal phosphate (p. 286) as a coenzyme.

$$
R{-}\underset{NH_2}{\underset{|}{C}H}{-}CO_2H \xrightarrow{-CO_2} R{-}CH_2{-}NH_2
$$

Almost all other preparative methods are specific syntheses of one or other of the three classes of amine.

Preparation of primary amines. Many compounds containing C—N bonds can be reduced to primary amines by catalytic methods (Ni/H_2) or metal-acid reducing agents ($Sn + HCl$). Examples of such compounds are:

$$\left.\begin{array}{l} R{-}NO_2 \\ \text{Nitroalkane} \\ R{-}C{\equiv}N \\ \text{Nitrile} \\ R_2C{=}NOH \\ \text{Oxime} \\ R_2C{=}N{-}NH_2 \\ \text{Hydrazone} \\ R_2C{=}N{-}NHC_6H_5 \\ \text{Phenylhydrazone} \end{array}\right\} \xrightarrow[\text{or Sn/HCl}]{\text{Ni/H}_2} \left\{\begin{array}{l} R{-}NH_2 \\ R{-}CH_2{-}NH_2 \\ R_2CH{-}NH_2 \\ R_2CH{-}NH_2 + NH_3 \\ R_2CH{-}NH_2 + H_2NC_6H_5 \end{array}\right.$$

Treatment of primary amides (i.e. $RCONH_2$) with alkali hypobromites gives the lower primary amine:

$$R{-}\overset{\overset{\displaystyle O}{\|}}{C}{-}NH_2 + KOBr \longrightarrow R{-}NH_2 + CO_2 + KBr$$

This reaction cannot be extended to amides having alkyl or aryl substituents on the nitrogen atom.

Preparation of secondary amines. Reduction of imines (Schiff's bases), i.e. compounds containing C=N— groups, produces secondary amines. The imines can be obtained from aldehydes or ketones (p. 100).

$$R{-}CH{=}N{-}R \xrightarrow{Pt/H_2} R{-}CH_2{-}NH{-}R$$

Preparation of tertiary amines. Pyrolysis or reduction of quaternary ammonium salts leads to the production of tertiary amines (p. 89).

Properties of amines. The lower amines are gases or liquids of characteristic ammoniacal or fishy odour, soluble in water and most organic solvents. The amines of higher relative molecular mass are progressively less water-soluble, and have repulsive odours, the evil smell of decaying meat being due to the production of putrescine, $H_2N(CH_2)_4NH_2$, and cadaverine, $H_2N(CH_2)_5NH_2$, by enzymic decomposition of protein (p. 267).

Reactions of amines

1. The most characteristic property of amines is their basicity. Like ammonia, they form salts with acids, utilising the lone pair of electrons on the

nitrogen atom to form a bond to a hydrogen ion. Their aqueous solutions are strongly alkaline.

$$C_2H_5NH_2 + H^+ \rightleftharpoons C_2H_5\overset{+}{N}H_3 \quad \text{Ethylammonium ion}$$
$$C_2H_5NH_2 + H_2O \rightleftharpoons C_2H_5\overset{+}{N}H_3 \; OH^-$$

Aliphatic amines are stronger bases than ammonia and the increased basicity is attributable to inductive electron displacement along the C—N bond in the cation, stabilising the positive charge on the nitrogen atom, in the same way that alkyl substituents stabilise carbonium ions (p. 32).

$$H_3C \overset{\delta+}{\longrightarrow} \overset{\delta-}{N^+}H_3$$

2. The lone pair of electrons on the nitrogen atom enables amines to react with many compounds by nucleophilic attack at electron-deficient centres. The formation of salts (nucleophilic attack on H^+ or H_3O^+) and the reaction with alkyl halides (see above and p. 51) are typical nucleophilic reactions. Amines, like ammonia, can also form complexes with metal ions by donation of the lone pair, e.g.

$$Cu^{++} + CH_3NH_2 \longrightarrow [Cu(NH_2CH_3)_4]^{++} \quad \text{cf. } [Cu(NH_3)_4]^{++}$$

Primary and secondary amines react with certain derivatives of carboxylic acids (i.e. acyl chlorides, acid anhydrides, esters, and thiol esters) to form amides, and these acylations are initiated by nucleophilic attack of the amine on the carbonyl group of the reagent. Examples of these reactions are:

$$\underset{\text{Acyl chloride}}{R{-}\overset{O}{\overset{\|}{C}}{-}Cl} + R'{-}NH_2 \longrightarrow \underset{\text{Amide}}{R{-}\overset{O}{\overset{\|}{C}}{-}NHR'} + R'{-}\overset{+}{N}H_3Cl^-$$

$$\underset{\text{Acid anhydride}}{R{-}\overset{O}{\overset{\|}{C}}{-}O{-}\overset{O}{\overset{\|}{C}}{-}R} + R'_2NH \longrightarrow R{-}\overset{O}{\overset{\|}{C}}{-}NR'_2 + R{-}\overset{O}{\overset{\|}{C}}{-}OH$$

$$\underset{\text{Ester}}{R{-}\overset{O}{\overset{\|}{C}}{-}OR''} + R'NH_2 \longrightarrow R{-}\overset{O}{\overset{\|}{C}}{-}NHR' + \underset{\text{Alcohol}}{R''OH}$$

$$\underset{\text{Thiol ester}}{R{-}\overset{O}{\overset{\|}{C}}{-}SR''} + R'_2NH \longrightarrow R{-}\overset{O}{\overset{\|}{C}}{-}NR'_2 + \underset{\text{Thiol}}{R''SH}$$

Similarly the chlorides of sulphonic acids (p. 78) form sulphonamides with primary and secondary amines:

$$C_6H_5SO_2Cl + R_2NH \longrightarrow C_6H_5SO_2NR_2 + R_2\overset{+}{N}H_2Cl^-$$

Tertiary amines, which have no replaceable hydrogen atoms, do not form amides. The mechanism of these reactions is described later (p. 142).

3. Primary amines combine with carbonyl compounds (aldehydes and ketones) eliminating water and forming imines (Schiff's bases) (p. 100).

$$R{-}NH_2 + O{=}C(R')_2 \longrightarrow R{-}N{=}C(R')_2$$

Imine

4. Primary and secondary aliphatic amines react with nitrous acid, usually produced *in situ* by addition of sodium nitrite solution to an ice-cold solution of the amine in excess dilute mineral acid. The effective reagent is the anhydride, N_2O_3, and this reacts with primary amines producing alcohols via a complex sequence of reactions:

$$2\,HNO_2 \rightleftharpoons H_2O + O{=}N{-}O{-}N{=}O$$

$$R{-}\ddot{N}H_2 + O{=}N{-}O{-}N{=}O \longrightarrow R{-}\overset{H}{\underset{+}{N}}{-}N(O^-){-}O{-}N{=}O \xrightarrow[-NO_2^-]{-H^+} R{-}\overset{H}{N}{-}N{=}O$$

Alkyl nitrosoamine

$$R{-}\overset{H}{N}{-}N{=}O \quad H^+ \longrightarrow R{-}N{=}N{-}O{-}H + H^+$$

Alkyl diazotic acid

$$\downarrow +H^+$$

$$R{-}\overset{+}{N}{\equiv}N: \xleftarrow{-H_2O} R{-}N{=}N{-}\overset{+}{O}H_2$$

Alkyldiazonium cation

$$H_2\ddot{O}: + R{-}\overset{+}{N}{\equiv}N: \xrightarrow{-N_2} H_2\overset{+}{O}{-}R \xrightarrow{-H^+} H{-}\ddot{O}{-}R$$

Alkyl oxonium cation

$$R{-}\overset{+}{N}{\equiv}N: \longrightarrow N_2 + R^+ \xrightarrow{H_2O}$$

Initially the primary amine is converted by the nitrous anhydride into a nitrosoamine, which, in the presence of acid, rearranges to an isomeric diazotic acid. Protonation of the oxygen atom of this intermediate by the mineral acid present, followed by elimination of water, gives an unstable alkyldiazonium ion, which rapidly decomposes to nitrogen and a carbonium ion. The carbonium ion instantly reacts with water to form an alcohol, via an alkyloxonium ion.* This is effectively an S_N1 reaction (p. 52) of the alkyldiazonium cation. Alternatively, the diazonium ion may undergo an S_N2 reaction (p. 51) with water, the expulsion of nitrogen and formation of the alkyloxonium ion being concerted. The overall reaction is:

$$RNH_2 + HNO_2 \longrightarrow ROH + H_2O + N_2$$

Secondary amines react with nitrous acid, but in this case the nitrosoamine produced is stable and can be isolated usually as a yellow oil or solid. Many nitrosoamines are carcinogenic.

$$R_2NH + N_2O_3 \longrightarrow \underset{\text{Dialkylnitrosoamine}}{R_2N{-}N{=}O} + HNO_2$$

Tertiary amines react with nitrous acid with loss of one alkyl group and formation of a nitrosoamine. The course of the reaction is thought to be:

$$R_2N{-}CHR'_2 \xrightarrow{N_2O_3} R_2\overset{\overset{\displaystyle NO}{|}}{\overset{+}{N}}{-}CHR'_2 \xrightarrow{-\text{HNO (nitroxyl)}}$$

$$R_2\overset{+}{N}{=}CR'_2 \xrightarrow{H_2O} R_2NH + O{=}CR'_2 \xrightarrow{N_2O_3} R_2NNO$$

5. Alkali hypochlorites react with primary and secondary amines, replacing the hydrogen atoms attached to nitrogen by chlorine. The resulting 'chloroamines' find some use as disinfectants, since in aqueous solution they are hydrolysed to the amine and hypochlorous acid.

$$RNH_2 + NaOCl \longrightarrow RNHCl \longrightarrow RNCl_2$$

$$R_2NH + NaOCl \longrightarrow R_2NCl$$

$$R_2NCl + H_2O \rightleftharpoons R_2NH + HOCl$$

6. Oxidation of amines leads to a variety of products depending on the circumstances. Primary amines are oxidised by alkaline permanganates to the oxidation products of the corresponding alcohol (N.B. The alcohol corres-

* The conversion of a primary amine into an alcohol by nitrous acid is a reaction which frequently goes in poor yield. Besides combining with water to give the alcohol, the carbonium ion R^+ may combine with other nucleophiles present, e.g. NO_2^-, or lose a proton to form an alkene, or undergo structural rearrangements.

ponding to a primary amine is *not* necessarily a primary alcohol, e.g. $(CH_3)_2CHNH_2$ and $(CH_3)_2CHOH$).

$$CH_3CH_2NH_2 \xrightarrow{KMnO_4/\bar{O}H} CH_3CHO \longrightarrow CH_3CO_2H$$

$$(CH_3)_2CHNH_2 \xrightarrow{KMnO_4/\bar{O}H} (CH_3)_2CO$$

Under these conditions, secondary amines are oxidised to tetra-alkyl derivatives of hydrazine.

$$(H_3C)_2NH \xrightarrow{KMnO_4/\bar{O}H} (H_3C)_2N{-}N(CH_3)_2$$

Tetramethylhydrazine

Tertiary amines are not readily oxidised by permanganates, but are converted by hydrogen peroxide into amine oxides (cf. sulphoxides, p. 76).

$$CH_3{-}N(CH_3)_2{:} \xrightarrow{H_2O_2} CH_3{-}N^+(CH_3)_2{-}O^-$$

Trimethylamine
N-oxide

Enzymic oxidation of amines is an important biological process, e.g. in the oxidative deamination of amino acids, initial dehydrogenation of the primary amino acid to an imino acid is followed by rapid hydrolysis to the corresponding ketone (an oxo acid) and ammonia.

$$\underset{\text{Amino acid}}{R{-}CH(NH_2){-}CO_2H} \xrightarrow[\text{oxidase}]{\text{amino acid}} \underset{\text{Imino acid}}{R{-}C({=}NH){-}CO_2H} \xrightarrow{H_2O} NH_3 + \underset{\text{Oxo acid}}{R{-}C({=}O){-}CO_2H}$$

Aliphatic quaternary ammonium salts

The only method available for the preparation of quaternary ammonium salts is the alkylation of tertiary amines with alkyl halides or other alkylating species, e.g. dimethyl sulphate (p. 159).

$$(C_2H_5)_3N + C_2H_5I \longrightarrow (C_2H_5)_4\overset{+}{N}\,I^-$$

The quaternary ammonium salts are ionic solids, very soluble in water, methanol, or ethanol, but insoluble in most non-hydroxylic solvents such as ether, petrol, and benzene. The hydroxides, which can be obtained by reaction of aqueous solutions of the halides with silver oxide or by use of an ion-exchange resin, are water-soluble, ionic solids, as strongly basic as sodium hydroxide. Aqueous solutions of

$$R_4\overset{+}{N}\,I^- + Ag_2O + H_2O \longrightarrow R_4\overset{+}{N}\,\overset{-}{O}H + AgI$$

quaternary ammonium hydroxides rapidly absorb carbon dioxide from the air forming carbonates.

Quaternary ammonium salts have few reactions. Heating the halides results in the formation of a tertiary amine and an alkyl halide by nucleophilic attack of the halide anion on a carbon atom adjacent to the positively charged nitrogen atom—the reverse of the reaction used to prepare the salts.

$$(CH_3)_4\overset{+}{N}\,I^- \xrightarrow{\text{heat}} CH_3I + (CH_3)_3N$$

Heating quaternary ammonium hydroxides affords alkenes and tertiary amines (p. 29, 84):

$$\underset{HO^-\ \ H}{CH_2{-}CH_2{-}\overset{+}{N}(C_2H_5)_3} \longrightarrow H_2C{=}CH_2 \quad :N(C_2H_5)_3 \quad H_2O$$

Quaternary ammonium salts react with lithium aluminium hydride, $Li^+(AlH_4)^-$, to give alkanes and tertiary amines. The reaction is a nucleophilic attack of hydride ion, H^-, on a carbon atom adjacent to the positively charged nitrogen (cf. the pyrolysis of halides above).

$$\bar{H}{:}\ CH_3{-}\overset{+}{N}(CH_3)_3 \longrightarrow CH_4 + :N(CH_3)_3$$

Aromatic amines

Aromatic amines have the amino group attached directly to one or more aromatic rings, e.g.

Aminobenzene
Aniline
Phenylamine
(Primary)

Diphenylamine

N-Methylaniline
N-Methylphenylamine
Methylaminobenzene

(Secondary)

Triphenylamine

N-Methyldiphenylamine (Tertiary)

N,N-Diethylaniline
N,N-Diethylphenylamine
Diethylaminobenzene

Primary aromatic amines are most readily obtained by reduction of the corresponding nitro-compounds, which can often be obtained by direct nitration of an aromatic compound, or by the reaction of alkali hypobromite with amides (p. 84).

$$Ar{-}H \xrightarrow[H_2SO_4]{HNO_3} Ar{-}NO_2 \xrightarrow[\text{or } Ni/H_2]{Sn/HCl} Ar{-}NH_2$$

$$Ar{-}CONH_2 \xrightarrow{KOBr} ArNH_2 + CO_2$$

Since aryl halides are inert to nucleophiles (p. 56) they do not react directly with ammonia, but at high temperatures in the presence of copper or copper salts amines are produced. This is not a simple nucleophilic substitution, like the reaction of alkyl halides with ammonia, but proceeds via complex copper-containing intermediates.

$$Ar{-}Br + NH_3 \xrightarrow{Cu} Ar{-}NH_2 + NH_4Br$$

Secondary and tertiary aromatic amines can be obtained either by alkylation of primary aromatic amines by alkyl halides (nucleophilic substitution, p. 51), or by reaction with aryl halides and copper at high temperatures (see above).

$$C_6H_5NH_2 \xrightarrow{+CH_3I} C_6H_5NHCH_3 \xrightarrow{+CH_3I} C_6H_5N(CH_3)_2$$

$$C_6H_5NHCH_3 \xrightarrow[150°C]{+C_6H_5I\ +Cu} (C_6H_5)_2NCH_3$$

$$C_6H_5NH_2 \xrightarrow[150°C]{+C_6H_5Br\ +Cu} (C_6H_5)_2NH \xrightarrow{+CH_3I} (C_6H_5)_2NCH_3$$

$$(C_6H_5)_2NH \xrightarrow[200°C]{+C_6H_5I\ +\ Cu} (C_6H_5)_3N$$

Aromatic amines are liquids of high boiling point, or low melting point solids, very sparingly soluble in water, and much weaker bases than ammonia (cf. aliphatic amines). Overlap of the orbital containing the lone pair on the nitrogen atom with the aromatic π orbitals (p. 11) results in resonance between the canonical structures:

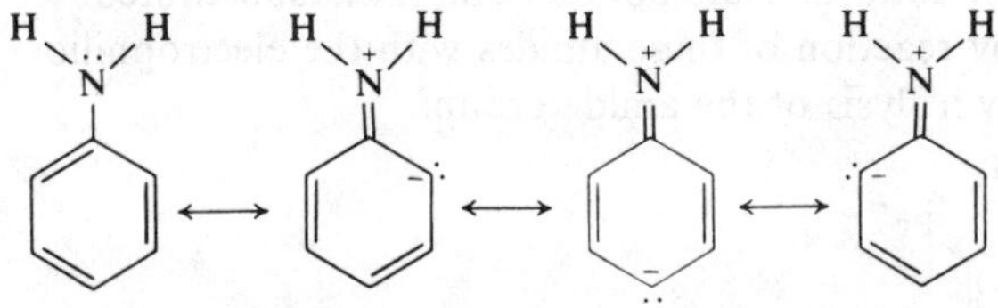

$$\text{PhNH}_2 \longleftrightarrow \text{(ortho carbanion, } \overset{+}{N}H_2\text{)} \longleftrightarrow \text{(para carbanion, } \overset{+}{N}H_2\text{)} \longleftrightarrow \text{(ortho' carbanion, } \overset{+}{N}H_2\text{)}$$

so that the lone pair is less readily available to form a bond to a hydrogen ion than in ammonia, where no such mesomeric interaction occurs.

Primary and secondary aromatic amines form salts and acyl derivatives just like the corresponding aliphatic amines. They also react with nitrous acid in a similar fashion (p. 86), primary aromatic amines giving phenols, and secondary aromatic amines forming nitrosoamines, but in the case of primary aromatic amines the intermediate diazonium cations are stable at 0°C, and diazonium salts can be isolated.

$$Ar{-}NH_2 \xrightarrow[0^\circ C]{NaNO_2/HCl} Ar{-}\overset{+}{N}{\equiv}N\ Cl^- \xrightarrow[H_2O]{T>5^\circ C} Ar{-}OH + N_2 + Cl^-$$

Arenediazonium salt

$$Ar_2NH \xrightarrow{NaNO_2/HCl} Ar_2N{-}N{=}O$$

Tertiary aromatic amines having two alkyl groups attached to the nitrogen atom react with nitrous acid to form compounds in which the aromatic ring has been nitrosated. These nitroso-derivatives are green-coloured bases, which form orange salts.

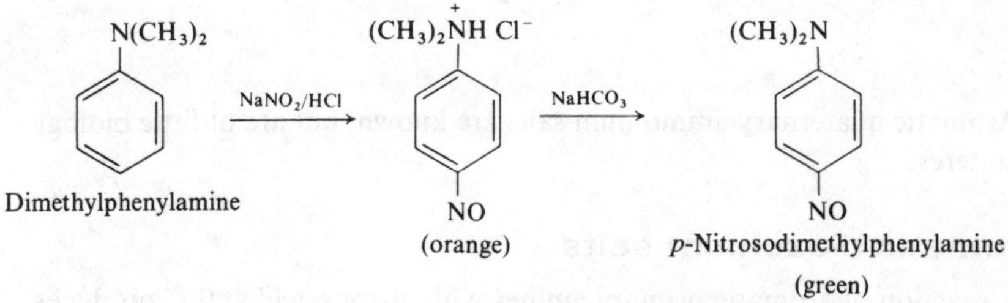

$$C_6H_5N(CH_3)_2 \xrightarrow{NaNO_2/HCl} p\text{-}ON{-}C_6H_4{-}(CH_3)_2\overset{+}{N}H\ Cl^- \xrightarrow{NaHCO_3} p\text{-}ON{-}C_6H_4{-}N(CH_3)_2$$

Dimethylphenylamine — (orange) — *p*-Nitrosodimethylphenylamine (green)

Electrophilic substitution of the aromatic ring of aryl amines occurs very readily in the *ortho*- and *para*-positions. The ease of substitution is indicated by the reaction of the tertiary amines with the feebly electrophilic nitrosonium ion, NO^+, to give nitroso-derivatives (see above and p. 71) (cf. nitronium ion, NO_2^+, p. 41). As is the case with phenols, direct halogenation or nitration cannot be controlled and leads to polysubstituted products (phenylamine and concentrated nitric acid react with explosive violence). However, acylation of the amino group of primary and secondary aromatic amines reduces the reactivity towards electrophiles, and monosubstituted products can be obtained by reaction of these amides with the electrophilic reagents, and subsequent hydrolysis of the amide group.

NH_2 Br_2/H_2O NH_2 Br Br Br

$(CH_3CO)_2O$

$NHCOCH_3$ Br_2 $NHCOCH_3$ Br Hydrolysis NH_2 Br

HNO_3 $NHCOCH_3$ NO_2 Hydrolysis NH_2 NO_2

$NHCOCH_3$ NO_2 Hydrolysis NH_2 NO_2

Aromatic quaternary ammonium salts are known, but are of little biological interest.

Aromatic diazonium salts

The reaction of aromatic primary amines with nitrous acid at 0°C produces diazonium salts, which are normally stable below 5°C. The pathway of this reaction is identical with that described for the reaction of aliphatic primary amines (p. 86).

$$C_6H_5NH_2 \xrightarrow[0°C]{NaNO_2/HCl} C_6H_5-\overset{+}{N}\equiv N \ Cl^-$$

These salts can be isolated as highly explosive, crystalline, ionic solids, but are generally prepared in solution and used without isolation. Although of considerable chemical significance, since the diazonium group can readily be replaced by many other functional groups, these compounds are of little biological interest, and will receive only brief attention here.

The reactions of aromatic diazonium salts can be divided into two groups, differing in the fate of the diazonium group.

Reactions of aromatic diazonium salts

A *In which nitrogen is formed.* The decomposition of diazonium salts in aqueous solution above 5°C leads to phenols with loss of nitrogen, but in the presence of halides and cuprous salts, the diazonium group is replaced by halogen. Similarly alkali cyanides with a cuprous cyanide catalyst will convert diazonium salts into aromatic nitriles.

CN + N_2 ← ($CN^-/CuCN$) — $C_6H_5N^+{\equiv}N$ — ($Cl^-/CuCl$) → Cl + N_2

(I^-) → I + N_2; (H_2O, T > 5°C) → OH + N_2; ($Br^-/CuBr$) → Br + N_2

B *In which the diazonium group is retained.*

1. Reduction of diazonium salts by stannous chloride and hydrochloric acid gives arylhydrazines.

$C_6H_5\overset{+}{N}{\equiv}N \xrightarrow{SnCl_2/HCl} C_6H_5NH{-}NH_2$

Phenylhydrazine

2. The diazonium cation is a weak electrophile, and will substitute reactive aromatic compounds. Phenols couple with diazonium salts in alkaline solution (the reaction is probably with the phenoxide anion) to give yellow or red 'azo-dyes', and coupling also occurs with tertiary aromatic amines. Primary and secondary aromatic amines react on the nitrogen atom, giving diaryltriazenes.

$$C_6H_5N_2^+ + C_6H_5O^- \longrightarrow C_6H_5N{=}NC_6H_4O^- \xrightarrow{\text{dil. acid}} C_6H_5N{=}NC_6H_4OH$$

An 'azo-dye'

$$C_6H_5N_2^+ + C_6H_5NR_2 \longrightarrow C_6H_5N{=}NC_6H_4NR_2$$

$$C_6H_5N_2^+ + C_6H_5NH_2 \longrightarrow C_6H_5N{=}N{-}NHC_6H_5$$

1,3-Diphenyltriazene

7

Simple carbonyl compounds

The carbonyl group, $>C{=}O$, is the common structural feature of a large number of functional groups, many of which are of great importance in naturally occurring substances, and some of which are illustrated and named below:

$R(H)C{=}O$	$R(R')C{=}O$	$R(HO)C{=}O$	$R(R'O)C{=}O$
Aldehyde	Ketone	Carboxylic Acid	Ester

$R(R'S)C{=}O$	$R(Cl)C{=}O$	$R(C{=}O)\text{–}O\text{–}(C{=}O)R$
Thioester	Acyl chloride	Carboxylic acid anhydride

$R(H_2N)C{=}O$	$R(R'HN)C{=}O$	$R(R'_2N)C{=}O$
Amides		

(R, R′ = Alkyl or aryl groups)

We shall consider the chemistry of several classes of carbonyl compounds, but, before doing so, some of the properties of the carbonyl group will be described, since it is the characteristics of this group which are responsible for much of the chemistry of carbonyl-containing functional groups.

The electronic structure of the carbonyl group follows from what has been described previously (p. 8). An sp^2 hybridised carbon atom and an oxygen atom, in which s and p orbitals are unmixed form a σ bond by overlap of an sp^2 orbital of carbon with a p orbital of oxygen. The π bond of the $\rangle C{=}O$ group is formed by overlap of p orbitals of carbon and oxygen, leaving two sp^2 orbitals on carbon available for bond formation, and an s and a p orbital on oxygen containing the two lone pairs.

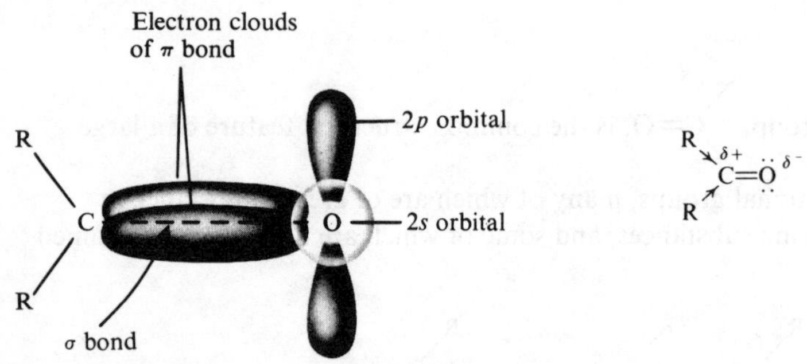

The electronic structure of a carbonyl group.

Because of the greater electronegativity of the oxygen atom, both C—O bonds are greatly polarised (p. 16), with the carbon atom bearing a partial positive charge. This positive charge in its turn causes an inductive displacement of electrons along the bonds joining the carbon atom to adjacent groups. These electronic factors are sufficient to explain many of the characteristic reactions of carbonyl compounds. The lone pairs of oxygen are the site of electrophilic attack; the carbon of the carbonyl group is the site of nucleophilic attack on account of its partial positive charge; and the inductive displacement along bonds between the carbon atom and neighbouring groups explains some of the characteristic properties of groups adjacent to carbonyl functions.

Aliphatic aldehydes and ketones

Nomenclature. The systematic nomenclature of organic compounds indicates the presence of an aldehyde either by the suffix 'al' to the name based on the parent hydrocarbon, or by describing the compound as a derivative of the corresponding carboxylic acid.

CH_3—CHO 'Ethanal' or 'Acetaldehyde' (from *acet*ic acid)

CH_3—CH(C_6H_5)—CHO '2-Phenylpropanal' or '2-Phenylpropionaldehyde' (from '2-*phenylpropion*ic acid)

C_6H_{11}—CHO 'Cyclohexanecarboxaldehyde' (from *cyclohexane carbox*ylic acid)

Ketones are described by the prefix 'oxo' or the suffix 'one'. In addition simple ketones are frequently given trivial names indicating the two groups attached to the carbonyl group.

CH_3—C(=O)—CH_2—CH_2—CH_3 '2-Oxopentane'
'Pentan-2-one'
(Methyl propyl ketone)

(cyclohexane ring with C=O) 'Oxocyclohexane'
'Cyclohexanone'

There are also many trivial names which persist in common usage, e.g. acetone for propan-2-one, acetophenone for methyl phenyl ketone.

Aliphatic aldehydes and ketones are the oxidation products of primary and secondary alcohols respectively. They are formally derived from alkanes by replacement of two hydrogen atoms of a methyl or methylene group by hydroxyl groups, followed by elimination of water from the '*gem*-diol'*

$$\text{R—CH}_3 \longrightarrow \text{R—CH(OH)}_2 \xrightarrow{-\text{H}_2\text{O}} \text{R—C(=O)—H} \quad \text{Aldehyde}$$

$$\underset{\text{Alkane}}{\text{R—CH}_2\text{—R}'} \longrightarrow \underset{gem\text{-Diol}}{\text{R—C(OH)}_2\text{—R}'} \xrightarrow{-\text{H}_2\text{O}} \underset{\text{Carbonyl compound}}{\text{R—C(=O)—R}'} \quad \text{Ketone}$$

(N.B. These schemes illustrate structural relationships only, and do not represent feasible synthetic routes for the preparation of carbonyl compounds.)

* '*Gem*'—disubstituted compounds have both functional groups attached to the same carbon atom (Latin *gemini*—twins).

Preparation of aldehydes and ketones

1. The oxidation of primary or secondary alcohols (p. 63) is the classical method of preparation of aldehydes and ketones. Chromic acid (acidified sodium dichromate) is frequently used in the laboratory, but industrially oxygen and catalysts are employed. When oxidising primary alcohols, care must be taken that the aldehyde produced is not oxidised further to a carboxylic acid. The oxidation of alcohols in biological systems uses complex coenzymes such as NAD^+ and $NADP^+$ (p. 299) as the oxidising agents.

$$R{-}CH_2OH \xrightarrow{CrO_3} R{-}CHO + H_2O$$

$$R{-}CHOH{-}R' \xrightarrow{CrO_3} R{-}CO{-}R' + H_2O$$

2. Hydrolysis of *gem*-dihaloalkanes gives aldehydes and ketones

$$R{-}CCl_2{-}R' \xrightarrow{H_2O} R{-}CO{-}R' + HCl$$

$$R{-}CHCl_2 \xrightarrow{H_2O} R{-}CHO + HCl$$

Acidic hydrolysis of a number of derivatives of aldehydes and ketones will also form the carbonyl compounds, e.g. oximes, hydrazones, etc. (p. 102).

3. Ozonolysis of alkenes (p. 33) affords aldehydes and/or ketones depending upon the structure of the alkene.

4. Oxidation of *vic*-diols* with periodic acid or lead tetra-acetate (p. 66) gives aldehydes or ketones depending upon the structure of the diol.

General preparations of ketones

5. Hydration of alkynes (p. 35) always produces ketones except for the hydration of ethyne.

6. Ketones can be synthesised from β-keto esters (p. 220).

Properties of aldehydes and ketones. Apart from methanal, HCHO, which is a gas, the lower aldehydes and ketones are neutral liquids, less soluble in water and of lower boiling point than the corresponding alcohols, and possessing characteristic pungent odours.

Reactions of aldehydes and ketones. The reactions of aliphatic aldehydes ar ketones can be divided into a number of sections, differing in the type of reaction and the group or atom initially involved. The sections which will be considered here are:

* Vicinally ('*vic*') disubstituted compounds have substituents on adjacent carbon atoms (Latin *vicinus*—neighbour).

A Nucleophilic attack on the carbonyl group.
B Electrophilic attack on the carbonyl group.
C Keto–enol tautomerism and related reactions.
D Oxidation of aldehydes and ketones.
E Miscellaneous reactions.

A *The reaction of aldehydes and ketones with nucleophiles.* Aldehydes and ketones form a series of characteristic derivatives by 'addition reactions' (i.e. reactions in which the molecular formula of the product is the sum of the molecular formulae of the reagents) and 'condensation reactions' (i.e. reactions in which combination of the reagents occurs with the elimination of water or an alcohol). Both of these types of reaction have a common initial pathway commencing with nucleophilic attack on the carbon of the carbonyl group.

The mechanism of addition of $H—\ddot{X}$ *to* $R—\overset{O}{\overset{\|}{C}}—R'$

$$RR'C^{\delta+}{=}\ddot{O}^{\delta-} + H—\ddot{X} \longrightarrow RR'C(\ddot{O}^{-})(\overset{+}{X}—H) \longrightarrow RR'C(\ddot{O}—H)(X:)$$

Overall reaction: R—CO—R′ + HX ⟶ R—C(OH)X—R′

The polarisation of the carbonyl group leaves the carbon atom with a partial positive charge. The nucleophilic group X of the reagent H—X donates a lone pair of electrons to the carbon atom, with simultaneous movement of the electrons of the π bond onto the oxygen atom, resulting in a doubly charged intermediate. The H—X bond in the intermediate will be highly polarised by the positive charge on X and the group will tend to lose H^+, whilst the negatively charged oxygen atom will be basic, and will tend to acquire a hydrogen ion. The result is a transfer of a hydrogen ion from X to oxygen, though in hydroxylic solvents (e.g. water or ethanol) the solvent may act as a proton acceptor and donor, so that it is not necessarily the identical proton lost from X which is attached to the oxygen atom.

(Alternatively, if H—X ionises to H^+ and X^- the initial nucleophilic attack may be by X^- rather than H—X, followed by addition of H^+ to the negatively charged oxygen atom.)

Addition reactions of aldehydes and ketones

$H—\ddot{X}$	+ $R(R')C=O \rightarrow$	*Product of Reaction*	
$H—C\equiv N$: (+KCN catalyst)		$R(R')C(OH)(CN)$	Cyanohydrin†
$H—\ddot{N}H_2$		$R(R')C(OH)(NH_2)$ *	
$H—\ddot{O}H$		$R(R')C(OH)(OH)$	*gem*-Diol *
$H—\ddot{O}R''$		$R(R')C(OH)(OR'')$	Hemiacetal *
$H—\ddot{S}—R''$		$R(R')C(OH)(SR'')$	Hemithioacetal *
$H—O—SO_2^- Na^+$		$R(R')C(OH)(SO_3^- Na^+)$	Sodium bisulphite adduct

R, R′ = alkyl or H
† Formation initiated by nucleophilic attack of CN^- (not HCN) on the carbonyl group.
* Usually impossible to isolate as pure compounds.

Addition reactions of this sort are frequently reversible, and some of the products mentioned above cannot be isolated. Ammonia and sodium bisulphite adducts are readily decomposed by dilute acid by the following mechanism:

$$R(R')C(\ddot{O}H)(X\!:) \xrightarrow{+H^+} R(R')C(\ddot{O}—H)(\overset{+}{X}—H) \xrightarrow{-HX} R(R')C=\overset{+}{O}—H \xrightarrow{-H^+} R(R')C=\ddot{O}$$

Condensation reactions occur between aldehydes or ketones and many compounds containing the $—NH_2$ group:

$$R(R')C=\ddot{O} + H_2N—Y \longrightarrow R(R')C=N—Y + H_2O$$

The mechanism of this reaction is:

$$R(R')C{=}\ddot{O} + H_2\ddot{N}{-}Y \longrightarrow R(R')C(\ddot{O}H)(\ddot{N}H{-}Y) \qquad \text{by the mechanism shown above}$$

$$R(R')C(\ddot{O}{-}H)(\ddot{N}(H){-}Y) \xrightarrow[(pH\sim 4{\cdot}5)]{+H^+} R(R')C(\overset{+}{O}H_2)(\ddot{N}(H){-}Y) \xrightarrow{-H_2O} R(R')C{=}\overset{+}{N}(H){-}Y \xrightarrow{-H^+} R(R')C{=}\ddot{N}{-}Y$$

Initial reaction of the carbonyl compound with $H_2N{-}Y$ forms the addition compound by the mechanism described above. Under weakly acidic conditions (e.g. an acetate buffer solution) protonation of the oxygen atom of the hydroxyl group is followed by the elimination of water and loss of a proton from the NH group, thereby forming the observed product. It may seem surprising that, in the step following the formation of the addition compound, protonation occurs on the oxygen atom rather than on nitrogen, when amines are well known to be more basic than alcohols. Protonation of nitrogen undoubtedly can occur, but the only reaction possibilities open to the resultant compound are loss of the newly acquired proton, or concerted elimination of

$$R(R')C(\ddot{O}{-}H)(\ddot{N}(H){-}Y) \underset{-H^+}{\overset{+H^+}{\rightleftharpoons}} R(R')C(\ddot{O}{-}H)(\overset{+}{N}H_2Y) \longrightarrow R(R')C{=}\ddot{O} \quad H^+ \quad H_2\ddot{N}{-}Y$$

a proton and $H_2N{-}Y$ to reform the original reagents. Although protonation of oxygen is not very favoured, it leads irreversibly (under these conditions) to the condensation compound.

Alternatively, condensation reactions may be catalysed by bases. The amino group of the addition compound can be deprotonated to form an anion, from which the hydroxyl group is subsequently expelled.

$$R(R')C(\ddot{O}{-}H)(\ddot{N}(H){-}Y) \underset{}{\overset{-H^+}{\rightleftharpoons}} R(R')C(\ddot{O}{-}H)(\ddot{\bar{N}}{-}Y) \xrightarrow{-\bar{O}H} R(R')C{=}\ddot{N}{-}Y$$

Typical condensation reactions of aldehydes and ketones

$H_2N{-}Y$		Product of Reaction
$H_2N{-}R$ Amine		$RR'C{=}N{-}R''$ Imine (Schiff's base)
$H_2N{-}OH$ Hydroxylamine		$RR'C{=}N{-}OH$ Oxime
$H_2N{-}NH{-}C(=O){-}NH_2$ Semicarbazide	$+\ RR'C{=}O \rightarrow$	$RR'C{=}N{-}NH{-}C(=O){-}NH_2$ Semicarbazone
$H_2N{-}NHC_6H_5$ Phenylhydrazine		$RR'C{=}N{-}NHC_6H_5$ Phenylhydrazone
$H_2N{-}NH{-}C_6H_3(NO_2)_2$ 2,4-Dinitrophenylhydrazine		$RR'C{=}N{-}NH{-}C_6H_3(NO_2)_2$ 2,4-Dinitrophenylhydrazone

$$\underset{\text{Hydrazine}}{H_2N{-}NH_2} + RR'C{=}O \longrightarrow \underset{\text{Hydrazone}}{RR'C{=}N{-}NH_2} \xrightarrow{+\ RR'C{=}O} \underset{\text{Azine}}{RR'C{=}N{-}N{=}CRR'}$$

As with some of the addition compounds, the products of these reactions can by hydrolysed by boiling with dilute mineral acid (e.g. 1M HCl, pH = 0). The mechanism of hydrolysis of these compounds is exactly the reverse of the formation.

$$R_2C{=}\ddot{N}{-}Y \xrightarrow{+H^+} RR'C{=}\overset{+}{N}H{-}Y \;(+\;H_2\ddot{O}) \longrightarrow RR'C(NH{-}Y)(\overset{+}{O}H_2) \xrightarrow{-H^+} RR'C(\ddot{N}H{-}Y)(\ddot{O}H)$$

$$RR'C(\ddot{N}H{-}Y)(\ddot{O}H) \xrightarrow{+H^+} RR'C(\overset{+}{N}H_2{-}Y)(\ddot{O}H) \longrightarrow RR'C{=}\ddot{O} + H_2\ddot{N}{-}Y + H^+$$

$$H_2\ddot{N}{-}Y + H^+ \rightleftharpoons H_3\overset{+}{N}{-}Y$$

Under these strongly acidic conditions (about 10^4 times more acidic than those used in the preparation) hydrolysis is initiated by protonation of the C=N group, and the succession of equilibria (all the stages illustrated are reversible) is displaced towards formation of the carbonyl compound by the final conversion of $H_2N{-}Y$ into its cation $H_3\overset{+}{N}{-}Y$, which, having no lone pair on the nitrogen atom, cannot initiate nucleophilic attack on the carbonyl group.

B *The reaction of aldehydes and ketones with electrophiles.* In the presence of catalytic amounts of strong acids, aldehydes polymerise to cyclic trimers and tetramers. The reaction, whose mechanism is given below, is initiated by protonation of the oxygen atom.

$$R{-}CH{=}\ddot{O} + H^+ \longrightarrow R{-}CH{=}\overset{+}{O}{-}H \longleftrightarrow R{-}\overset{+}{C}H{-}\ddot{O}{-}H$$

$$R{-}CH{=}\overset{+}{O}{-}H + \ddot{O}{=}CH{-}R \longrightarrow R{-}CH(OH){-}\overset{+}{O}{=}CH{-}R \longleftrightarrow R{-}CH(OH){-}O{-}\overset{+}{C}H{-}R$$

$$R{-}CH(OH){-}\overset{+}{O}{=}CH{-}R + O{=}CH{-}R \longrightarrow \text{cyclic } (R{-}CH{-}O)_3H^+ \longrightarrow \xrightarrow{-H^+} \text{cyclic trimer } (R{-}CH{-}O)_3$$

$$\text{i.e. } 3\,RCHO \xrightarrow{H^+} \begin{array}{c} RHC{-}O{-}CHR \\ | \qquad\quad | \\ O \qquad\quad O \\ \diagdown\ \diagup \\ CH \\ | \\ R \end{array}$$

$$\text{Similarly } 4\,RCHO \xrightarrow{H^+} \begin{array}{c} R \\ | \\ CH \\ \diagup\ \ \diagdown \\ O \qquad\quad O \\ \diagup \qquad\qquad \diagdown \\ R{-}CH \qquad\qquad CH{-}R \\ \diagdown \qquad\qquad \diagup \\ O \qquad\quad O \\ \diagdown\ \ \diagup \\ CH \\ | \\ R \end{array}$$

Thus ethanal (acetaldehyde) gives paraldehyde and metaldehyde by acid-catalysed polymerisation under different conditions:

$$\underset{\text{Metaldehyde}}{\begin{array}{c} CH_3 \\ | \\ CH \\ \diagup\ \ \diagdown \\ O \qquad\quad O \\ \diagup \qquad\qquad \diagdown \\ CH_3{-}CH \qquad\qquad CH{-}CH_3 \\ \diagdown \qquad\qquad \diagup \\ O \qquad\quad O \\ \diagdown\ \ \diagup \\ CH \\ | \\ CH_3 \end{array}} \xleftarrow[0°C]{HCl} CH_3CHO \xrightarrow[20°C]{H_2SO_4} \underset{\text{Paraldehyde}}{\begin{array}{c} H_3C \qquad\qquad CH_3 \\ \diagdown \qquad\quad \diagup \\ HC{-}O{-}CH \\ | \qquad\quad | \\ O \qquad\quad O \\ \diagdown\ \diagup \\ CH \\ | \\ CH_3 \end{array}}$$

These polymers are acetals (p. 115), and are readily reconverted into the parent aldehyde by heating with dilute mineral acid.

Ketones do not form polymers of this type, but on heating with concentrated sulphuric acid some ketones are converted into aromatic compounds, e.g. propanone (acetone) forms 1,3,5-trimethylbenzene (mesitylene).

$$3\,CH_3COCH_3 \xrightarrow{H_2SO_4} \text{1,3,5-}(CH_3)_3C_6H_3 + 3H_2O$$

It seems probable that the initial step in this condensation is the protonation of the oxygen atom of the carbonyl group.

C *Keto–enol tautomerism of aldehydes and ketones.* In the presence of bases, compounds containing the group

$$R'-\overset{\overset{R''}{|}}{\underset{\underset{H}{|}}{C}}-\overset{\overset{O}{\|}}{C}-R$$

are converted into anions by removal of a proton from the carbon atom adjacent to the carbonyl group.* The ease of removal of this proton is attributable to two causes, of which the latter is more important.

1. Inductive displacement along the C—CO bond produces a corresponding, weaker displacement along the C—H bond, thereby facilitating the removal of H^+ by a base.

$$-\underset{\underset{H}{\uparrow}}{\overset{|}{C}}\rightarrow\overset{|\delta+}{C}=\overset{\delta-}{O}$$

2. The negative charge on the resultant anion is not localised on the carbon atom, as the orbital containing the two electrons is conjugated (p. 11) to the π bond of the carbonyl group. The effect of this conjugation is to distribute the negative charge over more than one atom and in particular onto the electronegative oxygen atom, and such a mesomeric (i.e. resonance stabilised) anion is always much easier to produce than a corresponding anion in which the negative charge is localised on one atom.

$$R'-\overset{\overset{R''}{|}}{\underset{\underset{H}{|}}{C}}-\overset{\overset{R}{|}}{C}=\ddot{O}: \xrightarrow{-H^+} R'-\overset{\overset{R''}{|}}{\bar{C}}-\overset{\overset{R}{|}}{C}=\ddot{O} \longleftrightarrow R'-\overset{\overset{R''}{|}}{C}=\overset{\overset{R}{|}}{C}-\ddot{O}:^-$$

The anion produced in this way can be protonated in two ways, giving isomeric compounds, known as the 'keto' and 'enol' isomers, each of which can be reconverted into the common 'enolate' anion by loss of a proton.

$$-\overset{\gamma}{\overset{|}{C}}-\overset{\beta}{\overset{|}{C}}-\overset{\alpha}{\overset{|}{C}}-\overset{\overset{O}{\|}}{C}$$

* This is often referred to as the 'α position' from the old system of nomenclature.

$$\underset{\text{Keto isomer}}{R'-\overset{R''}{\underset{H}{C}}-\overset{R}{C}=\ddot{O}} \underset{+H^+}{\overset{-H^+}{\rightleftharpoons}} \left\{ \begin{array}{c} R'-\overset{R''}{\bar{C}}-\overset{R}{C}=\ddot{O} \\ \updownarrow \\ R'-\overset{R''}{C}=\overset{R}{C}-\ddot{O}:^- \end{array} \right\} \underset{-H^+}{\overset{+H^+}{\rightleftharpoons}} \underset{\text{Enol isomer}}{R'-\overset{R''}{C}=\overset{R}{C}-\ddot{O}-H}$$

Mesomeric enolate anion

Thus, in the presence of a base, any aldehyde or ketone with hydrogen atoms on the carbon atom adjacent to the carbonyl group will be in equilibrium with its enol isomer, conversion occurring via a low, equilibrium concentration of the enolate anion. In practice, although strong bases (e.g. $C_2H_5O^-$, $(CH_3)_3CO^-$) are needed to obtain high concentrations of enolate anions, even weak bases, such as the alkaline surface of glass, catalyse the interconversion of keto and enol isomers, so that any normal sample of an aldehyde or ketone will contain a low concentration of the enol, if such an isomer is possible. With simple ketones and aldehydes the proportion of enol is usually <1 per cent, but in special cases may rise to over 50 per cent.

In addition to the base-catalysed interconversion of keto and enol isomers, isomerisation is also catalysed by acid. The intermediate for acid-catalysed interconversion is the species produced by protonation of the carbonyl group. This cation can lose a proton from the oxygen atom to form the keto isomer, or from the α position to form the enol.

$$CH_3-\overset{:O:}{\overset{\|}{C}}-CH_3 \underset{-H^+}{\overset{+H^+}{\rightleftharpoons}} \left\{ \begin{array}{c} CH_3-\overset{H-:O^+}{\overset{\|}{C}}-CH_3 \\ \updownarrow \\ CH_3-\overset{H-\ddot{O}:}{\overset{|}{\overset{+}{C}}}-CH_3 \end{array} \right\} \underset{+H^+}{\overset{-H^+}{\rightleftharpoons}} CH_3-\overset{H-\ddot{O}:}{\overset{|}{C}}=CH_2$$

Some keto–enol equilibria

$$CH_3{-}CH{=}O \rightleftharpoons CH_2{=}CH{-}OH$$

$$CH_3CH_2{-}CH{=}O \rightleftharpoons CH_3CH{=}CH{-}OH$$

$$(CH_3)_2CH{-}CH{=}O \rightleftharpoons (CH_3)_2C{=}CH{-}OH$$

$$CH_3CH_2{-}C({=}O){-}CH_2CH_3 \rightleftharpoons CH_3CH{=}C(OH){-}CH_2CH_3$$

$$CH_3CH_2{-}C(OH){=}CH_2 \rightleftharpoons CH_3CH_2{-}C({=}O){-}CH_3 \rightleftharpoons CH_3CH{=}C(OH){-}CH_3$$

$$CH_3CH{=}C(OH){-}CH(CH_3)_2 \rightleftharpoons CH_3CH_2{-}C({=}O){-}CH(CH_3)_2 \rightleftharpoons CH_3CH_2{-}C(OH){=}C(CH_3)_2$$

$(CH_3)_3C{-}CHO$

$(CH_3)_3C{-}C({=}O){-}C(CH_3)_3$

No enol tautomer possible

Constitutional isomers which are rapidly interconverted are known as **'tautomers'**, and the phenomenon of tautomerism is most commonly encountered in compounds which have a 'mobile' hydrogen atom. Tautomerism is known to occur in other than carbonyl compounds, e.g. in the aliphatic nitro-compounds (p. 80), in which 'nitro' and 'aci' isomers are known, interconverted via a mesomeric anion:

$$\underset{\text{Nitro isomer}}{R_2CH{-}\overset{+}{N}(=\ddot{O})(\ddot{O}^-)} \underset{+H^+}{\overset{-H^+}{\rightleftharpoons}} \left\{ \begin{array}{c} R_2\bar{C}{-}\overset{+}{N}(=\ddot{O})(\ddot{O}^-) \\ \updownarrow \\ R_2C{=}\overset{+}{N}(\ddot{O}^-)(\ddot{O}^-) \end{array} \right\}_{\text{Mesomeric nitronate anion}} \underset{-H^+}{\overset{+H^+}{\rightleftharpoons}} \underset{\text{Aci isomer}}{R_2C{=}\overset{+}{N}(\ddot{O}{-}H)(\ddot{O}^-)}$$

It is important to distinguish between tautomerism and resonance. Tautomers are different compounds, which are normally rapidly interconverted, but which, in principle, are capable of isolation as distinct, demonstrably different compounds. (This separation has been achieved in a very few cases, p. 221).

Where canonical structures are written for a mesomeric compound, these are different, individually inadequate ways of describing a unique molecular species. Thus the keto and enol forms of propanone are different chemical compounds, which, on loss of a proton, both form the same mesomeric enolate anion, for which two canonical structures can be written.

The enolate anions of aldehydes and ketones are intermediates in a number of reactions. Two important examples are the halogenation of aldehydes and the aldol addition.

1. *Halogenation of aldehydes and ketones.* Ketones and aldehydes are readily substituted by halogens (including iodine) on the carbon atom adjacent to the carbonyl group. The rate of reaction is markedly accelerated by dilute alkali or acid, and kinetic studies have shown that the halogen reacts with the enol tautomer (or enolate anion), the catalytic effect of bases and acids being due to the formation of enolate anion and accelerated interconversion of keto and enol isomers respectively. The reaction of a halogen molecule with the enolate anion or enol is a normal electrophilic attack of a halogen on an alkene:

$$R{-}C(\text{:}\ddot{O}\text{:}^-){=}CH{-}R' \xleftarrow{\text{Base}} R{-}C(={:}O{:}){-}CH_2{-}R' \xrightarrow{\text{Acid}} R{-}C({:}O(H){:}){=}CH{-}R'$$

$$+\ Br{-}Br \qquad\qquad\qquad\qquad + \ Br{-}Br$$

$$\downarrow \qquad\qquad\qquad\qquad \downarrow$$

$$R{-}C(={:}O{:}){-}CH(Br){-}R' + Br^- \qquad\qquad R{-}C(={:}O{:}){-}CH(Br){-}R' + H^+ + Br^-$$

In the presence of excess halogen, all the hydrogen atoms in the α position will be replaced, whilst hydrocarbon groups remote from the carbonyl group

$$CH_3COCH_3 \xrightarrow[\text{Acetic acid}]{Cl_2} CCl_3COCCl_3 \text{ ultimately}$$

$$CH_3CHO \xrightarrow{I_2/NaOH} [CI_3CHO] \xrightarrow{H_2O/\bar{O}H} CHI_3 + HC\bar{O}_2$$

will not be substituted under these conditions, since enolisation has only local effect. To substitute these remote positions, radical conditions (i.e. halogen + light; p. 26) will be required.

2. *The aldol addition and related reactions.* In the presence of bases, most aldehydes and ketones are converted into dimers (compounds whose molecular formulae are twice that of the starting material—the monomer).

$$2\,CH_3-\overset{\displaystyle H}{\overset{|}{C}}=O \xrightleftharpoons[\text{or dil. NaOH}]{Na_2HPO_4} CH_3-\underset{\displaystyle OH}{\underset{|}{\overset{\displaystyle H}{\overset{|}{C}}}}-CH_2-\overset{\displaystyle H}{\overset{|}{C}}=O$$

3-Hydroxybutanal ('Aldol')

$$2\,CH_3-\overset{\displaystyle O}{\overset{\|}{C}}-CH_3 \xrightleftharpoons{Ba(OH)_2} CH_3-\underset{\displaystyle OH}{\underset{|}{\overset{\displaystyle CH_3}{\overset{|}{C}}}}-CH_2-\overset{\displaystyle O}{\overset{\|}{C}}-CH_3$$

2-Hydroxy-2-methylpentan-4-one
('Diacetone alcohol')

This reaction, known as the 'aldol addition'* after the trivial name of the ethanal dimer, occurs via the enolate anion, which, acting as a nucleophile, attacks the carbonyl group of another molecule of aldehyde or ketone.

$$R-CH_2-\overset{\displaystyle H}{\overset{|}{C}}=\ddot{O} \xrightleftharpoons[+H^+]{-H^+} R-\overset{-}{\ddot{C}}H-\overset{\displaystyle H}{\overset{|}{C}}=\ddot{O} \longleftrightarrow R-CH=\overset{\displaystyle H}{\overset{|}{C}}-\ddot{O}:^-$$

$$R-CH_2-\overset{\displaystyle H}{\overset{|}{C}}=\ddot{O} + R-\overset{-}{\ddot{C}}H-\overset{\displaystyle H}{\overset{|}{C}}=\ddot{O} \rightleftharpoons R-CH_2-\overset{\displaystyle H}{\overset{|}{C}}-\ddot{O}:^- \;(\text{bonded to } R-CH-\overset{\displaystyle H}{\overset{|}{C}}=\ddot{O})$$

$$\xrightleftharpoons[-H^+]{+H^+} R-CH_2-\underset{\displaystyle H}{\underset{|}{\overset{\displaystyle OH}{\overset{|}{C}}}}-\overset{\displaystyle R}{\overset{|}{C}H}-CH=O$$

$$\begin{array}{l} CH_2-O-PO_3H_2 \\ | \\ C=O \\ | \\ CH_2-OH \end{array} + \begin{array}{l} CH=O \\ | \\ H-C-OH \\ | \\ H-C-OH \\ | \\ CH_2-O-PO_3H_2 \end{array} \xrightleftharpoons{\text{aldolase}} \begin{array}{l} CH_2-O-PO_3H_2 \\ | \\ C=O \\ | \\ HO-C-H \\ | \\ H-C-OH \\ | \\ H-C-OH \\ | \\ H-C-OH \\ | \\ CH_2-O-PO_3H_2 \end{array}$$

Dihydroxyacetone phosphate   D-Erythrose-4-phosphate   D-Sedoheptulose-1,7-diphosphate

* Examples of the aldol addition are very important reactions in biochemistry, particularly in carbohydrate metabolism. One example is the combination of C_3 and C_4 sugars to give a C_7 sugar, a reversible reaction catalysed by the enzyme 'aldolase'.

The resultant alkoxide anion abstracts a proton from the solvent (usually water or ethanol) or from another molecule of the carbonyl compound. All the steps are reversible, and normally an equilibrium mixture of aldol and parent aldehyde or ketone is obtained. As in the case of halogenation, the reaction of simple aldehydes and ketones involves only carbon atoms adjacent to the carbonyl group, e.g.

$$2CH_3CH_2CH_2CHO \underset{\text{alkali}}{\overset{\text{weak}}{\rightleftharpoons}} CH_3CH_2CH_2CH(OH)CH(C_2H_5)CHO$$

Butanal — 2-Ethyl-3-hydroxyhexanal

When some aldehydes—but no ketones—are warmed with strong alkalies, dark, resinous products are obtained. This occurs only when the aldehyde has a methylene group adjacent to the carbonyl group, i.e. R—CH_2—CHO. The resin is formed via the aldol, which is converted into an enolate anion under the alkaline conditions. This enolate anion can lose a hydroxide ion, forming

$$R{-}CH_2{-}\overset{H}{\overset{|}{C}}{=}\ddot{O}\!: \overset{\text{alkali}}{\rightleftharpoons} R{-}CH_2{-}\overset{H}{\overset{|}{\underset{:\ddot{O}H}{\underset{|}{C}}}}{-}\overset{R}{\overset{|}{C}}H{-}\overset{H}{\overset{|}{C}}{=}\ddot{O}\!:$$

$$\Big\updownarrow \; +H^+ \;/\; -H^+$$

$$R{-}CH_2{-}\overset{H}{\overset{|}{\underset{:\ddot{O}H}{\underset{|}{C}}}}{-}\overset{R}{\overset{|}{C}}{=}\overset{H}{\overset{|}{C}}{-}\ddot{O}\!:^- \longleftrightarrow R{-}CH_2{-}\overset{H}{\overset{|}{\underset{:\ddot{O}H}{\underset{|}{C}}}}{-}\overset{R}{\overset{|}{\ddot{C}^-}}{-}\overset{H}{\overset{|}{C}}{=}\ddot{O}\!:$$

$$\Big\updownarrow \; +OH^- \;/\; -OH^-$$

$$R{-}CH_2{-}\overset{H}{\overset{|}{C}}{=}\overset{R}{\overset{|}{C}}{-}\overset{H}{\overset{|}{C}}{=}\ddot{O}\!:$$

an unsaturated aldehyde in which the newly formed double bond is conjugated with the carbonyl group. As a result of the conjugation of the C=C and C=O groups, the methylene group adjacent to the double bond can lose a proton to form an anion, since the mesomeric effect of the carbonyl group can be transmitted along the conjugated double bond (p. 11), and the anion produced by proton abstraction has a negative charge delocalised over five atoms, an even more favourable state than in the case of simple enolates in which delocalisation is over three atoms only.

$$R{-}CH_2{-}CH{=}C(R){-}C(H){=}O \xrightarrow{-H^+} R{-}\overset{..-}{C}H{-}CH{=}C(R){-}C(H){=}\ddot{O}$$

$$\updownarrow$$

$$R{-}CH{=}CH{-}C(R){=}C(H){-}\ddot{O}{:}^- \longleftrightarrow R{-}CH{=}CH{-}\overset{-}{C}(R){-}C(H){=}\ddot{O}$$

This anion, in its turn, can attack another molecule of the original aldehyde and form an aldol, which can lose water by a mechanism identical to that described previously.

$$R{-}CH_2{-}C(H){=}\ddot{O} + R{-}\overset{..-}{C}H{-}CH{=}C(R){-}C(H){=}\ddot{O}$$

$$\downarrow$$

$$R{-}CH_2{-}C(H)(:\ddot{O}:^-){-}CH(R){-}CH{=}C(R){-}C(H){=}\ddot{O}$$

$$\downarrow \; +H^+ \text{ (from the solvent)}$$

$$R{-}CH_2{-}C(H)(:\ddot{O}H){-}CH(R){-}CH{=}C(R){-}C(H){=}\ddot{O}$$

$$\downarrow \; -H^+$$

$$R{-}CH_2{-}C(H)(:\ddot{O}H){-}\overset{-}{C}(R){-}CH{=}C(R){-}C(H){=}\ddot{O} \longleftrightarrow R{-}CH_2{-}C(H)(:\ddot{O}H){-}C(R){=}CH{-}\overset{-}{C}(R){-}C(H){=}\ddot{O} \text{ etc.}$$

$$\downarrow \; -OH^-$$

$$R{-}CH_2{-}CH{=}C(R){-}CH{=}C(R){-}C(H){=}\ddot{O}$$

$$\downarrow \; \text{Repetition}$$

$$R{-}CH_2{-}\left({-}CH{=}\overset{R}{\overset{|}{C}}{-}\right)_n{-}CHO$$

An 'aldehyde resin'

Repetition of this process, via the conjugated enolates, produces a series of compounds with long chains of conjugated double bonds, responsible for the dark colours of the products.

Ketones possessing the $—CH_2—CO—$ group do not form resins with hot alkali, the position of the equilibrium in the initial aldol addition and subsequent steps being unfavourable for their formation. However, treatment with concentrated mineral acid results in the condensation of two or three molecules of ketone, the enol probably being the important intermediate in this case. Aromatic hydrocarbons are sometimes formed simultaneously (p. 104).

$$CH_3-\overset{O}{\overset{\|}{C}}-CH_3 \xrightarrow[\text{HCl}]{\text{conc.}} (H_3C)_2C{=}CH-\overset{O}{\overset{\|}{C}}-CH_3 + (H_3C)_2C{=}CH-\overset{O}{\overset{\|}{C}}-CH{=}C(CH_3)_2$$

$$(H_3C)_2C{=}\ddot{O} \underset{-H^+}{\overset{+H^+}{\rightleftharpoons}} (H_3C)_2C{=}\overset{..+}{O}-H \underset{+H^+}{\overset{-H^+}{\rightleftharpoons}} H_2C{=}C(CH_3)-\ddot{O}-H$$

$$(H_3C)_2C{=}\overset{+}{O}-H + H_2C{=}C(H_3C)-\ddot{O}-H \xrightarrow{-H^+} (H_3C)_2C(\ddot{O}-H)-CH_2-C(=\ddot{O})-CH_3 \xrightarrow{+H^+} (H_3C)_2C(\overset{+}{O}H_2)-CH(H)-C(=\ddot{O})-CH_3$$

$$\xrightarrow[-H^+]{-H_2O} (H_3C)_2C{=}CH-\overset{O}{\overset{\|}{C}}-CH_3$$

$$(H_3C)_2C{=}CH-\overset{O}{\overset{\|}{C}}-CH_3 \xrightarrow[(CH_3)_2C=O\ +\ H^+]{\text{Repeat}} (H_3C)_2C{=}CH-\overset{O}{\overset{\|}{C}}-CH{=}C(CH_3)_2$$

The formation of unsaturated, polymeric compounds by aldehydes and ketones is not possible in the absence of $—CH_2—CO—$ groups. Aldehydes and ketones containing only one α hydrogen atom are merely converted into the corresponding aldol by alkali. However aldehydes—but not ketones—which have no α hydrogen atom undergo the Cannizzaro reaction in the presence of concentrated alkali. This reaction results in mutual oxidation and reduction forming the carboxylate anion and alcohol corresponding to the aldehyde.

$$2\,R_3C{-}CHO + NaOH \longrightarrow R_3C{-}CH_2OH + R_3C{-}CO_2^-\,Na^+ \qquad (R \neq H)$$

The mechanism of this reaction involves initial nucleophilic attack of hydroxide ion on the carbonyl group, followed by transfer of a hydride ion (H:$^-$) to another molecule of the aldehyde. The resulting alkoxide anion and carboxylic acid exchange a proton to form the observed products. This reaction

$$R_3C{-}C({=}O){-}H + {:}\ddot{O}H^- \longrightarrow R_3C{-}C(O^-)(H){-}OH$$

$$R_3C{-}C(O^-)(H){-}OH + H{-}C({=}O){-}CR_3 \longrightarrow R_3C{-}C({=}O){-}OH + {}^-O{-}CH_2CR_3 \longrightarrow R_3C{-}CO_2^- + H{-}\ddot{O}{-}CH_2CR_3$$

is much slower than the aldol addition, and so is observed only when the rapid competing reaction is not possible.

The reactions of aldehydes under alkaline conditions can be summarised:

(a) $RCH_2CHO \xrightarrow[\text{alkali}]{\text{weak}} RCH_2CH(OH)CH(R)CHO \xrightarrow[\text{alkali}]{\text{strong}} RCH_2(CH{=}C(R))_nCHO$

(b) $R_2CHCHO \longrightarrow R_2CHCH(OH)CR_2CHO$

(c) $R_3CCHO \xrightarrow[\text{strong alkali}]{} R_3CCO_2^- + R_3CCH_2OH$

D *Oxidation of aldehydes and ketones*. Aldehydes are very readily oxidised to the corresponding carboxylic acids, in marked contrast to ketones, which are oxidised only with difficulty. As well as the usual powerful oxidising reagents such as chromic acid, potassium permanganate, and nitric acid, oxidation of aldehydes can be performed by a number of relatively weak oxidants such as alkaline solutions of copper(II) or silver(I) compounds. Fehling's solution (an alkaline solution of a copper(II) tartarate complex ion), Benedicts solution (alkaline copper(II) citrate complex), or Tollen's reagent (ammoniacal silver nitrate, i.e., *weakly* alkaline $[Ag(NH_3)_2]^+$ solution), will all oxidise simple aldehydes, being reduced to copper(I) oxide or metallic silver (the 'silver mirror' test). These reagents do not oxidise *simple* ketones (cf. carbohydrates, p. 248). (N.B. *Strongly* alkaline $[Ag(NH_3)_2]^+$ solution oxidises aldehydes very rapidly and many simple ketones more slowly, with the deposition of metallic silver.)

$$R{-}C(=O){-}H \xrightarrow{[O]} R{-}C(=O){-}OH$$

$$\text{e.g. } CH_3CHO + 2\,Cu^{++} + 3\,OH^- \longrightarrow CH_3CO_2^- + 2\,Cu^+ + 2\,H_2O$$

$$2\,Cu^+ + 2\,OH^- \longrightarrow Cu_2O\downarrow + H_2O$$

The oxidation of simple ketones is a much more difficult reaction. Propanone is only very slowly attacked by potassium permanganate or hot chromic acid. Prolonged, vigorous oxidation will, however, degrade a ketone by cleavage of the C—CO bond to give a mixture of carboxylic acids.

$$CH_3CH_2CH_2\vdots\overset{O}{\overset{\|}{C}}\vdots CH_2CH_3 \xrightarrow{CrO_3} \begin{cases} CH_3CH_2CH_2CO_2H + HO_2CCH_3 \\ CH_3CH_2CO_2H + HO_2CCH_2CH_3 \end{cases}$$

$$\text{cyclohexanone} \xrightarrow[\text{conc. } HNO_3]{\text{boiling}} \begin{matrix} H_2C{-}CH_2{-}CO_2H \\ | \\ H_2C{-}CH_2{-}CO_2H \end{matrix}$$

Adipic acid

E *Miscellaneous reactions*

1. *Reduction* of aldehydes or ketones can lead to either the parent alcohol or hydrocarbon, depending upon which method of reduction is employed. Reduction with lithium aluminium hydride or sodium borohydride ($NaBH_4$) leads to alcohols via nucleophilic attack of hydride ion on the carbonyl group.

$$RR'C{=}\ddot{O} + \bar{H}{:} \longrightarrow RR'C(H){-}\ddot{O}{:}^- \xrightarrow{+H_2O} RR'C(H){-}OH + \bar{O}H$$

Likewise metal–acid reduction (e.g. Zn/HCl, Sn/HCl) and catalytic hydrogenation (e.g. Ni/H_2) usually lead to the alcohol, but use of amalgamated zinc and hydrochloric acid (Clemmensen reduction) gives the hydrocarbon by a process which follows a totally different mechanism.

$$RR'C{=}O + Zn/Hg + HCl \longrightarrow RR'CH_2$$

An alternative method for conversion of aldehydes or ketones into the parent hydrocarbon is via the action of strong bases on the hydrazone (p. 102) (Wolff–Kishner reduction).

$$RR'C{=}\ddot{O} \xrightarrow{H_2NNH_2} RR'C{=}\ddot{N}{-}\ddot{N}H_2$$

$$RR'C{=}\ddot{N}{-}\ddot{N}H_2 \xrightarrow[-H^+]{KOH} RR'C{=}\ddot{N}{-}\bar{\ddot{N}}{-}H \longleftrightarrow RR'\bar{C}{-}\ddot{N}{=}\ddot{N}{-}H$$

$$\downarrow +H^+ \text{ (from the solvent)}$$

$$RR'C(H){-}\ddot{N}{=}\ddot{N}^- \xleftarrow{-H^+} RR'C(H){-}\ddot{N}{=}\ddot{N}{-}H$$

$$\downarrow$$

$$N_2 + RR'\bar{C}H \xrightarrow[\text{(from the solvent)}]{+H^+} RR'CH_2$$

$$\text{Overall reaction} \quad RR'C{=}NNH_2 \xrightarrow[\substack{(CH_2OH)_2 \\ 200°C}]{KOH \text{ in}} RR'CH_2 + N_2$$

2. *Formation of acetals*. The addition reaction of aldehydes and ketones with alcohols to form hemiacetals has already been described. In the presence of acid catalysts, these compounds react further with alcohols to form 'acetals', which are the dialkyl ethers of the diol:

$R_2C(OH)_2$	$R_2C(OR')(OH)$	$R_2C(OR')_2$
gem-Diol	Hemiacetal	Acetal

The mechanism of acetal formation:

$$R_2C{=}\ddot{O} + R'{-}\ddot{O}{-}H \rightleftharpoons R_2C(\ddot{O}{-}R')(\ddot{O}{-}H)$$

$$\xrightleftharpoons[-H^+]{+H^+} R_2C(\ddot{O}{-}R')(\overset{+}{\ddot{O}}H_2) \xrightleftharpoons[+H_2O]{-H_2O} R_2C{=}\overset{+}{\ddot{O}}{-}R' \;\; (+\; H{-}\ddot{O}{-}R')$$

$$\rightleftharpoons R_2C(\ddot{O}{-}R')(\overset{+}{\ddot{O}}(H){-}R') \xrightleftharpoons[+H^+]{-H^+} R_2C(\ddot{O}R')_2$$

Overall reaction $R_2C{=}O + 2R'OH \rightleftharpoons R_2C(OR')_2 + H_2O$

Initial formation of the hemiacetal (p. 99) is followed by protonation of the oxygen atom of the hydroxyl group and loss of water to form an oxonium ion. This cation reacts with a molecule of the alcohol to give a product which is transformed into the acetal by loss of a proton. It should be noted that all the steps are reversible, and that the reverse reaction is also acid-catalysed.

Whilst the equilibria favour acetal formation in the case of aldehydes, they are very unfavourable for ketones, which can be converted into their acetals only by removal of the water formed during the reaction and consequent displacement of the equilibria in the desired direction. Usually the orthoformic ester (p. 54) is used, which removes water by the irreversible, acid-catalysed reaction:

$$(R'O)_3CH + H_2O \xrightarrow{H^+} 2\,R'OH + R'OCHO$$

Trialkyl orthoformate; Alkyl formate / Alkyl methanoate

Acetals, though very readily hydrolysed by dilute acids, are stable to alkalis, being unaffected by prolonged boiling with dilute sodium hydroxide.

Thiols form thioacetals of aldehydes and ketones by an analogous sequence of reactions proceeding via the hemithioacetal. The usual catalyst employed is anhydrous zinc chloride, which also acts as a dehydrating agent.

$$R_2C{=}O + R'SH \longrightarrow R_2C(OH)(SR') \xrightarrow[R'SH]{ZnCl_2} R_2C(SR')_2$$

3. The oxygen atom of the carbonyl group of aldehydes and ketones can usually be replaced by chlorine atoms by reaction with phosphorus penta-chloride. This reaction provides a useful route to *gem*-dichloro compounds.

$$R_2C{=}O + PCl_5 \longrightarrow R_2CCl_2 + POCl_3$$

4. *Schiff's test*. Aldehydes, but not ketones, will restore the colour to solutions of magenta decolourised by sulphur dioxide. Although the chemistry of this reaction is obscure, it is known that the reaction is a specific test for the group —CHO, and is not given by the corresponding diol $—CH(OH)_2$, since in the few cases where both aldehyde and diol are obtainable as separate compounds (e.g. chloral, see below) only the free aldehyde gives this reaction. The reagent is sometimes used as a stain for histochemical purposes, polysaccharides (p. 249) being oxidised by sodium periodate (p. 66) to aldehydes, which are stained purple by the reagent.

Ethanal (acetaldehyde). CH_3CHO, is a sweet smelling, neutral liquid, b.p. 21°C, prepared commercially by the hydration of ethyne (acetylene) (p. 35). It is also produced in living systems as an intermediate in the enzymic oxidation of ethanol, or during the alcoholic fermentation of glucose. It has all the normal reactions of a typical aldehyde, and is oxidised by sodium hypohalites to the haloforms (e.g. chloroform). The cyclic trimer, paraldehyde (p. 104), is used as a sedative.

Chlorination of ethanal gives trichloroethanal, 'chloral', which is one of the few carbonyl compounds forming a stable diol, 'chloral hydrate'. The diol, which gives no Schiff's test, is a crystalline solid, m.p. 58°C,

$$CH_3CHO \xrightarrow{Cl_2} \underset{\text{Chloral}}{CCl_3CHO} \xrightarrow{H_2O} \underset{\text{Chloral hydrate}}{CCl_3CH(OH)_2}$$

whilst chloral itself is a liquid, b.p. 98°C. Chloral hydrate and several hemiacetals of chloral have been used as sedatives.

Methanal (formaldehyde). HCHO, is a pungent gas, b.p. −19°C, very soluble in water giving a solution known as 'formalin'. It is poisonous with powerful bactericidal action and consequently is used as a disinfectant and preservative. Although the simplest aldehyde, it is far from being typical of this class of compounds.

Methanal can be prepared by the oxidation of methanol with the usual reagents, or by catalytic oxidation with air over a silver or platinum catalyst.

$$CH_3OH + O_2 \xrightarrow[\text{catalyst}]{Ag} CH_2O + H_2O$$

It behaves normally in many of the addition and condensation reactions of aldehydes, but with ammonia forms a complex condensation compound.

$$6\,CH_2{=}O + 4\,NH_3 \longrightarrow (CH_2)_6N_4 + 6\,H_2O$$

Hexamethylenetetramine

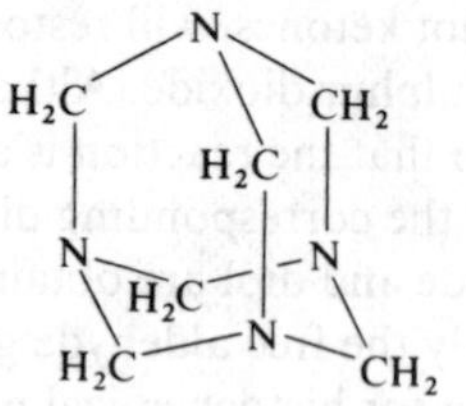

Methanal readily reacts with primary amines, converting them into methylene imines:

$$R{-}NH_2 + HCHO \longrightarrow [R{-}NH{-}CH_2OH] \xrightarrow{-H_2O} R{-}N{=}CH_2$$

Aqueous solutions of methanal slowly deposit a white polymeric solid, 'paraformaldehyde', which on warming with acid is converted into the normal cyclic trimer (p. 103):

$$H_2O + CH_2O \longrightarrow HO{-}CH_2{-}(O{-}CH_2)_n{-}OH$$

Paraformaldehyde

$$\downarrow H^+$$

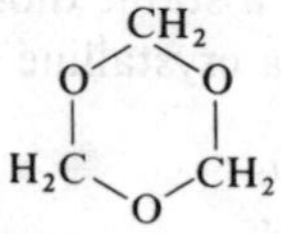

Trioxan

Because of the absence of any alkyl substituents on the carbonyl carbon atom, methanal cannot exist in an enol form (p. 105). Consequently, it is not polymerised by alkali, and undergoes the Cannizzaro reaction with concentrated alkali.

$$2\,CH_2O + KOH \longrightarrow HCO_2^-K^+ + CH_3OH$$

Methanal is a much more powerful reducing agent than other aldehydes. Fehling's solution is reduced to copper(I) oxide and metallic copper, and other weak oxidising agents can also be reduced, e.g. mercury(II) chloride is reduced to mercury(I) chloride. In these oxidations, methanal can be oxidised both to methanoic acid and thence to carbon dioxide and water.

$$HCHO \xrightarrow{[O]} HCO_2H \xrightarrow{[O]} CO_2 + H_2O$$

Propanone (acetone). CH_3COCH_3 is a neutral liquid, b.p. 56°C miscible with water. It occurs in the urine of diabetics. It is a typical ketone, and in addition to the normal reactions is converted into haloforms by sodium hypohalites, like all methyl ketones (p. 54). The sulphone derived from its diethylthioacetal has been used in the past as a hypnotic ('sulphonal').

$$(H_3C)_2C{=}O \xrightarrow[ZnCl_2]{C_2H_5SH} (H_3C)_2C(SC_2H_5)_2 \xrightarrow{KMnO_4} (H_3C)_2C(SO_2C_2H_5)_2$$

Sulphonal

Aromatic aldehydes and ketones

Aromatic aldehydes have the functional group —CHO attached directly to the aromatic ring, e.g.

Benzaldehyde — 4-Methylbenzaldehyde — 3-Nitrobenzaldehyde

They may be prepared by oxidation of the corresponding primary alcohol or hydrolysis of the *gem*-dichloro compound (pp. 47, 98). They are neutral liquids or solids, only very sparingly soluble in water, and of higher boiling point than the simple aliphatic aldehydes. The reactivity of the carbonyl groups is somewhat less than that of aliphatic aldehydes.

Aromatic aldehydes form most of the normal addition and condensation compounds of aliphatic aldehydes (p. 99), but, having no hydrogen atom on the carbon atom adjacent to the carbonyl group, they do not form enol tautomers (p. 105), and under strongly alkaline conditions undergo the Cannizzaro reaction (p. 112).

$$2\,ArCHO + KOH \longrightarrow ArCO_2^-K^+ + ArCH_2OH$$

Oxidation to the corresponding carboxylic acid proceeds readily with powerful oxidising agents (dilute nitric acid, chromic acid, potassium permanganate), but aromatic aldehydes do not always reduce Fehling's solution or ammoniacal silver nitrate (p. 113).

Reduction proceeds normally (p. 114), giving either the corresponding alcohol or hydrocarbon depending upon the reagent employed.

Electrophilic substitution of the aromatic nucleus (p. 42) occurs principally *meta* with respect to the carbonyl group.

Aromatic ketones may have either two aryl groups attached to the carbonyl group, or aryl and alkyl groups:

Methyl phenyl ketone (Acetophenone)

Diphenyl ketone (Benzophenone)

They may be prepared from the corresponding alcohol or *gem*-dihalo compound (p. 98) or by the Friedel–Crafts reaction of an acyl chloride with aromatic hydrocarbons (p. 40).

Alkyl aryl ketones behave as normal ketones, performing the usual addition and condensation reactions. Tautomerism is possible for these ketones with hydrogen atoms in the α position. Oxidation is difficult, and leads exclusively to the aromatic carboxylic acid:

$$C_6H_5CO{-}CH_2R \xrightarrow[KOH]{KMnO_4} C_6H_5CO_2^-K^+ + RCO_2^-K^+$$

Reduction to the corresponding secondary alcohol or hydrocarbon can be effected with the normal reagents (p. 114). Electrophilic substitution of the aromatic nucleus occurs predominantly *meta* with respect to the carbonyl group.

Diaryl ketones, though very much less reactive than dialkyl ketones, show the expected reactions of the carbonyl group.

Quinones

Quinones are the oxidation products of dihydric phenols, and they may be prepared by oxidation of the corresponding dihydroxybenzenes by weak oxidising agents such as iron(III) chloride, silver oxide, or even air.

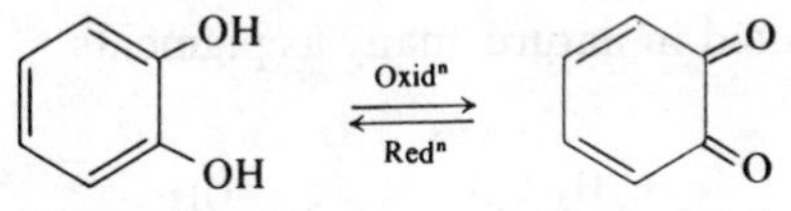

OH OH ⇌ (Oxid^n / Red^n) O O

1,2-Dihydroxybenzene (Catechol) — *ortho*-Benzoquinone

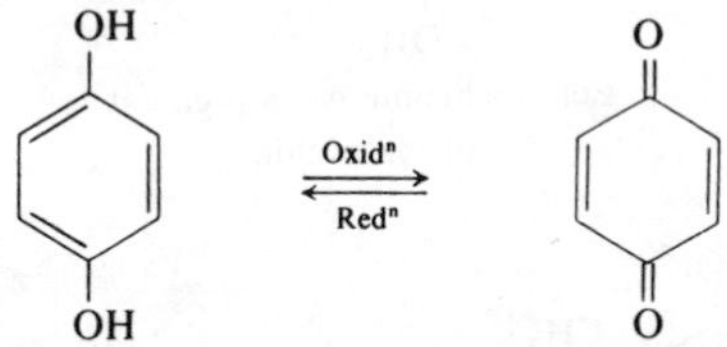

OH OH ⇌ (Oxid^n / Red^n) O O

1,4-Dihydroxybenzene (Quinol) — *para*-Benzoquinone

They have a very few ketonic properties. Oximes are formed by *para*-quinones, but scarcely any other of the normal derivatives can be prepared.

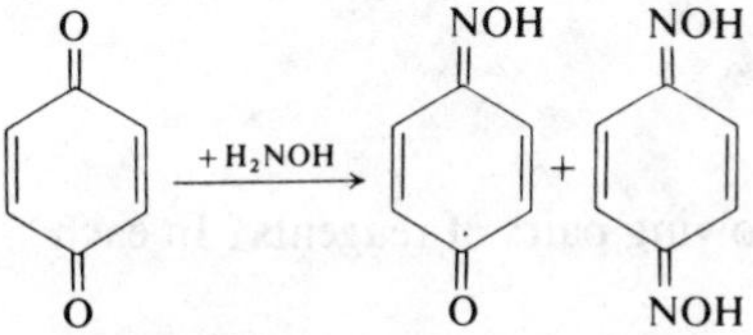

O O $\xrightarrow{+H_2NOH}$ NOH O + NOH NOH

Reduction by weak reducing agents, such as sulphurous acid, converts quinones into the corresponding dihydric phenols. Numerous biological compounds are known in which 'quinonoid' systems occur, and some of these are known to be involved in important oxidation or reduction steps in biochemical processes. In these cases a quinonoid structural feature is found, which can be reduced to the corresponding 'benzenoid' structure by an easily reversible process.

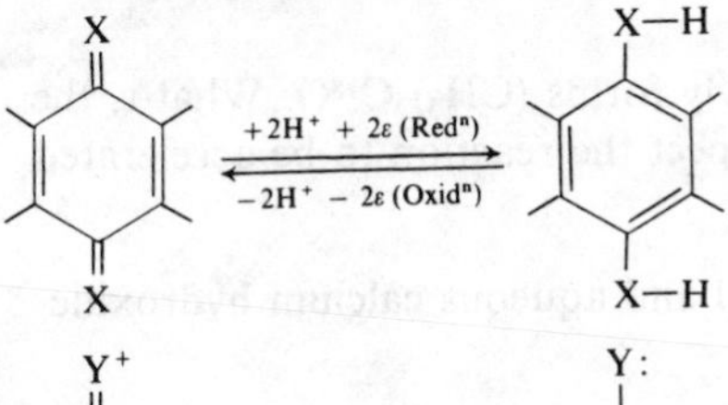

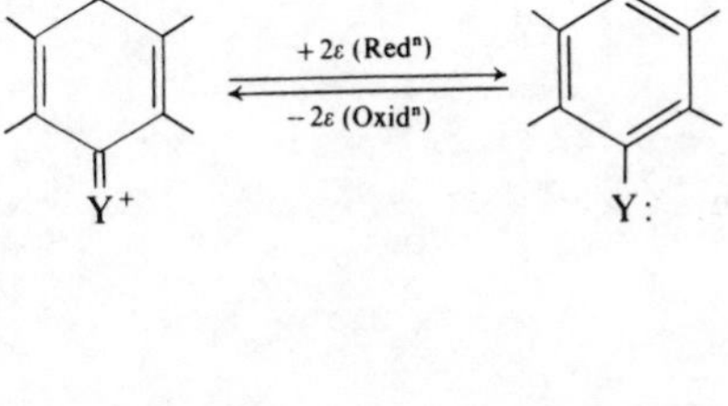

X X ⇌ (+2H$^+$ + 2ε (Red^n) / −2H$^+$ − 2ε (Oxid^n)) X—H X—H

Y$^+$ Y$^+$ ⇌ (+2ε (Red^n) / −2ε (Oxid^n)) Y: Y:

Complex quinones are widely distributed in nature, many as pigments:

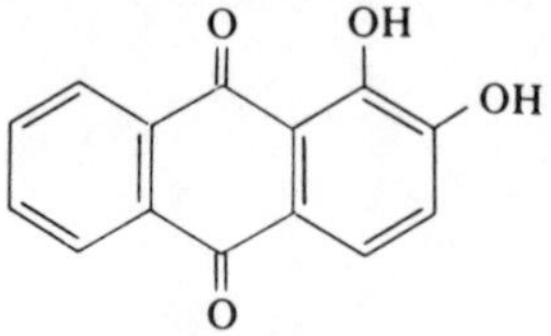

Alizarin—a plant pigment once used as a dye

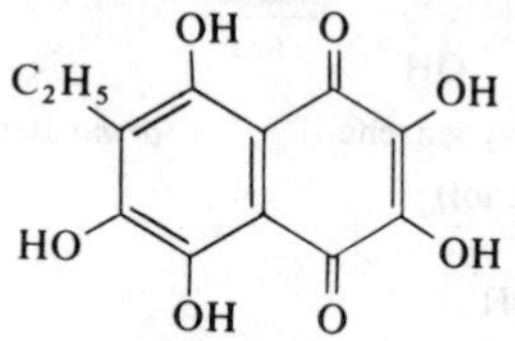

Echinochrome A—a pigment of sea-urchins

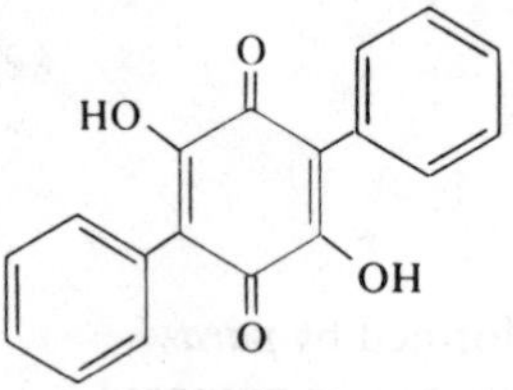

Polyporic acid —a fungal pigment

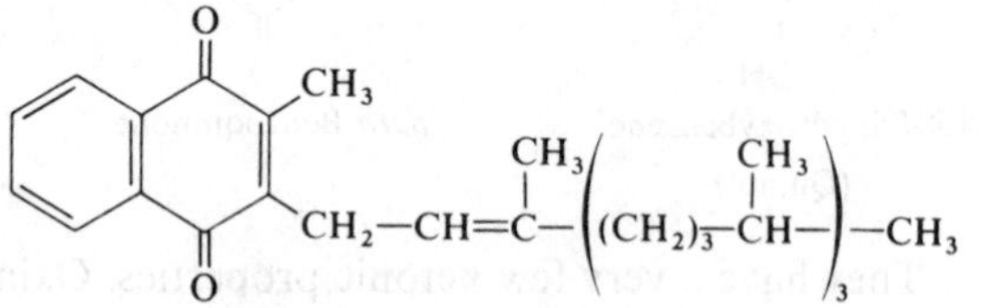

Vitamin K_1—a blood coagulation factor

Problems

1. What products are expected from the following pairs of reagents? In each case give the mechanism of reaction.
   $CH_3CH_2CHO + H_2NN(CH_3)_2$
   $(CH_3)_2CHCH_2CHO + NaOH$
   $CH_3CH_2COC_6H_5 + Br_2 + CH_3CO_2H$
   $(CH_3)_2CHCOCH_3 + C_2H_5SH + ZnCl_2$
2. A solution of ethanal in deuterium oxide, D_2O, (heavy water) slowly forms CH_2DCHO, CHD_2CHO, and CD_3CHO. This exchange is greatly accelerated by both acids and bases. How do these catalysts accelerate the isotopic exchange?
3. A solution of propanone in $H_2{}^{18}O$, slowly forms $(CH_3)_2C^{18}O$. What is the mechanism of this reaction? Do you expect the reaction to be accelerated by acid or base?
4. Treatment of nitroethane with methanal and aqueous calcium hydroxide gives the compound

   $$CH_3-C(CH_2OH)_2-NO_2$$

   What is the mechanism of this reaction?

5. Do you expect treatment of crotonaldehyde, $CH_3CH{=}CHCHO$, with iodine and sodium carbonate to give $CH_2ICH{=}CHCHO$? Explain.

6. Propanone and ethane-1,2-diol react in the presence of acid catalysts to give the compound

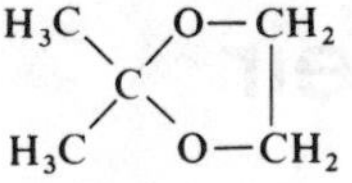

$$(H_3C)_2C\langle\begin{matrix}O-CH_2\\ |\\ O-CH_2\end{matrix}$$

What is the mechanism of this reaction?

7. By what mechanism are glyceraldehyde and dihydroxypropanone interconverted (p. 68)? Would the reaction be catalysed by acid or base?

8. Rationalise the following reactions mechanistically.

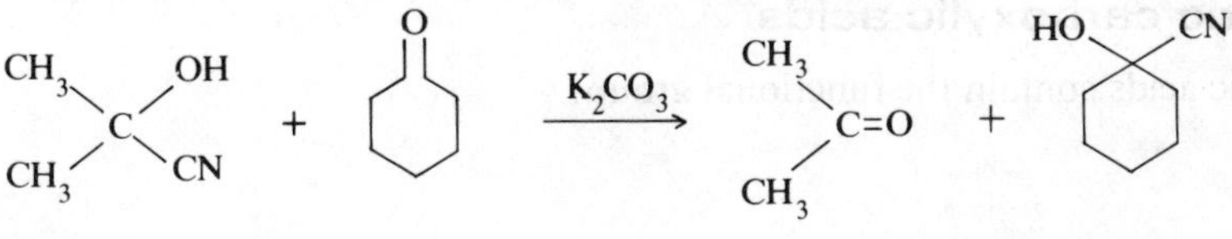

$$(CH_3)_2C(OH)CN + \text{cyclohexanone} \xrightarrow{K_2CO_3} (CH_3)_2C{=}O + \text{1-hydroxycyclohexanecarbonitrile (HO, CN)}$$

$$HC(OCH_3)_3 + H_2O \xrightarrow{H^+} HCO_2CH_3 + 2CH_3OH$$

$$C_6H_5CHO + CH_3NO_2 \xrightarrow{^-OH} C_6H_5CH{=}CHNO_2$$

9. If the cyclic ketone shown below is left to react with NaOD in D_2O for a prolonged period, which hydrogen atoms will be replaced by deuterium?

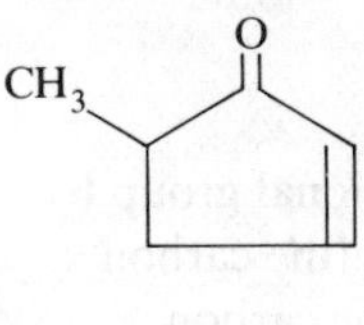

CH_3

8

Carboxylic acids and their derivatives

Aliphatic carboxylic acids

Carboxylic acids contain the functional group:

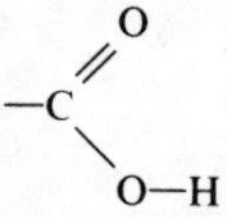

$$-C(=O)-O-H$$

which may formally be derived from a methyl group by substitution of all the hydrogen atoms by hydroxyl groups, followed by elimination of water from the ortho-acid so obtained.

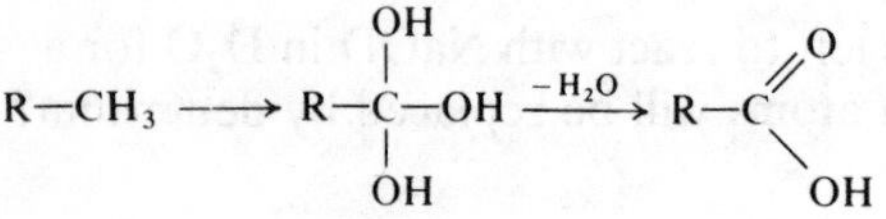

$$R-CH_3 \longrightarrow R-C(OH)_2-OH \xrightarrow{-H_2O} R-C(=O)-OH$$

Ortho-acid

Systematic nomenclature indicates the presence of this functional group by the suffix '-oic acid' attached to the name of the parent alkane (the carbon atom of the carboxylic acid group counting as one of the alkane carbon atoms), and the carbon chain is numbered from this functional group:

$CH_3(CH_2)_8CO_2H$

Decanoic acid

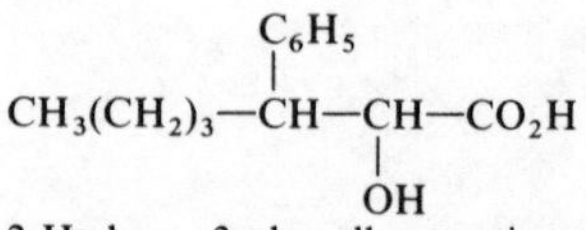

$CH_3(CH_2)_3-CH(C_6H_5)-CH(OH)-CO_2H$

2-Hydroxy-3-phenylheptanoic acid

Alternatively the carboxylic acid group may be regarded as a substituent, and described by the suffix 'carboxylic acid':

CO_2H

Cyclohexanecarboxylic acid

The anions derived from carboxylic acids are described by the suffixes '-oate' or '-carboxylate'.

$CH_3CH_2CH(CH_3)CO_2^-Na^+$
Sodium 2-methylbutanoate

$CO_2^- K^+$
Potassium cyclobutanecarboxylate

As with many other classes of compounds, long-established, trivial names persist in use for the first few members of the series:

	TRIVIAL NAME	SYSTEMATIC NAME
HCO_2H	Formic acid	Methanoic acid
CH_3CO_2H	Acetic acid	Ethanoic acid
$CH_3CH_2CO_2H$	Propionic acid	Propanoic acid
$CH_3CH_2CH_2CO_2H$	Butyric acid	Butanoic acid
$(CH_3)_2CHCO_2H$	Isobutyric acid	2-Methylpropanoic acid
$CH_3(CH_2)_3CO_2H$	Valeric acid	Pentanoic acid

and an archaic designation of the carbon atoms of the chain by Greek letters is still widely used:

$$\overset{\delta}{C}-\overset{\gamma}{C}-\overset{\beta}{C}-\overset{\alpha}{C}-CO_2H$$

e.g. $CH_3CHClCO_2H$ $\begin{cases}\text{2-Chloropropanoic acid}\\ \alpha\text{-Chloropropionic acid}\end{cases}$

(Note that, in this scheme, the α carbon atom is number 2 in the systematic method of numbering.)

Preparation of carboxylic acids

1. Carboxylic acids may be prepared by the oxidation of primary alcohols and aldehydes (pp. 63, 113); or by the more difficult oxidation of ketones (p. 114).

2. The hydrolysis of esters (p. 139), anhydrides, acyl chlorides, amides (p. 142), or nitriles (p. 149) leads to the corresponding carboxylic acids.

$$\left.\begin{array}{l} R{-}CO{-}OCH_3 \\ (R{-}CO)_2O \\ R{-}CO{-}Cl \\ R{-}CO{-}NH_2 \\ R{-}C{\equiv}N \end{array}\right\} \xrightarrow{H^+/H_2O} \left\{\begin{array}{l} R{-}CO_2H + CH_3OH \\ 2R{-}CO_2H \\ R{-}CO_2H + HCl \\ R{-}CO_2H + NH_4^+ \\ R{-}CO_2H + NH_4^+ \end{array}\right.$$

3. Carboxylic acids may be synthesised from diethyl malonate (p. 166) or ethyl acetoacetate (p. 220) by standard procedures.

Properties of carboxylic acids. The lower aliphatic carboxylic acids are pungent liquids of higher boiling point than the corresponding alcohols, and miscible with, or very soluble in water. The solubility of alkanoic acids in water decreases progressively as the non-polar hydrocarbon moiety increases in size the long-chain fatty acids (e.g. stearic acid, $CH_3(CH_2)_{16}CO_2H$) being waxy solids, insoluble in water.

Reactions of carboxylic acids

1. The most characteristic reaction of carboxylic acids is their ionisation:

$$R{-}C(=O){-}O{-}H \rightleftharpoons R{-}C(=O){-}O^- + H^+$$

The ease with which ionisation occurs is attributable to two causes, similar tc those cited for the very much more feebly acidic behaviour of ketones (p. 105

(a) The displacement of electrons along the double bond of the carbonyl group towards the oxygen atom leaves a partial positive charge on the carbon atom, which causes an inductive displacement along the C—OH and O—H bonds away from the hydrogen atom. The hydrogen atom can therefore be removed by interaction with a base and, in fact, ionisation of carboxylic acids is appreciable only in the presence of suitable proton acceptors (e.g. H_2O) and is negligible in hydrocarbon solvents.

$$R \rightarrow \overset{\delta+}{C}(=\overset{\delta-}{O}){-}O{\cdots}H$$

(b) The anion produced by loss of a proton is a resonance hybrid of two canonical structures. The delocalisation of the charge stabilises the anion, which is therefore more easily formed, e.g. in aqueous solution

$$R{-}C(=\ddot{O}{:}){-}\ddot{O}{-}H + {:}\ddot{O}(H){-}H \rightleftharpoons R{-}C(=\ddot{O}{:}){-}\ddot{O}{:}^- + H{-}\overset{+}{\ddot{O}}(H){-}H$$

$$\updownarrow$$

$$R{-}C(-\ddot{O}{:}^-)=\ddot{O}$$

The simple alkanoic acids are weak acids (p. 133) giving rise to stable salts with strong bases, e.g. $CH_3CO_2^-K^+$. They are stronger acids than carbonic acid, and will therefore liberate carbon dioxide from carbonates and bicarbonates.

2. In the presence of mineral acid catalysts, or very slowly in their absence, carboxylic acids react with alcohols to form esters.

$$R{-}C(=O){-}OH + H{-}O{-}R' \underset{}{\overset{H^+}{\rightleftharpoons}} R{-}C(=O){-}O{-}R' + H_2O$$

$$\text{e.g. } CH_3CO_2H + C_2H_5OH \rightleftharpoons CH_3CO_2C_2H_5 + H_2O$$

Acetic acid / Ethanoic acid; Ethanol; Ethyl acetate / Ethyl ethanoate

The mechanism of this important, reversible reaction is described later (p. 137). Esters can also be formed by the interaction of silver salts of carboxylic acids with alkyl halides in a typical nucleophilic displacement (p. 50).

$$RCO_2^-Ag^+ + R'I \longrightarrow RCO_2R' + AgI$$

3. When the ammonium salts of carboxylic acids are heated, amides are formed slowly,

$$RCO_2^-NH_4^+ \xrightarrow{\text{heat}} RCONH_2 + H_2O$$

The direct conversion of carboxylic acids into amides is normally achieved by heating a mixture of the ammonium salt and carboxylic acid, or by passing a stream of ammonia gas continuously into the heated acid.

4. Carboxylic acids react with non-metal halides to form acyl halides (cf. alcohols, p. 50).

$$R{-}C(=O)OH \xrightarrow[\text{or } SOCl_2]{+PCl_3,\ PCl_5} R{-}C(=O)Cl$$

$$R{-}C(=O)OH \xrightarrow{+PBr_3} R{-}C(=O)Br$$

All the above reactions are ones which formally involve the hydroxyl group alone. Carboxylic acids do not show the addition and condensation reactions of the carbonyl group, which are so characteristic of aldehydes and ketones, nor is the formation of enolate anions, analogous to those of aldehydes and ketones, observed with carboxylic acids. The basic conditions required to convert a ketone or aldehyde into its enolate convert a carboxylic acid into its carboxylate anion, and removal of a second proton would involve production

$$R{-}\overset{R}{\underset{\cdot\cdot}{C^-}}{-}\overset{O}{\overset{\|}{C}}{-}O{-}H \not\longleftarrow R{-}\overset{R}{\underset{H}{C}}{-}\overset{O}{\overset{\|}{C}}{-}O{-}H \xrightarrow{+\text{ base}} R{-}\overset{R}{\underset{H}{C}}{-}\overset{O}{\overset{\|}{C}}{-}O^-$$

of a second negative charge in close proximity to that already present, a very difficult process indeed.

5. Halogenation of aliphatic carboxylic acids can occur by radical attack on the alkyl group under the usual conditions, i.e. chlorine or bromine in the presence of light (p. 26). In this way ethanoic acid (acetic acid) can

$$CH_3CO_2H \xrightarrow[\text{light}]{Cl_2} CH_2ClCO_2H,\ CHCl_2CO_2H,\ CCl_3CO_2H$$

$$CH_3CH_2CH_2CO_2H \xrightarrow[\text{light}]{Cl_2} CH_3CH_2CHClCO_2H,\ CH_3CHClCH_2CO_2H$$
$$CH_2ClCH_2CH_2CO_2H,\ \text{etc.}$$

be smoothly converted into trichloroethanoic acid in three steps. However, with other alkanoic acids attack occurs randomly along the hydrocarbon chain leading to mixtures of products.

6. Reduction of carboxylic acids is very difficult and the normal metal–acid reducing systems are ineffective. Catalytic hydrogenation at high pressures (100 atm) with a copper chromite ($CuCrO_2$) catalyst reduces acids to alcohols, but carboxylic acids are inert to hydrogenation with the normal

metal catalysts (nickel, palladium, platinum). Lithium aluminium hydride reduces carboxylic acids smoothly to the corresponding alcohols.

$$R{-}CO_2H \xrightarrow[\text{or } H_2/CuCrO_2]{LiAlH_4} R{-}CH_2OH$$

7. The decarboxylation and electrolysis of salts of carboxylic acids are described elsewhere (pp. 222; 25).

Ethanoic acid (Acetic acid), CH_3CO_2H, is obtained commercially either by the catalytic oxidation of ethanol, or by hydration of ethyne to ethanal (p. 35) followed by oxidation to ethanoic acid.

$$C_2H_2 \xrightarrow[Hg^{++}]{H_2O/H^+} CH_3CHO \xrightarrow{[O]} CH_3CO_2H$$

It occurs naturally, e.g. in vinegar, being produced by bacterial oxidation of ethanol. The pure acid is a solid of m.p. 17°C (hence 'glacial' acetic acid), miscible with water and burning with difficulty. It is a typical aliphatic carboxylic acid, of $pK = 4{\cdot}74$.

Methanoic acid (Formic acid), HCO_2H, is unique in having no alkyl or aryl group attached to the carboxylic acid group. It may be prepared by the oxidation of methanol or methanal, or by hydrolysis of its esters, hydrocyanic acid (the nitrile of methanoic acid), or chloroform (p. 54). Commercially, methanoic acid is prepared by the catalytic hydration of carbon monoxide.

$$CO \xrightarrow[\text{catalyst}]{+H_2O} HCO_2H$$

Methanoic acid, $pK = 3{\cdot}62$, is a much stronger acid than the other simple aliphatic carboxylic acids. This is partly due to the absence of an alkyl group substituent on the carbonyl carbon atom. The H—C bond in methanoic acid is much less polarisable than the corresponding C—C bond in ethanoic acid, and consequently polarisation of the C—OH bond, and therefore of the O—H bond will be all the greater.

The pure acid is a colourless, sharp-smelling liquid, m.p. 8°C, b.p. 100°C, miscible with water. Its salts, esters, and amides are normal (R—NH—CHO is not an aldehyde, but an amide of methanoic acid). Methanoic anhydride and methanoyl chloride are both known but very unstable, the normal preparative methods give carbon monoxide. Warm concentrated sulphuric acid readily dehydrates methanoic acid giving carbon monoxide.

$$HCO_2H \xrightarrow{PCl_5} HCl + CO$$

$$HCO_2H \xrightarrow{H_2SO_4} CO + H_2O$$

The acid is a reducing agent, being readily oxidised to carbon dioxide and water, even by such mild oxidising reagents as Fehling's solution and ammoniacal silver nitrate (p. 113).

Aromatic carboxylic acids

Aromatic carboxylic acids have the functional group attached directly to an aromatic ring, e.g. Benzoic acid:

$$C_6H_5-C(=O)-OH$$

They may be prepared by the oxidation of the corresponding alcohols or aldehydes and in addition by the vigorous oxidation of aromatic ketones or even hydrocarbons (p. 46). Aromatic carboxylic acids are solids, sparingly soluble in water and slightly stronger than simple aliphatic acids. The chemistry of the carboxylic acid group is similar to that of the group in aliphatic acids, and esters, amides, and acyl chlorides are formed in the customary way. Reduction of the carboxylic acid function is achieved by lithium aluminium hydride without affecting the aromatic ring, whilst catalytic hydrogenation reduces the aromatic ring exclusively.

$$C_6H_5CH_2OH \xleftarrow{LiAlH_4} C_6H_5CO_2H \xrightarrow{H_2/Ni} C_6H_{11}CO_2H$$

Electrophilic substitution of the aromatic nucleus occurs predominantly in the *meta* position (p. 42), and to obtain aromatic acids substituted in the *ortho* and *para* positions, it is usually necessary to employ roundabout synthetic routes, e.g. in the case of the isomeric nitrobenzoic acids, only the *meta* isomer can be obtained directly, the *ortho* and *para* nitrobenzoic acids must be obtained via the corresponding nitrotoluenes.

$$C_6H_5CO_2H \xrightarrow[H_2SO_4]{HNO_3} m\text{-}NO_2C_6H_4CO_2H$$

CH_3 $\xrightarrow{HNO_3}$ CH_3, NO_2 + CH_3, NO_2

$KMnO_4/KOH$ $KMnO_4/KOH$

CO_2H, NO_2 CO_2H, NO_2

*p*H and *pK*

Acid–base systems are of the greatest importance in biological reactions, and it is necessary to be able to describe both the acidity (or alkalinity) of solutions and the relative strengths of acids and bases by a convenient quantitative scale. *p*H and *pK* are the functions used for this purpose.

*p*H. Pure water dissociates to a minute extent $H_2O \rightleftharpoons HO^- + H^+$ and the equilibrium constant for this dissociation is given by the expression:

$$K_{e_{H_2O}} = \frac{[H^+][OH^-]}{[H_2O]}$$

where$[H^+]$, etc., is the concentration in mol dm^{-3}. Since in pure water and *dilute* aqueous solutions $[H_2O]$ is very much larger than $[H^+]$ or $[OH^-]$ and virtually constant, the expression can be rewritten

$$K_{e_{H_2O}}[H_2O] = [H^+][OH^-] = \text{ionic product}$$

and for pure water and dilute aqueous solutions at 25°C

$$[H^+][OH^-] = 10^{-14}\ (\text{mol dm}^{-3})^2*$$

Thus for pure water $[H^+] = [OH^-] = 10^{-7}$ mol dm^{-3}, and in dilute solutions of acid or alkali the ratio of $[H^+]$ and $[OH^-]$ is fixed by the ionic product.

* This is not the dissociation constant (K_e) for water, which is given by

$$K_{e_{H_2O}} = \frac{10^{-14}}{1000/18} = 1{\cdot}8 \times 10^{-16}\ \text{mol dm}^{-3}$$

since one litre of water contains 1000/18 mol water.

In 1M strong monobasic acid (i.e. completely dissociated) $[H^+] = 1$ and $[OH^-] = 10^{-14}$ mol dm^{-3}, whilst in 1M strong monacid base $[OH^-] = 1$ and therefore $[H^+] = 10^{-14}$ mol dm^{-3}. Many reactions occur between these limits, and the description of acidity must cover values of $[H^+]$ stretching over fourteen powers of ten. In order to contract this enormous scale into workable figures the concentration of hydrogen ion is expressed as a logarithmic term, *p*H, which is defined by:

$$pH = -\log_{10}[H^+]$$

the negative sign ensuring that in the relevant range of acidity, *p*H values will be positive. As $[H^+]$ changes from 1 to 10^{-14} mol dm^{-3} *p*H changes from 0 to 14. Thus for a 0·1M solution of strong monobasic acid $[H^+] = 0.1$ mol dm^{-3} whence $pH = -\log_{10}(0{\cdot}1) = 1$. Similarly 0·001M strong monobasic acid has a $pH = 3$, whilst for a 0·01M solution of a strong monacid base, $[OH^-] = 0{\cdot}01$, therefore $[H^+] = 10^{-12}$ mol dm^{-3}, whence $pH = 12$. The *p*H of pure water is 7, this being synonymous with neutrality.

Strictly, a 10^{-n}M solution of a strong monobasic acid has $[H^+] = 10^{-n} + x$ and $[OH^-] = x$ mol dm^{-3} where x is the concentration of ions produced by the dissociation of water molecules. Then:

$$[H^+][OH^-] = (10^{-n} + x)x = 10^{-n}x + x^2 = 10^{-14}$$

The correct value of $[H^+]$ can be found only by solving this quadratic equation, but where $n < 6$, $10^{-n}x \gg x^2$, and x^2 can be ignored in the expression above, which is equivalent to ignoring the concentration of hydrogen ion produced by dissociation of water. However if $n > 6$, this factor must be taken into account, and *p*H values can be obtained only by solution of the quadratic equation, otherwise the results would be nonsensical, e.g. 10^{-8}M monobasic acid would have a $pH = 8$, which is weakly alkaline.

The pH of very dilute solutions of strong monobasic acid

CONC. OF STRONG MONOBASIC ACID	$[H^+]$ FROM IONISATION OF STRONG ACID (mol dm^{-3})	$[H^+]$ FROM DISSOCIATION OF WATER (mol dm^{-3})	TOTAL $[H^+]$ (mol dm^{-3})	*p*H
10^{-5}M	$1{\cdot}0 \times 10^{-5}$	$0{\cdot}00010 \times 10^{-5}$	$1{\cdot}00010 \times 10^{-5}$	4·99996
10^{-6}M	$1{\cdot}0 \times 10^{-6}$	$0{\cdot}00990 \times 10^{-6}$	$1{\cdot}00990 \times 10^{-6}$	5·99572
10^{-7}M	$1{\cdot}0 \times 10^{-7}$	$0{\cdot}61804 \times 10^{-7}$	$1{\cdot}61804 \times 10^{-7}$	6·79101
10^{-8}M	$1{\cdot}0 \times 10^{-8}$	$0{\cdot}95125 \times 10^{-7}$	$1{\cdot}05125 \times 10^{-7}$	6·97829
10^{-9}M	$1{\cdot}0 \times 10^{-9}$	$0{\cdot}995025 \times 10^{-7}$	$1{\cdot}005025 \times 10^{-7}$	6·99782

Note how the proportion of $[H^+]$ arising from the dissociation of water becomes progressively more important as the concentration of strong acid decreases.

pK of acids. The equilibrium constant for the dissociation of an acid HA

$$HA \rightleftharpoons H^+ + A^-$$

is given by

$$K_{HA} = \frac{[H^+][A^-]}{[HA]}$$

and the stronger the acid, the greater the degree of dissociation, and the larger the value of K_{HA}. The relative strengths of acids can be compared by use of K values, but it is frequently more convenient to use the term pK, which is defined by:

$$pK = -\log_{10} K_{HA} = -\log_{10} \frac{[H^+][A^-]}{[HA]}$$

Polybasic acids, with several dissociation steps have pK_1, pK_2, pK_3, etc., corresponding to the first, second, and third ionisations.

The relationships between pH and pK can be seen by considering the situation when the acid HA is exactly half dissociated. In these circumstances $[HA] = [A^-]$, whence $K_{HA} = [H^+]$ and

$$-\log_{10} K_{HA} = -\log_{10} [H^+], \text{ i.e. } pK = p\text{H}$$

Thus the physical significance of the pK value for an acid is that it is the pH at which the acid is half ionised. Since pK, like pH, is a logarithmic term, a difference of unity means a tenfold difference in K_{HA}, e.g. an acid of pK = 2·5 is ten times as strong as one of pK = 3·5 and a hundred times as strong as one of pK = 4·5. The strong mineral acids, which are completely ionised even in fairly concentrated solutions, have infinitely large K_{HA} values, but the weak acids commonly encountered in organic compounds have pK values between 0 and 14.

The pK values for a series of acids, phenols, alcohols, and thiols are listed below.

pK_a values of some acids

	ACID	CONJUGATE BASE*	pK_a
1	HCO_2H	HCO_2^-	3·75
2	CH_3CO_2H	$CH_3CO_2^-$	4·75
3	$CH_3CH_2CO_2H$	$CH_3CH_2CO_2^-$	4·87
4	$CH_3(CH_2)_2CO_2H$	$CH_3(CH_2)_2CO_2^-$	4·81

* The anion of an acid is termed the 'conjugate base' of the acid, and likewise the protonated cation of a base is known as the 'conjugate acid' of the base.

	ACID	CONJUGATE BASE	pK_a
5	$(CH_3)_2CHCO_2H$	$(CH_3)_2CHCO_2^-$	4·84
6	$ClCH_2CO_2H$	$ClCH_2CO_2^-$	2·85
7	Cl_2CHCO_2H	$Cl_2CHCO_2^-$	1·48
8	Cl_3CCO_2H	$Cl_3CCO_2^-$	0·70
9	$CH_3CHClCO_2H$	$CH_3CHClCO_2^-$	2·83
10	$CH_2ClCH_2CO_2H$	$CH_2ClCH_2CO_2^-$	3·98
11	$CH_3CH_2CHClCO_2H$	$CH_3CH_2CHClCO_2^-$	2·86
12	$CH_3CHClCH_2CO_2H$	$CH_3CHClCH_2CO_2^-$	4·05
13	$CH_2Cl(CH_2)_2CO_2H$	$CH_2Cl(CH_2)_2CO_2^-$	4·52
14	$HOCH_2CO_2H$	$HOCH_2CO_2^-$	3·12
15	$CH_3CH(OH)CO_2H$	$CH_3CH(OH)CO_2^-$	3·83
16	$C_6H_5CO_2H$	$C_6H_5CO_2^-$	4·19
17	$2\text{-}NO_2C_6H_4CO_2H$	$2\text{-}NO_2C_6H_4CO_2^-$	2·16
18	$3\text{-}NO_2C_6H_4CO_2H$	$3\text{-}NO_2C_6H_4CO_2^-$	3·47
19	$4\text{-}NO_2C_6H_4CO_2H$	$4\text{-}NO_2C_6H_4CO_2^-$	3·41
20	$2\text{-}ClC_6H_4CO_2H$	$2\text{-}ClC_6H_4CO_2^-$	2·92
21	$3\text{-}ClC_6H_4CO_2H$	$3\text{-}ClC_6H_4CO_2^-$	3·82
22	$4\text{-}ClC_6H_4CO_2H$	$4\text{-}ClC_6H_4CO_2^-$	3·98
23	CH_3OH	CH_3O^-	*ca.* 18
24	C_2H_5OH	$C_2H_5O^-$	*ca.* 18
25	Glycerol	Monoanion	14·15
26	C_2H_5SH	$C_2H_5S^-$	10·64
27	$CH_3(CH_2)_2SH$	$CH_3(CH_2)_2S^-$	10·83
28	C_6H_5OH	$C_6H_5O^-$	9·89*
29	C_6H_5SH	$C_6H_5S^-$	7·47
30	NH_4^+	NH_3	9·25
31	$CH_3\overset{+}{N}H_3$	CH_3NH_2	10·64
32	$(CH_3)_2\overset{+}{N}H_2$	$(CH_3)_2NH$	10·72
33	$(CH_3)_3\overset{+}{N}H$	$(CH_3)_3N$	9·74
34	$C_2H_5\overset{+}{N}H_3$	$C_2H_5NH_2$	10·75
35	$(C_2H_5)_2\overset{+}{N}H_2$	$(C_2H_5)_2NH$	10·98
36	$(C_2H_5)_3\overset{+}{N}H$	$(C_2H_5)_3N$	10·76
37	$C_6H_5CH_2\overset{+}{N}H_3$	$C_6H_5CH_2NH_2$	9·37
38	$C_6H_5\overset{+}{N}H_3$	$C_6H_5NH_2$	4·58
39	$(C_6H_5)_2\overset{+}{N}H_2$	$(C_6H_5)_2NH$	0·88
40		$(C_6H_5)_3N$	Non-basic
41	$C_6H_5\overset{+}{N}H_2CH_3$	$C_6H_5NHCH_3$	4·70

* For the pK_a values for nitrophenols, see p. 70.

	ACID	CONJUGATE BASE	pK_a
42	$C_6H_5\overset{+}{N}H(CH_3)_2$	$C_6H_5N(CH_3)_2$	5·16
43	$2\text{-}NO_2C_6H_4\overset{+}{N}H_3$	$2\text{-}NO_2C_6H_4NH_2$	−0·26
44	$3\text{-}NO_2C_6H_4\overset{+}{N}H_3$	$3\text{-}NO_2C_6H_4NH_2$	2·5
45	$4\text{-}NO_2C_6H_4\overset{+}{N}H_3$	$4\text{-}NO_2C_6H_4NH_2$	1·0
46	H_2O	HO^-	15·75
47	C_2H_2	C_2H^-	*ca.* 22
48	NH_3	NH_2^-	*ca.* 35
49	$C_6H_5NH_2$	$C_6H_5\bar{N}H$	*ca.* 27
50	CH_3COCH_3	$CH_3CO\bar{C}H_2$	*ca.* 20
51	C_2H_4	$C_2H_3^-$	*ca.* 40
52	C_2H_6	$C_2H_5^-$	> 40

In the table above note the following trends:

(i) the large increase in pK_a on going from methanoic acid to ethanoic acid, and the relatively small change thereafter on extension of the side chain (1–5);

(ii) the large decreases in pK_a with progressive substitution of the methyl group of ethanoic acid by chlorine (2, 6, 7, 8);

(iii) the rapidly diminishing effect of a chlorine substituent in the side chain of an aliphatic acid as the position of substitution moves away from the carboxylic acid group (9–13);

(iv) the (inductive) effect of electronegative substituents on the pK_a of benzoic acid, and the variation of this effect with the position of substitution (16–22);

(v) the comparative acidity of thiol and hydroxyl groups in monofunctional compounds (23–29);

(vi) the comparative basicities of alkylamines, arylamines, and ammonia (the more strongly basic the amine, the less acidic the protonated cation) (30–42).

pK of bases. There are two common ways of describing and comparing the strengths of bases on a logarithmic scale, differing in the criterion used to measure basicity.

The strength of a base is inversely related to the degree of dissociation of the protonated cation (the 'conjugate acid') derived from the base. If a base B gives rise to a cation $(BH)^+$, then this cation can be regarded as a weak acid:

$$(BH)^+ \rightleftharpoons B + H^+$$

for which a dissociation constant can be written

$$K_a = \frac{[B][H^+]}{[(BH)^+]}$$

$$\text{whence } pK_a = -\log_{10} K_a = -\log_{10} \frac{[B][H^+]}{[(BH)^+]}$$

Note the similarity of this expression to that of the pK of an acid.

Alternatively the strength of the base can be related to the concentration of hydroxide anions generated in a dilute aqueous solution of the base.

$$B + H_2O \rightleftharpoons (BH)^+ + OH^-$$

for which

$$K_b = \frac{[(BH)^+][OH^-]}{[B]}$$

(if $[H_2O]$ is regarded as constant). We can therefore define another logarithmic scale:

$$pK_b = -\log_{10} K_b = -\log_{10} \frac{[(BH)^+][OH^-]}{[B]}$$

and since $[H^+][OH^-] = 10^{-14}$ it is easily demonstrated that $pK_b = 14 - pK_a$.

Of the two scales the former is more convenient, as the pK_a for a base, which is numerically equal to the pH at which the base is half in the protonated form (i.e. the cation is half dissociated) is directly comparable with the pK scale for acids, whereas pK_b is not.

The pK_a values of some common bases are listed in the table above. Note the general increase in basic strength (increase in pK_a) with progressive substitution of the hydrogen atoms of ammonia by alkyl groups, and the very marked decrease in basicity when the substituent is an aromatic group (p. 90).

Esters

Esters are the *O*-alkyl or *O*-aryl derivatives of carboxylic acids, and their systematic names are derived either from a combination of those of the alkyl or aryl group and the carboxylic acid, or by describing the ester group or acyl group as a substituent.

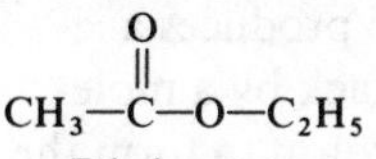

$$CH_3-\overset{O}{\overset{\|}{C}}-O-C_2H_5$$

Ethyl acetate
Ethyl ethanoate

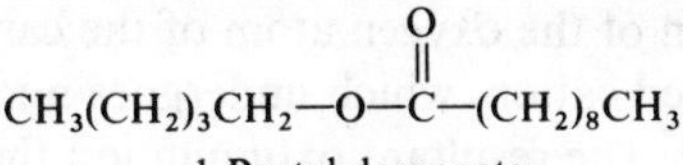

$$CH_3(CH_2)_3CH_2-O-\overset{O}{\overset{\|}{C}}-(CH_2)_8CH_3$$

1-Pentyl decanoate

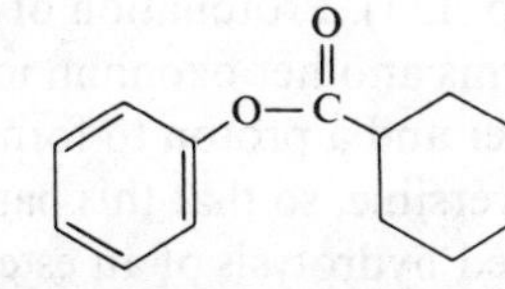

Phenyl cyclohexanecarboxylate

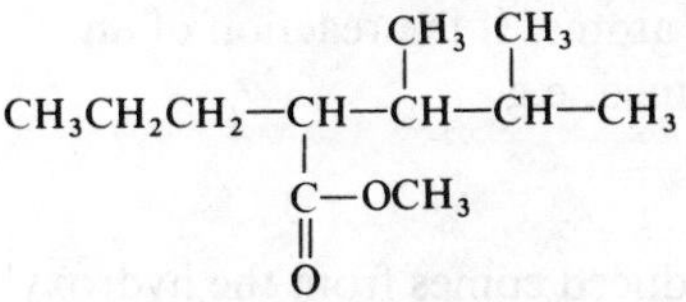

2,3-Dimethyl-4-methoxycarbonylheptane

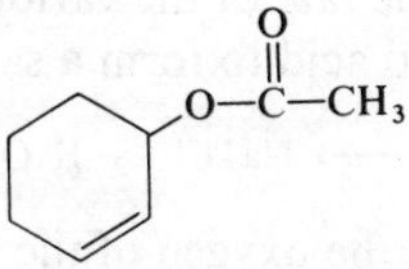

3-Acetoxycyclohexene

Preparation of esters

1. Esters of alcohols can be prepared by the acid-catalysed reaction of carboxylic acids with alcohols.

$$RCO_2H + R'OH \underset{}{\overset{H^+}{\rightleftharpoons}} RCO_2R' + H_2O$$

The mechanism of this reaction is:

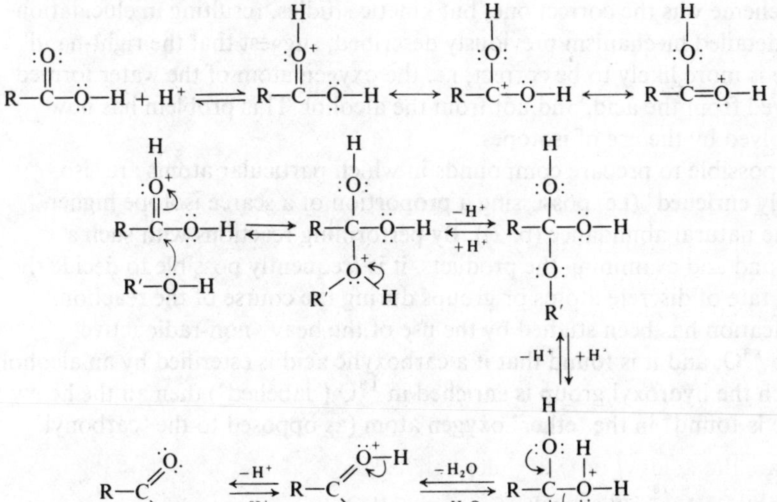

$$R-\overset{:O:}{\overset{\|}{C}}-\ddot{O}-H + H^+ \rightleftharpoons R-\overset{H-\overset{+}{O}}{\overset{\|}{C}}-\ddot{O}-H \longleftrightarrow R-\overset{H-\ddot{O}:}{\overset{|}{\overset{+}{C}}}-\ddot{O}-H \longleftrightarrow R-\overset{H-\ddot{O}:}{\overset{|}{C}}=\overset{+}{O}-H$$

$$R-\overset{H-\overset{+}{O}:}{\overset{\|}{C}}-\ddot{O}-H + R'-\ddot{O}-H \rightleftharpoons R-\overset{H-\ddot{O}:}{\overset{|}{C}}(-\overset{+}{O}HR')-\ddot{O}-H \underset{+H^+}{\overset{-H^+}{\rightleftharpoons}} R-\overset{H-\ddot{O}:}{\overset{|}{C}}(-\ddot{O}-R')-\ddot{O}-H$$

$$\underset{+H^+}{\overset{-H^+}{\rightleftharpoons}}$$

$$R-C(=\ddot{O})-\ddot{O}-R' \underset{+H^+}{\overset{-H^+}{\rightleftharpoons}} R-C(=\overset{+}{O}-H)-\ddot{O}-R' \underset{+H_2O}{\overset{-H_2O}{\rightleftharpoons}} R-\overset{H-\ddot{O}:}{\overset{|}{C}}(-\ddot{O}-R')-\overset{+}{O}H_2$$

Initial protonation of the oxygen atom of the carbonyl group produces a resonance stabilised cation, which undergoes nucleophilic attack by a molecule of an alcohol. The resultant oxonium ion then loses a proton to form the monoalkyl derivative of an ortho-acid (p. 124). Protonation of one of the hydroxyl groups of this intermediate forms another oxonium ion, which can successively eliminate a molecule of water and a proton to form the ester. It should be noted that all the steps are reversible, so that this pathway also describes the mechanism of acid-catalysed hydrolysis of an ester.

One important conclusion coming from the mechanistic study of this reaction concerns the fate of the various oxygen atoms. If the reaction of an inorganic base and acid to form a salt is examined, e.g.

$$Na^+OH^- + H^+Cl^- \longrightarrow Na^+Cl^- + H_2O$$

it is apparent that the oxygen of the water produced comes from the hydroxyl group of the base. However, in esterification, water can be schematically removed from the alcohol and carboxylic acid by two methods represented below by the archaic 'box-reactions':

$$R{-}C({=}O){-}O{-}\boxed{H \quad H{-}O}{-}R' \qquad\qquad R{-}C({=}O){-}\boxed{O{-}H \quad H}{-}O{-}R'$$

On the basis of the inorganic analogy, it was originally thought that the left-hand scheme was the correct one, but kinetic studies, resulting in elucidation of the detailed mechanism previously described, suggest that the right-hand scheme is more likely to be correct, i.e. the oxygen atom of the water formed is derived from the acid, and not from the alcohol. This problem has now been solved by the use of isotopes.

It is possible to prepare compounds in which particular atoms are 'isotopically enriched' (i.e. possessing a proportion of a scarce isotope higher than the natural abundance (p. 2)). By performing reactions with such a compound and examining the products, it is frequently possible to decide the precise fate of discrete atoms or groups during the course of the reaction. Esterification has been studied by the use of the heavy non-radioactive isotope ^{18}O, and it is found that if a carboxylic acid is esterified by an alcohol in which the hydroxyl group is enriched in ^{18}O ('labelled') then all the heavy isotope is found* in the 'ether' oxygen atom (as opposed to the 'carbonyl'

$$R{-}C({=}O){-}O{-}H + H{-}\overset{18}{O}{-}R' \rightleftharpoons R{-}C({=}O){-}\overset{18}{O}{-}R' + H_2O$$

* By the technique of mass spectrometry.

oxygen atom) of the ester, and none is found in the water produced. Because of the dynamic equilibrium between the covalent and ionised forms of a carboxylic acid, it is not possible to label the oxygen atoms of the hydroxyl and carbonyl groups of a carboxylic acid independently, and the label is said to be 'scrambled'. However, if such an isotopically labelled carboxylic acid is

$$\mathrm{R{-}C({=}O){-}{}^{18}O{-}H \rightleftharpoons H^+ + \left[\mathrm{R{-}C(O^-){=}{}^{18}O} \leftrightarrow \mathrm{R{-}C({=}O){-}{}^{18}O^-}\right] \rightleftharpoons R{-}C(O{-}H){=}{}^{18}O}$$

esterified, then half of the heavy isotope is found in the water formed, and half is found in the carbonyl oxygen atom of the ester. These results can be

$$\mathrm{R{-}C({=}{}^{18}O){-}{}^{18}O{-}H + H{-}O{-}R' \rightleftharpoons R{-}C({=}{}^{18}O){-}O{-}R' + H_2{}^{18}O}$$

seen to coincide exactly with predictions based on the esterification mechanism described above.

2. Esters may also be prepared by the reaction of alcohols with acyl chlorides (p. 144) and acid anhydrides (p. 145), by acid-catalysed reaction with amides (p. 148) and nitriles, or by transesterification.

3. The reaction of silver salts of carboxylic acids with alkyl halides gives esters (p. 127).

Properties of esters. The lower aliphatic esters are neutral, pleasant-smelling liquids, generally insoluble in water, and of lower boiling point than the corresponding carboxylic acids. They are soluble in most organic solvents, and some find use as solvents (e.g. ethyl acetate; 'amyl' acetate = 3-methyl-1-butyl ethanoate).

Reactions of esters

1. *Hydrolysis.* Esters react with water, forming the corresponding alcohol and carboxylic acid. This hydrolysis is slow under neutral conditions, but greatly accelerated by mineral acids, following exactly the reverse of the mechanism of esterification (p. 137). Esters also react with aqueous alkalies, ultimately forming the alcohol and the carboxylate anion. The mechanism of this reaction is:

$$R-\overset{\overset{:O:}{\|}}{C}-\ddot{O}-R' + {}^{-}:\ddot{O}-H \rightleftharpoons R-\underset{:\ddot{O}-H}{\overset{:\ddot{O}:^{-}}{\underset{|}{\overset{|}{C}}}}-\ddot{O}-R' \longrightarrow R-C\overset{\ddot{O}:}{\underset{\ddot{O}-H}{\lessgtr}} + {}^{-}:\ddot{O}-R' \longrightarrow R-C\overset{\ddot{O}:}{\underset{\ddot{O}:^{-}}{\lessgtr}} + H-\ddot{O}-R'$$

$$\text{Overall reaction } RCO_2R' + OH^- \longrightarrow RCO_2^- + R'OH$$

Nucleophilic attack on the carbonyl group by the hydroxide anion gives an intermediate anion which can eliminate either a hydroxide ion, reforming the starting materials, or an alkoxide anion. Loss of an alkoxide anion results in the formation of a carboxylic acid, which reacts with the powerfully basic alkoxide ion to form the carboxylate anion and a molecule of the alcohol. Though the final step is, in principle, reversible, the equilibrium lies virtually entirely on the side of the alcohol and carboxylate anion, thus the alkaline hydrolysis (or 'saponification') of esters, unlike acidic hydrolysis, leads quantitatively to the carboxylate salts. It should be noted that the metal cation of the alkali has absolutely no role in this reaction, merely serving to preserve electrical neutrality.

2. *Transesterification.* Just as an ester can undergo acidic hydrolysis, so an ester of one alcohol can undergo an acid-catalysed reaction with a second alcohol resulting in an equilibrium mixture of the two possible esters. Similarly an ester of one alcohol will react with the alkoxide anion derived from

$$R-C\overset{O}{\underset{O-R'}{\lessgtr}} + R''-O-H \overset{H^+}{\rightleftharpoons} R-C\overset{O}{\underset{O-R''}{\lessgtr}} + R'-O-H$$

another alcohol, again producing an equilibrium mixture of esters. The mechanisms of these reactions are closely similar to those of acid-catalysed and alkaline hydrolysis, and conform to the general mechanisms I and III (p. 142) of carboxylic acid derivatives.

$$R-C\overset{O}{\underset{O-R'}{\lessgtr}} + R''-O^- \rightleftharpoons R-C\overset{O}{\underset{O-R''}{\lessgtr}} + R'-O^-$$

3. Esters react slowly with ammonia or primary and secondary amines to give the corresponding amides (reaction mechanism II, p. 142).

$$R-\overset{\overset{O}{\|}}{C}-OR' + NH_3 \longrightarrow R-\overset{\overset{O}{\|}}{C}-NH_2 + R'OH$$

4. Esters react with hydroxylamine in the presence of strong bases (e.g. alcoholic potassium hydroxide) to give hydroxamic acids. These compounds

form deep purple complexes with iron(III) ion, and the reaction provides a colorimetric method of estimating ester groups or their equivalent (e.g. amides) in biological material such as fats.

$$H_2NOH + C_2H_5O^- \rightleftharpoons C_2H_5OH + H\bar{N}OH$$

$$R{-}\overset{O}{\overset{\|}{C}}{-}OR' + H\bar{N}OH \longrightarrow \underset{\text{Hydroxamic acid}}{R{-}\overset{O}{\overset{\|}{C}}{-}NHOH} + R'O^-$$

$$R{-}\overset{O}{\overset{\|}{C}}{-}NHOH + Fe^{+3} \longrightarrow \left[R{-}C\begin{matrix}\overset{+}{O}{-}\bar{F}e \\ \| \qquad | \\ N{-}O \\ H\end{matrix}\right]^{++} + H^+$$

5. *Reduction.* Esters, like carboxylic acids, are resistant to reduction by most of the common reducing agents, but reduction to the corresponding lithium alkoxides can be achieved by use of lithium aluminium hydride, which may be regarded as a source of hydride anions:

$$R{-}\overset{:O:}{\overset{\|}{C}}{-}\ddot{O}{-}R' \;(+\; :H^-) \longrightarrow R{-}\underset{H}{\overset{:\ddot{O}:^-}{\overset{|}{\underset{|}{C}}}}{-}\ddot{O}{-}R' \longrightarrow R{-}\underset{H}{C}{=}\ddot{O} + :\ddot{O}^-{-}R'$$

$$R{-}\underset{H}{C}{=}\ddot{O} + H:^- \longrightarrow R{-}\underset{H}{\overset{H}{\overset{|}{\underset{|}{C}}}}{-}\ddot{O}:^- \xrightarrow{+H_2O} R{-}CH_2OH$$

Initial attack of hydride ion and loss of alkoxide anion results in the formation of an aldehyde, which promptly reacts with a hydride ion to form a second alkoxide anion. Subsequently, water is added during the work-up of the reaction mixture, when the alkoxides are converted into the alcohols.

$$(CH_3)_2CHCO_2CH_3 \xrightarrow[\text{(ii) } H_2O]{\text{(i) } LiAlH_4/(C_2H_5)_2O} (CH_3)_2CHCH_2OH + CH_3OH$$

6. *Enolate anions* can be formed by esters having a hydrogen atom in the α position. However, very much stronger bases are necessary than in the case of aldehydes and ketones (p. 105). In an ester the carbonyl group is involved in resonance of the type:

$$R{-}CH_2{-}\overset{:O:}{\overset{\|}{C}}{-}\ddot{O}{-}R' \longleftrightarrow R{-}CH_2{-}\overset{:\ddot{O}:^-}{\overset{|}{C}}{=}\overset{+}{\dot{O}}{-}R'$$

and is less able to stabilise an adjacent negative charge than in the case of aldehydes and ketones.

The mechanisms of some general reactions of carboxylic acid derivatives

Most of the numerous reactions of derivatives of carboxylic acids proceed by one of three mechanisms:

(i) Acid-catalysed reaction of R—CO—X with H—Y:
(ii) Uncatalysed reaction of R—CO—X with H—Y:
(iii) Reaction of R—CO—X with an anion Z^-:

I *Acid-catalysed reaction of* R—CO—X *with* H—Y:

$$R{-}C({=}O){-}X \xrightarrow{+H^+} R{-}C({=}\overset{+}{O}H){-}X \;(+\;\ddot{Y}{-}H) \longrightarrow R{-}C(OH)(X)(\overset{+}{Y}H) \xrightarrow{-H^+} R{-}C(OH)(X)(Y) \xrightarrow{+H^+} R{-}C(OH)(\overset{+}{X}H)(Y) \xrightarrow{-H{-}X} R{-}C({=}\overset{+}{O}H){-}Y \xrightarrow{-H^+} R{-}C({=}O){-}Y$$

For example, esterification of a carboxylic acid (—X = —OH, —Y = —OR); transesterification (—X = —OR′, —Y = —OR″); acid-catalysed hydrolysis (—X = —OR, —Y = —OH).

II *Uncatalysed reaction of* R—CO—X *with* H—Y:

$$R{-}C({=}O){-}X \;(+\;\ddot{Y}{-}H) \longrightarrow R{-}C(O^-)(X)(\overset{+}{Y}H) \longrightarrow R{-}C({=}O){-}Y + :X^- + H^+ \longrightarrow R{-}C({=}O){-}Y + H{-}X$$

For example, the reaction of esters with amines (—X = —OR, —Y =—NR_2') or acyl chlorides with alcohols or water (—X =—Cl, —Y = —OR or —OH).

III *Reaction of* R—CO—X *with an anion* Z^-:

$$R{-}C({=}O){-}X \;(+\;:Z^-) \longrightarrow R{-}C(O^-)(X)(Z) \longrightarrow R{-}C({=}O){-}Z + :X^-$$

For example, saponification of an ester by alkali (—X = —OR, Z = OH^-) base-catalysed transesterification (—X = —OR, Z = OR′);

formation of hydroxamic acid (p. 141) (—X = —OR, $Z^- = H\bar{N}OH$); or the first step in the reduction of an ester by $LiAlH_4$ (—X = —OR, $Z^- = \bar{H}$).

Of these alternatives, the basic mechanism is type II involving the unassisted attack of a nucleophile on the carbonyl group. This mechanism is usually employed only where either the carbonyl group is highly electrophilic (i.e. X is very electronegative) or HY is a very good nucleophile. Where the carbonyl group is only weakly electrophilic and HY is a poor nucleophile the reactivity of the system can be increased either by protonation of the carbonyl group to enhance its electrophilicity (type I), or by deprotonation of HY or HZ to generate a more powerful nucleophile (type III). In fluid solution it is clearly not possible to have both these processes operating simultaneously. However, where reactions of these types are catalysed by adsorption of the reagents on a surface, e.g. of an enzyme, it is possible to have both proton-donating (acidic) and proton-abstracting (basic) groups simultaneously catalysing the process.

Aromatic esters

Aromatic esters may either be derived from aromatic carboxylic acids, or from phenols, or from both groups of compound, e.g.

C_6H_5—C(=O)—O—CH_3 C_6H_5—O—C(=O)—CH_3 C_6H_5—O—C(=O)—C_6H_5

Methyl benzoate   Phenyl acetate
Phenyl ethanoate   Phenyl benzoate

All of these esters may be obtained by the standard methods (p. 137) and all series of esters exhibit the normal reactions of the ester group, but the esters of aromatic carboxylic acids are less reactive than the corresponding aliphatic compounds. Electrophilic substitution of the aromatic nucleus occurs normally (p. 40).

Acyl halides

Acyl halides bear the same relationship to carboxylic acids as that between alkyl halides and alcohols, i.e. replacement of a hydroxyl function by a halogen atom. Of the acyl halides the chlorides are most commonly encountered, and their chemistry is typical of the group.

The systematic nomenclature of acyl halides describes them either as derivatives of the carboxylic acid (i.e. alkanoyl halide) or by describing the functional group as a substituent, e.g.

$CH_3CH_2CH_2COCl$
Butanoyl chloride

H_3C–C_6H_{10}–COBr
4-Methylcyclohexanecarbonyl bromide

Preparation and properties of acyl chlorides. The only important preparation of acyl chlorides is the reaction of carboxylic acids with non-metal chlorides (p. 128). Phosphorus trichloride or pentachloride can be used, but thionyl chloride ($SOCl_2$) is more convenient, since the reagent is volatile and the by-products are gases:

$$CH_3CH_2CO_2H + SOCl_2 \longrightarrow CH_3CH_2COCl + SO_2 + HCl$$

Aliphatic acyl chlorides are very reactive liquids, of lower boiling point than the corresponding acid. Like many reactive halogen compounds, they are lachrymators and cause severe burns in contact with the skin. In air they emit fumes of hydrogen chloride on account of their reaction with water vapour.

Reaction of acyl chlorides. Acyl chlorides react very readily with nucleophiles, giving products similar to those obtained by the corresponding reactions of esters (p. 139). The reactivity is, however, very much greater than that of the

$$R{-}\overset{\overset{\displaystyle O^{\delta-}}{\|}}{\underset{\delta+}{C}}\overset{\delta+}{\rightarrow}\underset{\delta-}{Cl}$$

corresponding esters, since the powerfully electronegative halogen atom polarises the C—Cl bond, leaving the carbonyl carbon atom even more electron depleted than in the ester. As a result, acyl chlorides react readily with weak nucleophiles, such as water or alcohols, without acid catalysts. The reactions summarised below all proceed by the mechanism II (p. 142).

$$R{-}\overset{\overset{\displaystyle O}{\|}}{C}{-}Cl + \left.\begin{matrix} H_2O \\ R'OH \\ RCO_2^- \\ NH_3 \\ R'NH_2 \\ R'_2NH \\ R'SH \end{matrix}\right\} \longrightarrow \left\{\begin{matrix} RCO_2H + HCl \\ RCO_2R' + HCl \\ (RCO)_2O + Cl^- \\ RCONH_2 + \overset{+}{N}H_4Cl^- \\ RCONHR' + R'\overset{+}{N}H_3Cl^- \\ RCONR'_2 + R'_2\overset{+}{N}H_2Cl^- \\ RCOSR' + HCl \end{matrix}\right.$$

It should be noted that the reaction with ammonia, or primary and secondary amines, results in the formation of the amide and the ammonium halide (cf.

acid anhydrides). In addition to these ester-like reactions, acyl chlorides react with a number of simple anions with replacement of the halogen atoms, e.g.

$$RCOCl + \left.\begin{matrix} CN^- \\ Br^- \\ I^- \end{matrix}\right\} \longrightarrow \left\{\begin{matrix} RCOCN + Cl^- \\ RCOBr + Cl^- \\ RCOI + Cl^- \end{matrix}\right.$$

Aromatic acyl chlorides, e.g. benzoyl chloride C_6H_5COCl, are prepared by the same methods as aliphatic acyl chlorides. They are solids or liquids, of high boiling point, with extremely irritant vapours. They exhibit similar reactions with nucleophiles to those of aliphatic acyl chlorides but with greatly reduced reactivity, e.g. ethanoyl (acetyl) chloride reacts almost instantaneously with water or dilute ammonia, whilst benzoyl chloride reacts with water over a period of hours, and over a period of minutes with dilute aqueous ammonia.

Aliphatic acid anhydrides

Preparation and properties. Although acid anhydrides are formally obtained by the removal of the elements of water from two molecules of a carboxylic acid, direct dehydration of the carboxylic acid is scarcely ever possible. The anhydrides may be obtained by the reaction of an acyl chloride and a carboxylate salt:

$$CH_3COCl + CH_3CO_2^-Na^+ \longrightarrow (CH_3CO)_2O + NaCl$$

Acetic anhydride
Ethanoic anhydride

This method can give mixed anhydrides if the anion and the acyl chloride are derived from different carboxylic acids (cf. Williamson synthesis of ethers, p. 73).

$$CH_3COCl + CH_3CH_2CO_2^- \longrightarrow CH_3COOCOCH_2CH_3 + Cl^-$$

The lower aliphatic acid anhydrides are sharp-smelling liquids, sparingly soluble in water.

Reactions of acid anhydrides. Acid anhydrides react with the same range of nucleophiles as esters, but with reactivity intermediate between that of esters and acyl chlorides. Thus acid anhydrides react slowly with water or alcohols, but much more rapidly in the presence of acid catalysts. Similarly, amines are acylated by acid anhydrides much more rapidly than by esters, but without the violence of the reaction with acyl chlorides. The range of reactions is summarised below; it should be noted that on account of the weakness of carboxylic acids and consequent dissociation of their salts with amines, acylation of amines goes completely to the amide and carboxylic acid (cf. p. 144).

$$\begin{matrix} R-C\overset{\displaystyle O}{\diagup\!\!\diagup} \\ \quad\diagdown \\ \qquad O \\ \quad\diagup \\ R-C\underset{\displaystyle O}{\diagdown\!\!\diagdown} \end{matrix} + \left.\begin{matrix} H_2O \\ R'OH \\ NH_3 \\ R'NH_2 \\ R'_2NH \\ R'SH \end{matrix}\right\} \longrightarrow \left\{\begin{matrix} 2\,RCO_2H \\ RCO_2R' + RCO_2H \\ RCONH_2 + RCO_2H \\ RCONHR' + RCO_2H \\ RCONR'_2 + RCO_2H \\ RCOSR' + RCO_2H \end{matrix}\right.$$

Aromatic acid anhydrides, e.g. benzoic anhydride, $(C_6H_5CO)_2O$, are known but are of little interest. They may be prepared in the same way as aliphatic acid anhydrides, which they resemble in the scope of their reactions.

Thioesters

Thioesters are the acyl derivatives of thiols, and contain a C—S—C linkage. They may be prepared by the acid-catalysed reaction of thiols with carboxylic acids, or by the reaction of thiols with acyl chlorides (p. 144) or acid anhydrides.

$$CH_3CO_2H + HSC_2H_5 \overset{H^+}{\rightleftharpoons} CH_3COSC_2H_5 + H_2O$$

Ethyl thioacetate
Ethyl thioethanoate

The reactions of thioesters resemble those of acid anhydrides and normal esters, but with reactivity approaching that of acid anhydrides. The importance of thioesters in biochemical reactions is partly due to their enhanced reactivity, e.g. they readily acylate amines:

$$RCOSR' + R''NH_2 \longrightarrow RCONHR'' + HSR$$

Carboxylic acids are frequently incorporated into metabolic processes in the cell via the formation of their esters of coenzyme A, a complex thiol (p. 300).

Amides

Amides, the acyl derivatives of amines, can be named either as derivatives of the corresponding acid or by use of 'aminocarbonyl-' to describe the functional group:

$CH_3(CH_2)_4CONH_2$
Hexanamide (from *hexan*oic acid)

There are three types of amide known as primary, secondary, and tertiary, depending on the extent of substitution on the nitrogen atom.

$RCONH_2$	$RCONHR'$	$RCONR'_2$
Primary	Secondary	Tertiary

Most of the preparations and reactions are common to all groups, but primary amides also exhibit a few special reactions.

Preparation of amides. The commonest general methods of preparation of amides are by heating the appropriate amine salt of the carboxylic acid (p. 127), or by the action of ammonia or amines on a derivative of the carboxylic acid, e.g. ester, acyl chloride, or anhydride. Primary amides can also be prepared by the controlled hydrolysis of nitriles (p. 149).

Properties of amides. Amides are either solids or high-boiling liquids, neutral to indicators. The lower aliphatic amides are very soluble in water.

Reactions of amides

1. Although they contain the amino group, amides are only very feebly basic, on account of the mesomeric interaction between the carbonyl double bond and the lone pair on the nitrogen atom.

$$\mathrm{R{-}C(=\ddot{O}{:})\ddot{N}R'_2} \longleftrightarrow \mathrm{R{-}C(-{:}\ddot{O}{:}^-)={\overset{+}{N}}R'_2}$$

The resultant partial positive charge on the nitrogen atom decreases the basicity very greatly, and protonation of amides, which is significant only under very strongly acidic conditions, occurs on the oxygen atom, as the charge on the resultant cation is delocalised, unlike the alternative nitrogen-protonated cation, in which delocalisation is impossible.

$$\mathrm{R{-}C(=\ddot{O})\ddot{N}R'_2} + \mathrm{H^+} \not\longrightarrow \mathrm{R{-}C(=\ddot{O})\overset{+}{N}R_2{-}H}$$

$$\Updownarrow$$

$$\mathrm{R{-}C(=\overset{+}{\ddot{O}}{-}H)\ddot{N}R'_2} \longleftrightarrow \mathrm{R{-}\overset{+}{C}(\ddot{O}{-}H)\ddot{N}R'_2} \longleftrightarrow \mathrm{R{-}C(\ddot{O}{-}H)={\overset{+}{N}}R'_2}$$

The very low basicity of simple amides (pK_a of the protonated amide ~ −1) means that in aqueous media the cations are virtually completely dissociated.

Primary and secondary amides can also act as very weak acids (pK_a ~ 16) losing a proton under strongly basic conditions.

$$R{-}C(=\ddot{O})\ddot{N}R{-}H \xrightarrow[(K^+(CH_3)_3CO^-)]{-H^+} R{-}C(=\ddot{O})\ddot{N}^-R' \longleftrightarrow R{-}C(\ddot{O}^-)=\ddot{N}R'$$

Here again, the charge on the anion is delocalised as in the case of enolates (p. 105) and carboxylates (p. 127).

2. Amides can be hydrolysed under acidic or alkaline conditions (mechanisms I and III, p. 142) acidic hydrolysis being the more rapid.

$$RCONR'_2 + H_2O \xrightarrow{+H^+} RCO_2H + H_2\overset{+}{N}R'_2$$

$$RCONR'_2 + H_2O \xrightarrow{+OH^-} RCO_2^- + HNR'_2$$

3. Amides may be converted directly into esters by reaction with acidified alcohols:

$$RCONH_2 + R'OH + H^+ \longrightarrow RCO_2R' + \overset{+}{N}H_4$$

4. Amides cannot be reduced by metal–acid or catalytic methods, but lithium aluminium hydride converts them into amines (p. 83)

Special reactions of primary amides

5. Primary amides can be dehydrated by heating with phosphorus pentoxide, forming nitriles:

$$RCONH_2 \xrightarrow[-H_2O]{P_2O_5} R{-}C{\equiv}N$$

6. Primary amides react with nitrous acid, forming the carboxylic acid and nitrogen (cf. primary amines, p. 86).

$$RCONH_2 + HNO_2 \longrightarrow RCO_2H + N_2 + H_2O$$

7. Primary amides react with alkali hypohalites (e.g. KBrO) to form the lower amine (p. 84).

Nitriles

Nitriles may either be regarded as derivatives of carboxylic acids, being at a comparable level of oxidation, or as esters of hydrocyanic acid, or as cyano-derivatives of hydrocarbons:

$CH_3CH_2CH_2CN$
Butyronitrile
1-Propyl cyanide
1-Cyanopropane

C_6H_5–C≡N
Benzonitrile
Phenyl cyanide
Cyanobenzene

Preparation and properties of nitriles. Aliphatic nitriles may be prepared from alkyl halides by reaction with alkali cyanides, or by the dehydration of primary amides or oximes (of aldehydes).

$$RI + CN^- \longrightarrow RCN + I^-$$

$$\left.\begin{array}{l} RCONH_2 \\ RCH{=}NOH \end{array}\right\} \xrightarrow[-H_2O]{+P_2O_5} RCN$$

Aromatic nitriles can be obtained only by the last two methods.

Nitriles are neutral liquids or solids, the lower aliphatic nitriles being miscible with water. They are covalent compounds and do not exhibit the high toxicity of ionic cyanides, since no hydrocyanic acid is produced on hydrolysis.

Reactions of nitriles

1. Acidic hydrolysis of nitriles gives first the amide, and subsequently the carboxylic acid.

$$R{-}C{\equiv}N \xrightarrow{H^+/H_2O} R{-}CONH_2 \xrightarrow{H^+/H_2O} R{-}CO_2H + \overset{+}{N}H_4$$

$$R{-}C{\equiv}N{:} + H^+ \rightleftharpoons R{-}C{\equiv}\overset{+}{N}{-}H \longleftrightarrow R{-}\overset{+}{C}{=}\ddot{N}{-}H$$

$$R{-}C{\equiv}\overset{+}{N}{-}H + H{-}\ddot{O}{-}H \longrightarrow R{-}C(=\ddot{N}{-}H)(\overset{+}{\ddot{O}}H{-}H) \xrightarrow{-H^+} R{-}C(=\ddot{N}{-}H)(\ddot{O}{-}H)$$

Tautomerisation

$$R{-}C(\ddot{N}H_2)(=\ddot{O}) \xrightarrow[\text{(Mechanism I, p. 142)}]{H^+/H_2O} RCO_2H + \overset{+}{N}H_4$$

2. Nitriles can be reduced by a variety of reagents, forming primary amines (p. 84).

Fats

Fats are naturally occurring esters of glycerol (p. 67). Although some glycerol esters ('acylglycerols' or 'glycerides') of short-chain fatty acids are known (e.g. glyceryl tributyrate in cows' milk-fat), animal fats and vegetable oils usually contain many complex acylglycerols of long-chain saturated and unsaturated acids.

Some of the commoner long-chain acids found as glycerol esters are:

Stearic acid	$CH_3(CH_2)_{16}CO_2H$	
Palmitic acid	$CH_3(CH_2)_{14}CO_2H$	
Oleic acid	$CH_3(CH_2)_7CH{=}CH(CH_2)_7CO_2H$	*cis (Z)* isomer
Linoleic acid	$CH_3(CH_2)_4CH{=}CHCH_2CH{=}CH(CH_2)_7CO_2H$	*cis, cis (Z, Z)* isomer
Linolenic acid	$CH_3CH_2(CH{=}CHCH_2)_3(CH_2)_6CO_2H$	

A trivial system of nomenclature for acylglycerols describes the esters by adding 'in' to the root of the name of the acid, with a prefix indicating how many hydroxyl groups are esterified, e.g. 'tristearin' is the triester of stearic acid, 'diacetin' is a diester of acetic acid. However, natural fats, which are biologically important both as food reserves and structural material of living tissues, frequently consist of glycerol esterified by more than one type of acid, and are usually very complex mixtures. In addition to the esters of carboxylic acids, mixed esters of glycerol, fatty acids, and phosphoric acid ('glycerol phosphatides') are widely distributed in living tissue (p. 306).

The reactions of acylglycerols are those expected of esters. The alkaline hydrolysis of fats (saponification) is an old established process for making soaps, the alkali metal salts of long-chain fatty acids (p. 67)

Acetylcholine

Acetylcholine, $(CH_3)_3\overset{+}{N}CH_2CH_2OCOCH_3\ OH^-$, is the ethanoyl (acetyl) ester of choline, $(CH_3)_3\overset{+}{N}CH_2CH_2OH\ OH^-$, which is a basic compound, widely distributed in nature. Acetylcholine is of great physiological importance, since it is a neurohormone responsible for the transmission of impulses from nerve endings to muscle fibres. Its lifetime *in vivo* is comparatively short, as it is catalytically hydrolysed to choline and acetic acid by the enzyme choline esterase. The very poisonous 'nerve gases' (e.g. di-isopropyl fluorophosphonate $((CH_3)_2CHO)_2POF$) owe their toxicity to inhibition of choline esterase.

Urea

Urea is the diamide of carbonic acid, $(HO)_2CO$, and can be prepared by the reaction of ammonia with the corresponding acyl chloride or ester.

$$\underset{\text{Phosgene}}{COCl_2} + NH_3 \longrightarrow \underset{\text{Urea}}{CO(NH_2)_2} + \overset{+}{N}H_4Cl^-$$

$$\underset{\text{Diethyl carbonate}}{CO(OC_2H_5)_2} + NH_3 \longrightarrow CO(NH_2)_2 + C_2H_5OH$$

The preparation by the thermal isomerisation of ammonium cyanate, $\overset{+}{N}H_4CNO^-$, is of historical significance only.

Urea is one of the major end products of nitrogen metabolism in animals, and is present in human urine in relatively large amounts. It is a crystalline solid, m.p. 133°C, very soluble in water, soluble in ethanol, but insoluble in non-hydroxylic solvents such as benzene, ether, or chloroform. Although a purely covalent structure can be written for urea, its properties are more consistent with the resonant ionic canonical structures (structures with both positive and negative charges on the same molecule are known as '**zwitterions**'):

$$\ddot{O}{=}C(\ddot{N}H_2)_2 \longleftrightarrow {}^-\!:\ddot{O}{-}\overset{+}{C}(\ddot{N}H_2)_2 \longleftrightarrow {}^-\!:\ddot{O}{-}C(=\overset{+}{N}H_2)(\ddot{N}H_2) \longleftrightarrow {}^-\!:\ddot{O}{-}C(\ddot{N}H_2)(=\overset{+}{N}H_2)$$

Reaction of urea

1. Urea is a feeble monacid base, (pK_a of protonated urea = 0·18). Sparingly soluble salts of nitric and oxalic acid are known, and contain the oxygen-protonated cation $[HOC(NH_2)_2]^+$, in which the positive charge is delocalised over several atoms (cf. amides, p. 147):

$$H{-}\overset{+}{\ddot{O}}{=}C(\ddot{N}H_2)_2 \longleftrightarrow H{-}\ddot{O}{-}\overset{+}{C}(\ddot{N}H_2)_2 \longleftrightarrow H{-}\ddot{O}{-}C(=\overset{+}{N}H_2)(\ddot{N}H_2) \longleftrightarrow H{-}\ddot{O}{-}C(\ddot{N}H_2)(=\overset{+}{N}H_2)$$

2. Urea, like other amides, is hydrolysed under both acidic or alkaline conditions to ammonia and carbonic acid.

$$CO(NH_2)_2 \xrightarrow{H^+/H_2O} CO_2 + \overset{+}{N}H_4$$

$$CO(NH_2)_2 \xrightarrow{\bar{O}H/H_2O} CO_3^{--} + NH_3$$

3. Urea reacts with nitrous acid to give nitrogen, like other primary amides, and it is oxidised to nitrogen by hypohalites.

$$CO(NH_2)_2 + HNO_2 \longrightarrow CO_2 + H_2O + N_2$$

$$CO(NH_2)_2 + KBrO \longrightarrow CO_2 + KBr + N_2$$

The latter reaction has been used to estimate urea by measuring the volume of nitrogen evolved when a test sample (e.g. of urine) is mixed with excess potassium hypobromite.

4. Urea reacts with acylating agents giving *N*-acylureas, 'ureides':

$$(RCO)_2O + H_2NCONH_2 \longrightarrow \underset{\text{A ureide}}{RCONHCONH_2} + RCO_2H$$

Guanidine

Guanidine, $HNC(NH_2)_2$, is the imide (i.e. $>C{=}NH$ compound) corresponding to urea. It occurs naturally, both in the free state and as a structural feature in more complex compounds (e.g. arginine, p. 263). It is a strong monoacid base (pK_a of protonated guanidine = 13·65), very soluble in water. Its basic strength is due to the extensive delocalisation of the positive charge on the protonated cation.

$$H\ddot{N}{=}C(\ddot{N}H_2)_2 + H^+ \rightleftharpoons H_2\overset{+}{N}{=}C(\ddot{N}H_2)_2 \longleftrightarrow H_2\ddot{N}{-}\overset{+}{C}(\ddot{N}H_2)_2$$

$$\updownarrow \qquad\qquad\qquad \updownarrow$$

$$H_2\ddot{N}{-}C(=\overset{+}{N}H_2)(\ddot{N}H_2) \longleftrightarrow H_2\ddot{N}{-}C(\ddot{N}H_2)(=\overset{+}{N}H_2)$$

Problems

1. When chloroethanoic acid ($pK_a = 2{\cdot}85$) is added to a buffer solution of pH 3·00 what proportion of the chloroethanoic acid remains undissociated?

2. A dibasic acid of H_2Y is dissolved in a buffer solution of pH 4·5. Physical measurements show that 5 per cent of the acid remains undissociated and 10 per cent is in the form of the dianion Y^{-2}. Calculate the two pK_a values for H_2Y.

3. If hydrolysis of the ester $CH_3CO_2C_2H_5$ with potassium hydroxide and

$H_2{}^{18}O$ is interrupted before the reaction has gone to completion, the recovered ester is found to have incorporated some of the ^{18}O isotope. Explain this observation.

4. By what mechanisms do the following reactions occur?
   (a) The acid-catalysed reaction
   $CH_3CH_2CO_2C_6H_5 + CH_3OH \rightarrow CH_3CH_2CO_2CH_3 + C_6H_5OH$
   (b) The base-catalysed reaction
   $(CH_3)_2NH + CH_3CO_2C_2H_5 \rightarrow CH_3CON(CH_3)_2 + C_2H_5OH.$

5. When ammonium acetate is heated it undergoes two reactions:
   (a) $NH_4^+CH_3CO_2^- \rightleftharpoons NH_3 + CH_3CO_2H$
   (b) $NH_4^+CH_3CO_2^- \rightarrow CH_3CONH_2 + H_2O$

   At one time it was thought that these were two independent reactions. Why must the dissociation reaction (a) be the first step in the formation of ethanamide (acetamide)? What is the mechanism of formation of ethanamide?

6. Treatment of C_2H_5CN with a solution of hydrogen chloride in dry methanol gives a compound *A*, C_4H_9NO, which, on treatment with dilute aqueous acid, forms methyl propionate and ammonia. Formulate this reaction sequence mechanistically, and deduce the structure of compound *A*.

7. If a mixture of benzoic acid, ethanoic anhydride, and a trace of sulphuric acid is heated, ethanoic acid can be distilled from the reaction mixture and benzoic anhydride is formed. By what mechanism does this reaction proceed?

8. The pK_a values for CH_3CN, $CH_2(CN)_2$, and $CH(CN)_3$ are approximately 25, 12, and 0 respectively. Explain why the acidity increases with increasing substitution by cyano groups.

9

Hydrogen bonds in organic compounds

It is normally found that, in any series of compounds of similar chemical structure, the boiling points increase with increasing relative molecular mass. If the intermolecular forces in the liquid phase are similar for structurally similar compounds, then a heavy molecule will need a greater kinetic energy ($\frac{1}{2}mV^2$) than is required by a light molecule to achieve the 'escape velocity' necessary to carry it from the liquid into the vapour phase.

Examination of the covalent hydrides of the elements of groups IV–VII of the periodic table shows that, with three remarkable exceptions, the relationship between relative molecular mass and boiling point is observed within groups, and to some extent along periods. The exceptions are ammonia, water, and hydrogen fluoride, all of which have boiling points much in excess of the value expected by comparison with the other hydrides of the groups. Extrapolation from hydrogen sulphide would predict a boiling point of approximately −80°C for water.

The boiling points of simple hydrides

GROUP:	IV	V	VI	VII
	CH_4 (−161°C)	NH_3 (−33°C)	H_2O (+100°C)	HF (+20°C)
	SiH_4 (−112°C)	Ph_3 (−87°C)	H_2S (−59°C)	HCl (−85°C)
	GeH_4 (−90°C)	AsH_3 (−54°C)	H_2Se (−41°C)	HBr (−66°C)
		SbH_3 (−18°C)	H_2Te (0°C)	HI (−35°C)

The high boiling points of these three hydrides indicate that in the liquid phase there exists some type of intermolecular force which is absent from the other simple hydrides. Many cases also exist in organic chemistry of groups of compounds having unexpectedly high boiling points, and the great majority of these involve compounds containing OH or NH groups. Since the phenomenon is restricted to compounds containing hydrogen atoms in particular molecular environments, the name 'hydrogen bond' has been given to the intermolecular force.

Hydrogen bonding is found to occur in organic compounds when a molecule, in which a hydrogen atom is joined to an electronegative atom or group, is in close proximity to a molecule containing an oxygen atom or nitrogen atom with a lone pair of electrons:

$X^{\delta-}$——$^{\delta+}$H··········:Y X = Electronegative atom or group
'Hydrogen bond' Y = O or N atom

In these circumstances, polarisation of the X—H bond reduces the electron density around the hydrogen nucleus, which then interacts electrostatically with the neighbouring lone pair. The hydrogen bond is, in fact, a dipolar interaction, and is very much weaker than a normal covalent bond, being never more than one-tenth and usually only one-twentieth the strength of normal C—H, O—H, or N—H bonds. The absence of similar effects with atoms other than hydrogen is attributable to the screening effect of the completed inner shells of electrons, which prevent any pronounced electrostatic interactions with the nucleus. Only in the hydrogen atom is there no such completed inner shell.

In the simple hydrides H_3N, H_2O, and HF, the atoms N, O, and F are very much more electronegative than hydrogen, and also possess lone pairs. As a result, these molecules can form a sort of loosely bound polymer, which

```
                                  H        H
                                  |        |
          H       H           H—N—H····:N—H
          |       |       ··       :       |
H     H—N:····H—N:                 :       H
  \ ·     |       |                H
   N      H       H····:N/
  /  \      ·           | \
H     H      ·     H    H   H
               ·  /
                N
              /   \
             H     H
```

can be represented by a structure in which hydrogen bonds are represented by dotted lines. In the liquid phase these hydrogen bonds are being continually broken and reformed by the thermal motions of the molecules, so that no discrete polymeric species can be isolated. It should be noted that in these polymeric species, the hydrogen-bonded hydrogen atom is in no way 'shared' between two other nuclei, remaining covalently bound to one atom at all times with only a weak attachment to the other.

The same type of interaction occurs in organic molecules, showing similar effects. By comparing the boiling points of a series of isomeric organic compounds, it can be seen that where —OH and —NH groups occur, e.g. in simple alcohols and amines, the boiling points of these compounds are considerably higher than for isomeric compounds lacking these features.

The boiling points of isomeric alcohols and ethers

CH_3OH	65°C		
CH_3CH_2OH	78°C	$(CH_3)_2O$	−25°C
$CH_3CH_2CH_2OH$	97°C	$CH_3OCH_2CH_3$	8°C
$(CH_3)_2CHOH$	82°C		
$CH_3CH_2CH_2CH_2OH$	118°C	$CH_3OCH_2CH_2CH_3$	39°C
$CH_3CH_2CHOHCH_3$	99°C	$CH_3OCH(CH_3)_2$	32°C
$(CH_3)_2CHCH_2OH$	107°C	$(C_2H_5)_2O$	35°C
$(CH_3)_3COH$	83°C		

The boiling points of amines

CH_3NH_2	−7°C				
$CH_3CH_2NH_2$	16°C	$(CH_3)_2NH$	7°C		
$CH_3CH_2CH_2NH_2$	49°C	$CH_3NHCH_2CH_3$	36°C	$(CH_3)_3N$	3°C
$(CH_3)_2CHNH_2$	33°C				
$CH_3CH_2CH_2CH_2NH_2$	78°C	$CH_3NHCH_2CH_2CH_3$	62°C	$(CH_3)_2NCH_2CH_3$	36°C
$CH_3CH_2CHNH_2CH_3$	63·5°C	$CH_3NHCH(CH_3)_2$	50°C		
$(CH_3)_2CHCH_2NH_2$	67·5°C	$CH_3CH_2NHCH_2CH_3$	56°C		
$(CH_3)_3CNH_2$	44·5°C				

An even more dramatic effect can be seen by comparing the boiling points of a series of compounds with increasing substitution by hydroxyl groups. Even allowing for the increasing relative molecular mass, it is clear that increasing hydroxyl substitution has a very pronounced effect on the intermolecular forces in the liquid.

CH_3CH_3	(−89°C)	CH_3CH_2OH	(78°C)	CH_2OHCH_2OH	(197°C)		
$CH_3CH_2CH_3$	(−42°C)	$CH_3CH_2CH_2OH$	(97°C)	$CH_3CHOHCH_2OH$	(189°C)	CH_2OH–$CHOH$–CH_2OH	(290°C)
		$CH_3CHOHCH_3$	(82°C)	$CH_2OHCH_2CH_2OH$	(210°C)		

Sulphur compounds, on the other hand, show little evidence of hydrogen bonding, the boiling points of isomeric thiols and thioethers differing only slightly.

The boiling points of isomeric sulphur compounds

CH_3SH	6°C		
CH_3CH_2SH	35°C	$(CH_3)_2S$	37°C
$CH_3CH_2CH_2SH$	68°C	$CH_3SCH_2CH_3$	67°C
$(CH_3)_2CHSH$	59°C		
$CH_3CH_2CH_2CH_2SH$	98°C	$CH_3SCH_2CH_2CH_3$	95·5°C
$CH_3CH_2CHSHCH_3$	85°C	$CH_3SCH(CH_3)_2$	94°C
$(CH_3)_2CHCH_2SH$	88°C	$(CH_3CH_2)_2S$	92°C
$(CH_3)_3CSH$	64°C		

In carboxylic acids, hydrogen bonding is responsible for the dimerisation observed in solutions in non-hydroxylic solvents. In benzene solution, ethanoic acids exists predominantly as the hydrogen-bonded dimer:

```
        O····H—O
       //        \
CH3—C             C—CH3
       \         //
        O—H····O
```

and these dimers also exist in the gaseous state. Primary and secondary amides can form hydrogen bonds, and in these compounds the carbonyl group is the lone pair donor:

```
      R' R          R' R         R' R
      |  |          |  |         |  |
···H—N—C=O···H—N—C=O···H—N—C=O···
```

The effect of this can be seen on both the melting points and boiling points of a series of isomeric amides, in which progressive substitution on the nitrogen atom reduces the possibility of hydrogen bonding:

The melting and boiling points of some isomeric amides

CH_3CONH_2 (82°C, 222°C)	$HCONHCH_3$ (−5°C, 131°C)	
$CH_3CH_2CONH_2$ (79°C, 222°C)	$CH_3CONHCH_3$ (28°C, 203°C)	$HCON(CH_3)_2$ (−61°C, 153°C)
$CH_3CH_2CH_2CONH_2$ (115°C, 216°C)	$CH_3CH_2CONHCH_3$ (−43°C, 220°C)	$CH_3CON(CH_3)_2$ (−20°C, 165°C)
$(CH_3)_2CHCONH_2$ (128°C, 216°C)	$CH_3CONHC_2H_5$ (—. 205°C)	
	$HCONHCH(CH_3)_2$ (— , 220°C)	$HCON(CH_3)C_2H_5$(—, 170°C)

Although the melting point of a compound is decided partly by factors such as molecular size and shape, which determine how the molecules pack into the regular crystal lattice, it is noticeable that the melting points of the tertiary amides are much lower than those of the corresponding primary amides, probably on account of hydrogen bonds in the crystals of the latter. Another property of crystals affected by hydrogen bonding is that of hardness. Since the forces between adjacent organic molecules are usually weak, unlike those between adjacent ions in an inorganic ionic compound, the crystals of organic compounds are usually soft. However, in some cases, such as polyhydroxy compounds (e.g. sugars) crystals of exceptional hardness are formed, presumably on account of hydrogen bonding.

Hydrogen bonding is also responsible for the limited solubility of many hydroxylic compounds in non-hydroxylic solvents. Thus, whilst ethanol is miscible with ether, ethane-1,2-diol is only sparingly soluble in ether (like water), and glycerol is virtually insoluble in ether, petrol, benzene, or chloroform. However, ethanol, water, ethane-1,2-diol, and glycerol are all mutually miscible in all proportions, since all have the hydrogen bonding hydroxyl group.

Although all the cases of hydrogen bonding mentioned so far are of compounds in which the hydrogen bonding hydrogen atoms are attached to oxygen or nitrogen atoms, cases are known in which other electronegative groups are responsible for polarisation of the covalent bond to hydrogen. Thus hydrogen bonding is found in mixtures of chloroform with tertiary amines or propanone, and the abnormally high solubility of ethyne in propanone is probably also due to hydrogen bonding.

$$Cl_3C^{\delta-}\text{—}^{\delta+}H\cdots\cdots:NR_3 \qquad Cl_3C^{\delta-}\text{—}^{\delta+}H\cdots\cdots:\ddot{O}{=}C(CH_3)_2$$

$$(CH_3)_2C{=}\ddot{O}:\cdots\cdot H\text{—}C{\equiv}C\text{—}H\cdots\cdot:\ddot{O}{=}C(CH_3)_2$$

In addition to the many examples of hydrogen bonding in simple organic compounds, hydrogen bonds are of great importance in many biological molecules (pp. 272, 294). In proteins, for example, the biological functions are intimately associated with particular conformations of these large molecules, and maintenance of the biochemically correct shape, which is of truly vital importance, is frequently the result of extensive intramolecular hydrogen bonding.

Problem

In the list of boiling points of isomeric alcohols and amines (p. 156) it can be seen that *t*-butyl alcohol has a boiling point significantly lower than the isomeric alcohols, and similarly for *t*-butylamine. Comment upon this observation. (Molecular models may be helpful.)

Aliphatic esters of mineral acids

Sulphate esters

Sulphuric acid can give rise to two series of esters:

$$R{-}O{-}S(=O)_2{-}O{-}H$$

Alkyl hydrogen sulphates

$$R{-}O{-}S(=O)_2{-}O{-}R$$

Dialkyl sulphates

Primary alcohols and concentrated sulphuric acid react together on warming to form the monoalkyl esters, which on vacuum distillation disproportionate to sulphuric acid and the dialkyl ester.

$$CH_3OH + H_2SO_4 \xrightarrow{60°C} CH_3OSO_3H \xrightarrow[\text{in vacuo}]{\text{heat}} (CH_3O)_2SO_2 + H_2SO_4$$

The alkyl hydrogen sulphates are strongly acidic and form a series of metal salts, but their chemistry is of little interest. The dialkyl sulphates are oily liquids of high boiling point, which are very poisonous.

The only reaction of dialkyl sulphates which will be considered here is the reaction with nucleophiles. Unlike the esters of carboxylic acids, which are acylating reagents, dialkyl sulphates behave like alkyl halides and are alkylating agents, since nucleophiles attack at the carbon atom of the alkyl group rather than at the sulphur atom of the sulphate moiety. Usually only one of the alkyl groups of the sulphate ester is lost, the alkyl hydrogen sulphates and their anions reacting very much less readily than the dialkyl esters.

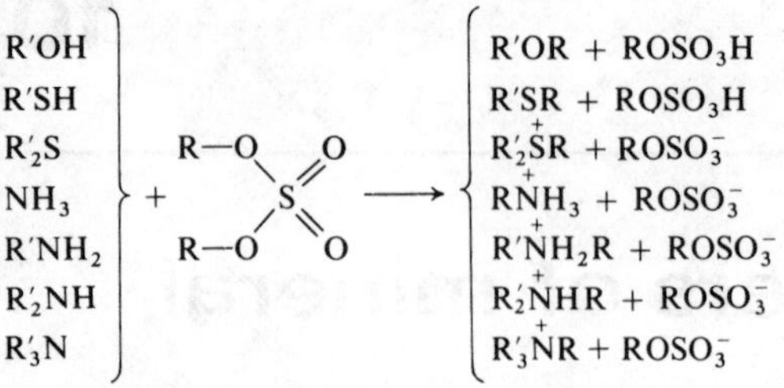

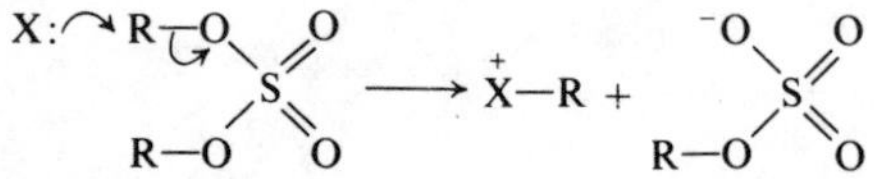

$$\left.\begin{array}{l} R'OH \\ R'SH \\ R'_2S \\ NH_3 \\ R'NH_2 \\ R'_2NH \\ R'_3N \end{array}\right\} + (RO)_2SO_2 \longrightarrow \left\{\begin{array}{l} R'OR + ROSO_3H \\ R'SR + ROSO_3H \\ R'_2\overset{+}{S}R + ROSO_3^- \\ R\overset{+}{N}H_3 + ROSO_3^- \\ R'\overset{+}{N}H_2R + ROSO_3^- \\ R'_2\overset{+}{N}HR + ROSO_3^- \\ R'_3\overset{+}{N}R + ROSO_3^- \end{array}\right.$$

$$X{:} + R{-}O{-}SO_2{-}O{-}R \longrightarrow \overset{+}{X}{-}R + {}^-O{-}SO_2{-}O{-}R$$

Nitrite esters

Esters of nitrous acid can be obtained by the reaction of an ice-cold mixture of aqueous sodium nitrite, sulphuric acid, and the appropriate alcohol:

$$ROH + HNO_2 \xrightarrow{H^+} R{-}O{-}N{=}O + H_2O$$

or by the reaction of alkali nitrites with alkyl halides (p. 51), when some of the isomeric nitroalkanes may also be produced.

The lower aliphatic nitrite esters are volatile liquids of fruity smell, which are extremely readily hydrolysed by water, dilute alkali or dilute acid, regenerating nitrous acid. The use of 'amyl nitrite' ($(CH_3)_2CHCH_2CH_2ONO$) as a vasodilator depends on this rapid hydrolysis, the nitrous acid formed from the inhaled vapour of the ester being the physiologically active species. Nitrite esters also find use in organic chemistry as sources of nitrous acid in non-aqueous media (e.g. in diazotisation of aromatic amines when the crystalline diazonium salt is required).

Phosphate esters

There are several phosphorus oxyacids of which esters are known, but those of biological interest are derived from orthophosphoric acid, and only these will be considered here. Orthophosphoric acid, H_3PO_4, is a tribasic acid and can form three series of aliphatic esters.

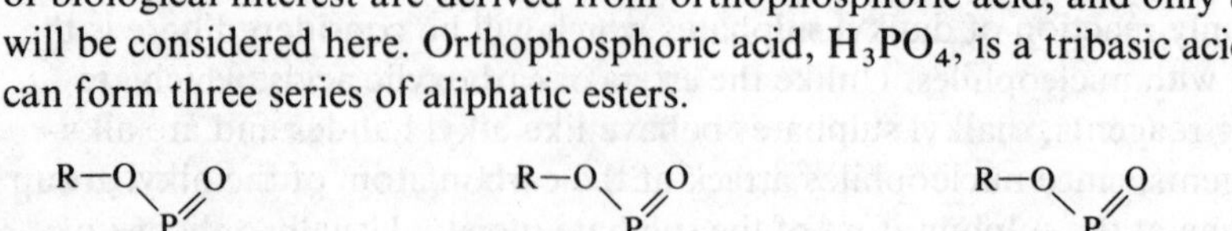

$$(RO)(HO)P(=O)OH \qquad (RO)_2P(=O)OH \qquad (RO)_2P(=O)OR$$

Monoalkyl phosphates — Dialkyl phosphates — Trialkyl phosphates

The trialkyl esters can be prepared by the reaction of phosphoryl chloride with alcohols in the presence of a tertiary amine:

$$C_2H_5OH + POCl_3 + (C_2H_5)_3N \longrightarrow (C_2H_5O)_3PO + (C_2H_5)_3\overset{+}{N}H\ Cl^-$$

(cf. the predominant reaction in the absence of base.

$$C_2H_5OH + POCl_3 \longrightarrow C_2H_5Cl + H_3PO_4)$$

The lower esters can be obtained by controlled hydrolysis of the trialkyl phosphates.

$$(C_2H_5O)_3PO \xrightarrow{H_2O} (C_2H_5O)_2PO.OH \xrightarrow{H_2O} C_2H_5OPO(OH)_2$$

The monoalkyl and dialkyl phosphates are acids, giving rise to a series of anions:

$$(RO)_2PO_2^- \qquad (RO)(HO)PO_2^- \qquad (RO)PO_3^{2-}$$

The *p*Ks of the two ionisation steps of monoalkyl phosphates are of the order $pK_1 = 1\text{–}2$, $pK_2 = 6\text{–}7$, i.e. the first dissociation is that of a strong acid, whilst the second is that of a very weak acid. Dialkyl phosphates are also strongly acidic.

In their reactions with nucleophiles, phosphate esters can behave both as acylating and alkylating agents. The two possible mechanisms for attack of a nucleophile X^- on a phosphate ester are:

(i) Phosphorylation (cf. carboxylate esters, p. 142)

$$R\text{—}\ddot{O}\text{—}P(=O)(OR)_2 + \ddot{X}^- \longrightarrow R\text{—}\ddot{O}\text{—}P(O^-)(X)(OR)_2 \longrightarrow R\text{—}\ddot{O}:^- + X\text{—}P(=O)(OR)_2$$

(ii) Alkylation (cf. sulphate esters, p. 160)

$$\bar{X}: + R\text{—}\ddot{O}\text{—}P(=O)(OR)_2 \longrightarrow X\text{—}R + :\ddot{O}^-\text{—}P(=O)(OR)_2$$

Examples of both types of process are known, and the pathway followed in any particular reaction depends on the nature of the nucleophile (e.g. H_2O, HO^-, NH_3, etc.), the nature of the phosphate ester (mono-, di-, or tri-alkyl ester), the structure of the alkyl group, and the *p*H of the reaction mixture, which determines whether it is the neutral or protonated ester, monoanion or dianion which reacts. Numerous examples of phosphate esters behaving both as alkylating and acylating reagents are known in biochemistry.

Esters of polyphosphoric acids

Unlike most of the inorganic oxyacids, phosphoric acid gives rise to a series of stable polyacids, some esters of which are encountered in biological systems. The commonest are the esters of pyrophosphoric acid (diphosphoric acid) and triphosphoric acid.

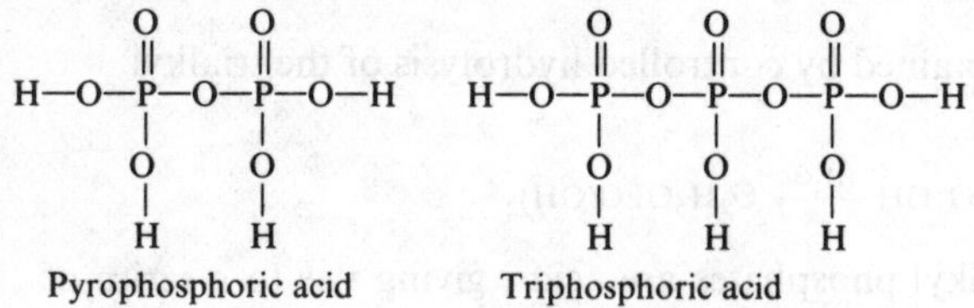

As in the case of orthophosphate esters, the polyphosphate esters can act as phosphorylating or alkylating reagents in cell reactions (see p. 295, 297).

11

Aliphatic dicarboxylic acids

General preparations of dicarboxylic acids. Aliphatic dicarboxylic acids can, in general, be obtained by any of the normal methods available for preparation of aliphatic monocarboxylic acids (p. 125), but starting from a suitable bifunctional compound. Thus, the oxidation of suitable diols, dialdehydes, hydroxy-acids, etc., will form the corresponding dicarboxylic acids:

$$\left.\begin{array}{l} HOCH_2(CH_2)_nCH_2OH \\ HOCH_2(CH_2)_nCHO \\ HOCH_2(CH_2)_nCO_2H \\ OHC(CH_2)_nCHO \\ OHC(CH_2)_nCO_2H \end{array}\right\} \xrightarrow{\text{Oxidation}} HO_2C(CH_2)_nCO_2H$$

Likewise the hydrolysis of cyano acids, $NC(CH_2)_nCO_2H$, or of dicyano compounds, $NC(CH_2)_nCN$, is also a feasible synthetic route to these compounds.

General reactions of dicarboxylic acids. The dicarboxylic acids have all the normal reactions of carboxylic acids, forming salts, esters, acyl chlorides, and amides, with occasional cases of unusual reactions, particularly in the case of anhydride formation. The presence of two functional groups in the molecule, which can react independently, leads to a more complex series of derivatives. Depending upon whether or not the two carboxylic acid groups are in identical molecular environment, two or three series of esters can be formed, e.g.:

$$\begin{array}{l} CH_2CO_2CH_3 \\ | \\ CH_3CHCO_2H \end{array} \qquad \begin{array}{l} CH_2CO_2H \\ | \\ CH_3CHCO_2CH_3 \end{array} \qquad \begin{array}{l} CH_2CO_2CH_3 \\ | \\ CH_3CHCO_2CH_3 \end{array}$$

Mixed derivatives are also possible:

$$(CH_2)_n\begin{matrix}CO_2C_2H_5\\CONH_2\end{matrix} \qquad (CH_2)_n\begin{matrix}CO_2CH_3\\COCl\end{matrix} \qquad (CH_2)_n\begin{matrix}CON(CH_3)_2\\CO_2H\end{matrix} \qquad (CH_2)_n\begin{matrix}CO_2C_2H_5\\CO_2^-Na^+\end{matrix}$$

Dicarboxylic acids will have two dissociation constants, corresponding to the ionisation of the two functional groups. A comparison of the values pK_1 and pK_2 for a series of dicarboxylic acids, in which the carboxyl groups are

$$(CH_2)_n\begin{matrix}CO_2H\\CO_2H\end{matrix} \xrightarrow{-H^+} (CH_2)_n\begin{matrix}CO_2^-\\CO_2H\end{matrix} \xrightarrow{-H^+} (CH_2)_n\begin{matrix}CO_2^-\\CO_2^-\end{matrix}$$

separated by an increasing number of carbon atoms, shows that with widely separated groups, the two *pK*s become identical and approach the *pK* of aliphatic monocarboxylic acids. However, when the two carboxylic acid groups are close together in the molecule, pK_1 and pK_2 differ appreciably. In these cases pK_1 is low as one undissociated carboxylic acid group acts as an electronegative substituent, increasing the acidity of the other group (p. 135), and pK_2 is much higher than pK_1 as the negative charge on the monoanion inhibits the second stage of ionisation, which produces a second negatively charged group in close proximity to the first.

pK values for some dicarboxylic acids, $HO_2C(CH_2)_nCO_2H$

n	NAME	pK_1	pK_2
0	Oxalic acid	1·19	4·21
1	Malonic acid	2·85	6·10
2	Succinic acid	4·19	5·57
3	Glutaric acid	4·34	$pK_1 \approx pK_2$
4	Adipic acid	4·48	
5	Pimelic acid	4·31	
	cf. Butyric acid	4·82	

Oxalic acid (Ethanedioic acid). $(CO_2H)_2$, occurs naturally in a number of plants, e.g. sorrel and rhubarb. It may be obtained by the oxidation of ethane-1,2-diol, or hydrolysis of its nitrile, cyanogen, $(CN)_2$. Oxidation of sugars or polysaccharides (p. 237), with nitric acid results in the formation of oxalic acid, and the commercial source is via the pyrolysis of sodium methanoate (formate):

$$2\overset{+}{Na}\overset{-}{O}_2CH \xrightarrow{\text{heat}} (\overset{-}{CO_2}\overset{+}{Na})_2 + H_2$$

Oxalic acid is a crystalline solid, sparingly soluble in water, and very poisonous. It is one of the strongest organic acids known. It forms normal salts, esters and amides, and the acid chloride, $(COCl)_2$, can be produced by reaction of the acid with PCl_5. No anhydride is known, and reactions with dehydrating agents form only carbon monoxide and carbon dioxide.

$$(CO_2H)_2 \xrightarrow[H_2SO_4]{\text{hot conc.}} CO + CO_2 + H_2O$$

Malonic acid (Propanedioic acid). $CH_2(CO_2H)_2$, is obtained from chloroethanoic acid via cyanoethanoic acid, or by oxidation of malic acid:

$$ClCH_2CO_2K \xrightarrow{KCN} NCCH_2CO_2K \xrightarrow{H^+/H_2O} CH_2(CO_2H)_2$$

$$\underset{\text{Malic acid}}{\begin{matrix} CH(OH)CO_2H \\ | \\ CH_2CO_2H \end{matrix}} \xrightarrow{CrO_3} \underset{\text{Malonic acid}}{CH_2(CO_2H)_2}$$

Malonic acid is a crystalline solid, m.p. 135°C, soluble in water and a fairly strong acid. It forms normal salts, esters, acyl chloride, and amides. Attempts to form an anhydride result in production of an amorphous polymeric anhydride,

$$\left(-O-\overset{O}{\overset{\|}{C}}-CH_2-\overset{O}{\overset{\|}{C}}- \right)_n$$

The most important characteristic reaction of malonic acid is its easy thermal decarboxylation. Malonic acid, and all acids of the type $RR'C(CO_2H)_2$ lose carbon dioxide when heated above the melting point:

$$CH_2(CO_2H)_2 \xrightarrow{\text{heat}} CH_3CO_2H + CO_2$$

$$RR'C(CO_2H)_2 \xrightarrow{\text{heat}} RR'CHCO_2H + CO_2$$

Diethyl malonate. $CH_2(CO_2C_2H_5)_2$, is a reagent of great importance in organic chemistry. It possesses a methylene group situated between two carbonyl groups, and readily loses a proton under basic conditions to form an enolate anion (p. 105), which is stabilised by resonance:

$$CH_2(CO_2C_2H_5)_2 + C_2H_5O^- \rightleftharpoons \bar{C}H(CO_2C_2H_5)_2 + C_2H_5OH$$

$$C_2H_5O{-}\overset{\overset{:O:}{\|}}{C}{-}\underset{\underset{H}{|}}{\overset{..-}{C}}{-}\overset{\overset{:O:}{\|}}{C}{-}OC_2H_5 \longleftrightarrow C_2H_5O{-}\overset{\overset{:\ddot{O}:^-}{|}}{C}{=}\underset{\underset{H}{|}}{C}{-}\overset{\overset{:O:}{\|}}{C}{-}OC_2H_5$$

$$\longleftrightarrow C_2H_5O{-}\overset{\overset{:O:}{\|}}{C}{-}\underset{\underset{H}{|}}{C}{=}\overset{\overset{:\ddot{O}:^-}{|}}{C}{-}OC_2H_5$$

The pK of this dissociation is 13·3, so that even in 1 M aqueous strong monacid base, the ester is more than half in the form of its enolate anion.

The synthetic importance of diethyl malonate lies in the possibility of utilising the readily formed enolate as a nucleophile, which can be alkylated by alkyl halides. The products of this reaction are the esters of alkyl malonic acids, and these can be hydrolysed to the parent dicarboxylic acids, which, on heating, decarboxylate to form the corresponding monocarboxylic acids. By control of the conditions and proportions of reagents, either one or both of the methylene hydrogen atoms can be substituted by alkyl groups in this way. The synthetic route is summarised by the reaction sequence:

$$CH_2(CO_2C_2H_5)_2 \xrightarrow{Na^+\bar{O}C_2H_5} Na^+\ \bar{C}H(CO_2C_2H_5)_2 \xrightarrow{R\text{–}I} RCH(CO_2C_2H_5)_2$$

$$RCH(CO_2C_2H_5)_2 \xrightarrow{H^+/H_2O} RCH(CO_2H)_2 \xrightarrow[-CO_2]{heat} RCH_2CO_2H$$

$$RCH(CO_2C_2H_5)_2 \xrightarrow{Na^+\bar{O}C_2H_5} Na^+R\bar{C}(CO_2C_2H_5)_2 \xrightarrow{R'\text{—}I} RR'C(CO_2C_2H_5)_2$$

$$RR'C(CO_2C_2H_5)_2 \xrightarrow{H^+/H_2O} RR'C(CO_2H)_2 \xrightarrow[-CO_2]{heat} RR'CHCO_2H$$

Barbituric acid and its derivatives. If diethyl malonate is heated with urea in the presence of sodium ethoxide, a cyclic ureide (p. 152), barbituric acid, is formed:

$$H_2C\begin{matrix}/CO_2C_2H_5\\ \backslash CO_2C_2H_5\end{matrix} + \begin{matrix}H_2N\backslash\\ H_2N/\end{matrix}C{=}O \xrightarrow{+Na^+\bar{O}C_2H_5} H_2C\begin{matrix}/\overset{O}{\overset{\|}{C}}{-}NH\backslash\\ \backslash\underset{\underset{O}{\|}}{C}{-}NH/\end{matrix}C{=}O$$

This compound is a remarkably strong acid, $pK_1 = 3{\cdot}98$, the acidic protons being those of the methylene groups, the negative charge of the mono-anion being delocalised over adjacent carbonyl groups. Barbituric acid is the parent compound of an extensive series of sedative drugs, of which examples are:

```
              CH3       O
               |        ||
CH3CH2CH2CH        C—NH
             \    /     \
              C          C=O
             /    \     /
   CH2CHCH2        C—NH
                   ||
                   O
```

Seconal

```
       CH3    O
        |     ||
CH3CH2CH2CH   C—NH
          \  /    \
           C       C=S
          / \     /
     H5C2    C—NH
             ||
             O
```

Pentothal (from thiourea, $(H_2N)_2CS$)

```
              O
              ||
      H5C6    C—NH
          \  /    \
           C       C=O
          / \     /
      H5C2   C—NH
              ||
              O
```

Phenobarbitone

Succinic acid. $(CH_2CO_2H)_2$, is widely distributed in living tissue, being an intermediate in the tricarboxylic acid cycle (p. 234). It may be synthesised by the two routes outlined below, both of which establish its structure unambiguously.

$$\begin{matrix} CH_2 \\ \| \\ CH_2 \end{matrix} \xrightarrow{+Br_2} \begin{matrix} CH_2Br \\ | \\ CH_2Br \end{matrix} \xrightarrow{KCN} \begin{matrix} CH_2CN \\ | \\ CH_2CN \end{matrix} \xrightarrow{H^+/H_2O} \begin{matrix} CH_2CO_2H \\ | \\ CH_2CO_2H \end{matrix}$$

Succinic acid

$$CH_2(CO_2C_2H_5)_2 \xrightarrow{Na^+\bar{O}C_2H_5} Na^+\bar{C}H(CO_2C_2H_5)_2 \xrightarrow{I_2} \begin{matrix} CH(CO_2C_2H_5)_2 \\ | \\ CH(CO_2C_2H_5)_2 \end{matrix}$$

$$\downarrow H^+/H_2O$$

$$\begin{matrix} CHCO_2H \\ \| \\ CHCO_2H \end{matrix} \xrightarrow{Ni/H_2} \begin{matrix} CH_2CO_2H \\ | \\ CH_2CO_2H \end{matrix} \xleftarrow{heat} \begin{matrix} CH(CO_2H)_2 \\ | \\ CH(CO_2H)_2 \end{matrix}$$

It may also be obtained by hydrogenation of the corresponding unsaturated acids, maleic and fumaric acids (p. 232).

Succinic acid is a crystalline, water-soluble solid, m.p. 185°C. It gives a normal series of esters, salts, amides, and acyl chloride, but is unusual in that on heating to the boiling point (235°C) it forms a cyclic anhydride.

$$\begin{array}{l} CH_2-CO_2H \\ | \\ CH_2-CO_2H \end{array} \xrightarrow[-H_2O]{\text{heat}} \text{Succinic anhydride } (H_2C-C(=O)-O-C(=O)-CH_2)$$

Succinic anhydride

This ease of dehydration is also shown by many dicarboxylic acids, whose carboxyl groups are separated by two or three carbon atoms (cf. glutaric acid) and is, in part, due to the ease of formation of rings of five or six atoms. The amide is also readily converted into a cyclic imide on heating.

$$\begin{array}{l} CH_2-CONH_2 \\ | \\ CH_2-CONH_2 \end{array} \xrightarrow[-NH_3]{\text{heat}} \text{Succinimide } (H_2C-C(=O)-NH-C(=O)-CH_2)$$

Succinimide

Glutaric acid. $CH_2(CH_2CO_2H)_2$, occurs naturally, and can be synthesised from diethyl malonate.

$$CH_2(CO_2C_2H_5)_2 \xrightarrow{Na^+\bar{O}C_2H_5} Na^+\bar{C}H(CO_2C_2H_5)_2 \xrightarrow{CH_2I_2} \begin{array}{l} CH(CO_2C_2H_5)_2 \\ | \\ CH_2 \\ | \\ CH(CO_2C_2H_5)_2 \end{array}$$

$$\xrightarrow{H^+/H_2O} \begin{array}{l} CH(CO_2H)_2 \\ | \\ CH_2 \\ | \\ CH(CO_2H)_2 \end{array} \xrightarrow{\text{heat}} \begin{array}{l} CH_2CO_2H \\ | \\ CH_2 \\ | \\ CH_2CO_2H \end{array}$$

Glutaric acid

Glutaric acid gives the normal derivatives of a carboxylic acid, and, like succinic acid, forms a cyclic anhydride on heating.

$$H_2C\begin{array}{l} \diagup CH_2-CO_2H \\ \diagdown CH_2-CO_2H \end{array} \xrightarrow[-H_2O]{\text{heat}} \text{Glutaric anhydride } (H_2C-C(=O)-O-C(=O)-CH_2-CH_2, \text{ six-membered ring})$$

Glutaric anhydride

Adipic acid. $(CH_2CH_2CO_2H)_2$, occurs in beet-juice. It can be prepared by the oxidation of cyclohexanone with hot concentrated nitric acid (p. 114), and is a normal dicarboxylic acid which does not form a cyclic anhydride on heating.

Problems

1. Starting from diethyl malonate, devise schemes for the synthesis of the following compounds:

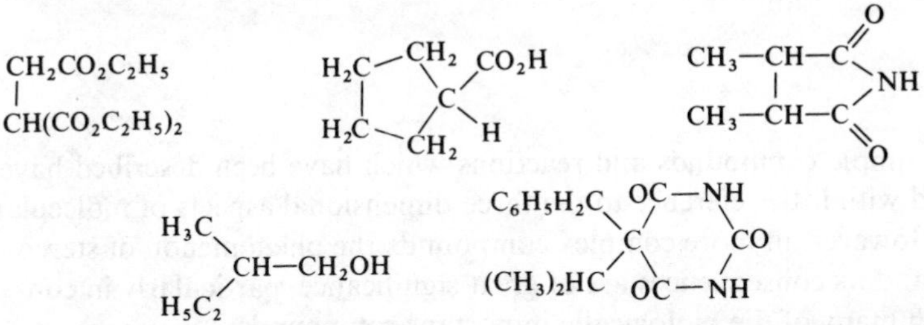

$CH_2CO_2C_2H_5$
|
$CH(CO_2C_2H_5)_2$

H_2C—CH_2 CO_2H
| C
H_2C—CH_2 H

O
CH_3—CH—C
| NH
CH_3—CH—C
O

H_3C
CH—CH_2OH
H_5C_2

$C_6H_5H_2C$ OC—NH
C CO
$(CH_3)_2HC$ OC—NH

2. Draw all the canonical structures for all the mono-anions which can be derived from barbituric acid (p. 166). What products might be formed by treatment of barbituric acid with one molar proportion of sodium ethoxide followed by 1-bromobutane?

3. Which hydrogen atom(s) in succinimide (p. 168) do you expect to undergo/isotopic exchange most readily in KOD/D_2O?

12

Stereochemistry part I.
Enantiomerism

So far, the simple compounds and reactions which have been described have been treated with little reference to the three-dimensional aspects of molecular structure. However, in more complex compounds the phenomenon of stereoisomerism and its consequences are of great significance, particularly in connection with many of the biologically important compounds.

Stereoisomerism is a form of isomerism, in which compounds of the same constitutional formulae differ in the spatial arrangements of the functional groups. Stereoisomerism in simple molecules can be divided into **optical isomerism (enantiomerism)** and **geometrical isomerism (*cis, trans* isomerism)**, but in complex molecules the distinction between these types may often be less clear-cut.

Optical isomerism arises from the possibility of having a three-dimensional structure which is not superposable on its mirror image. Such structures are said to have the property of **'chirality'**, or to be **'chiral'**, and are characterised by the lack of a plane of symmetry (referring, of course, to the three-dimensional structure, and not to the representation on paper).

The simplest case of a chiral structure is that of an 'asymmetrically substituted carbon atom', i.e. a carbon atom bearing four, different, covalently bound substituents. Such a structure has two non-superposable isomers, **enantiomers**, which are related as object to mirror-image, e.g. (I) and (II).

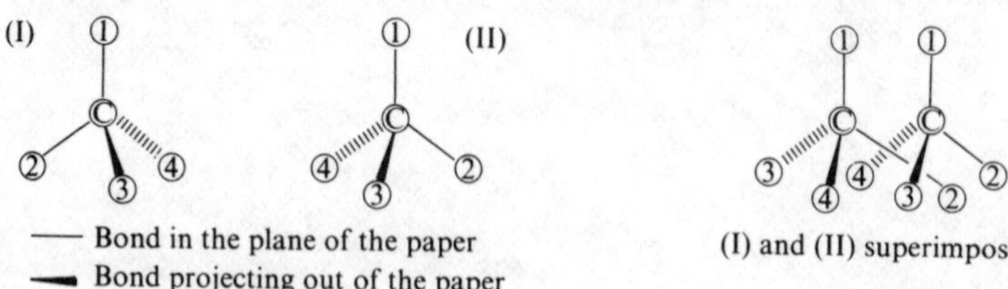

(I) and (II) superimposed

However, if only two of the substituents are the same, this structure acquires a plane of symmetry, and becomes superposable on its mirror image. In the diagrams (I) and (II), if the substituents ③ and ④ are identical, then the structures will have planes of symmetry passing through ①, Ⓒ, and ②.*

Numerous other types of chiral structure are known to chemists. However, in the compounds we shall consider in this and later chapters, the overwhelming majority of cases of optical isomerism will be concerned with structures in which the chirality is directly attributable to asymmetric substitution of a tetracovalent atom.

Polarised light. Light is an electromagnetic radiation composed of oscillating electric and magnetic fields, which are mutually perpendicular to each other and to the direction of propagation of the light beam.

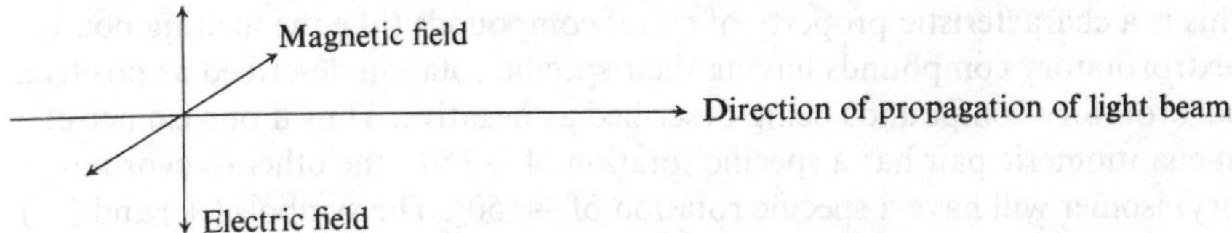

Normal light consists of multiple beams, which have random relative orientation of electric vectors. In plane-polarised light (usually referred to simply as 'polarised light') all the beams have their electric fields aligned parallel, with all the magnetic fields oscillating in the perpendicular plane.

Polarised light is unaffected by solutions of compounds whose structures have symmetry which precludes enantiomerism. However, it is found that if a beam of plane-polarised light is passed through a solution of one enantiomer of a compound with a chiral structure, the plane of polarisation of the light is rotated either clockwise or anticlockwise, and an equal and opposite rotation of the plane of polarisation occurs on passing through an equimolar solution of the other enantiomer.† The isomer whose solution rotates the plane of polarisation in a clockwise direction (with the observer facing the light source) is described as **dextrorotatory**, that which rotates the plane of polarisation in an anticlockwise direction is described as **laevorotatory**.

If solutions of enantiomeric compounds, which separately are dextro- and laevorotatory, are mixed to produce a solution containing equal concentrations of the two isomers, then this mixed solution is optically inactive. Such a mixture of enantiomers is known as a **racemic mixture**.

* It is recommended that students should use simple molecular models ('ball and stick' type) to study these structures and others mentioned in this chapter.

† These observations are made with an instrument known as a polarimeter. It is beyond the scope of this text to describe its method of operation, which, if required, should be sought in a textbook of physics.

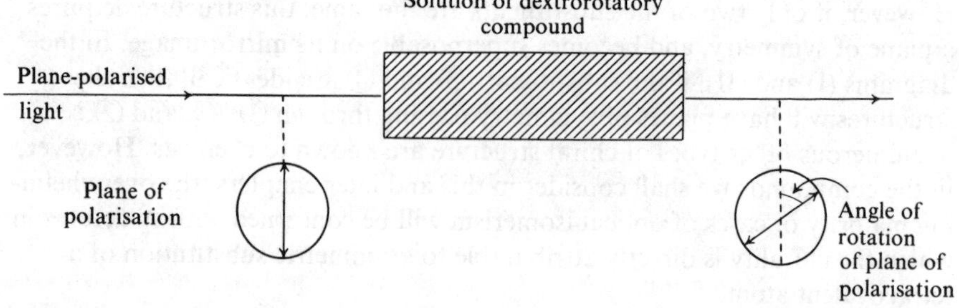

The specific rotation (α) of a substance is defined as the angle (in degrees) through which the plane of polarisation is rotated when plane-polarised light is passed through a 10 cm length of a solution of concentration 1 g/ml. This is a characteristic property of chiral compounds (like the melting point), dextrorotatory compounds having their specific rotation described as positive, laevorotatory compounds being described as negative. Thus if one isomer of an enantiomeric pair has a specific rotation of $+150°$, the other (laevorotatory) isomer will have a specific rotation of $-150°$. The symbols (+) and (−) are used in nomenclature to distinguish between two enantiomers, thus the two isomers of 2-bromobutane, $CH_3CH_2CHBrCH_3$, can be described as (+)-2-bromobutane (i.e. dextrorotatory) and (−)-2-bromobutane (laevorotatory), and the racemic mixture of these as (±)-2-bromobutane.*

Although enantiomers have markedly different effects on polarised light, all the other physical properties of the separate enantiomers, and all those chemical properties, which do not involve other chiral molecules, are identical. Thus two enantiomeric carboxylic acids (e.g. (+) and (−) $C_6H_5CH(C_2H_5)CO_2H$) will have the same melting and boiling points, the same refractive index, density, solubility, and viscosity. Their *pK*s will be identical, and they will form esters with, say, methanol or ethanol, whose properties, other than optical rotation, will likewise be identical. However, they will react at different rates with, say, (+)-butan-2-ol, and the two esters will not now be identical (p. 173).

The synthesis of a compound with a chiral structure usually produces both enantiomers in equal proportions (i.e. a racemic product), e.g.

$$CH_3CH_2COCH_3 \xrightarrow{LiAlH_4} CH_3H_2C{-}C(CH_3)(H)(OH) + CH_3H_2C{-}C(CH_3)(OH)(H)$$

* The archaic equivalents *d*, *l*, and *dl* are sometimes encountered, and refer solely to the direction of rotation of the plane of polarised light.

and the close similarity of the properties of the two enantiomers makes their separation difficult. No simple physical separation by techniques such as fractional distillation or recrystallisation is possible. However, the following methods have been employed to **resolve** (i.e. separate) racemic mixtures.

A *Mechanical separation*. If a solution of a racemic mixture is allowed to crystallise, two types of product may be obtained. Either one type of crystal (a **racemate**) will separate, in which the crystal lattice is built up of equal numbers of molecules of each of the enantiomers, or the solution will deposit a mixture of two types of crystal, one composed solely of the (+) enantiomer, the second containing solely the (−) enantiomer. In the latter case, if the two types of crystal can be distinguished (e.g. by the mirror-image relationship of the disposition of minor facets of the crystals), and if the individual crystals are big enough, then they can be separated by hand-picking. This very laborious and unsatisfactory method, which is rarely possible, is now only of historical interest, being the way in which Pasteur first resolved sodium ammonium (±)-tartarate. Many racemic mixtures crystallise as racemates and cannot, therefore, be resolved by this method. It may be noted in passing that the racemate, having a different crystal structure, may have a widely different melting point and solubility from those of the separate enantiomers, and cases are known where mixing the saturated solutions of the enantiomers produces a precipitate of the less soluble racemate. These differences correspond to the relative ease of packing 'right-handed' and 'left-handed' molecules alternately into the crystal lattice, compared with forming a lattice from right- or left-handed molecules alone.

B *Resolution via diastereoisomeric compounds.* If a racemic mixture of the two enantiomers of a chiral reagent is allowed to react with an optically inactive compound, the enantiomers will react separately, producing two enantiomeric products. This can be illustrated by the analogy of two pieces of rod with left-and right-hand screw threads (representing the enantiomers) being joined to another piece of rod which has no such chiral pattern. However, if the racemic mixture reacts with *one enantiomer of a second chiral reagent,* then the two products are no longer enantiomeric (i.e. related as object and mirror image), although they are stereoisomeric. Stereoisomers which are not mirror images are known as **diastereoisomers**, and diastereoisomers have physical properties which may differ widely, permitting separation by techniques such as fractional recrystallisation, distillation, or chromatography.

A familiar illustration of the formation of diastereoisomers is in the insertion of left and right hands (which have a mirror image relationship) into a right-

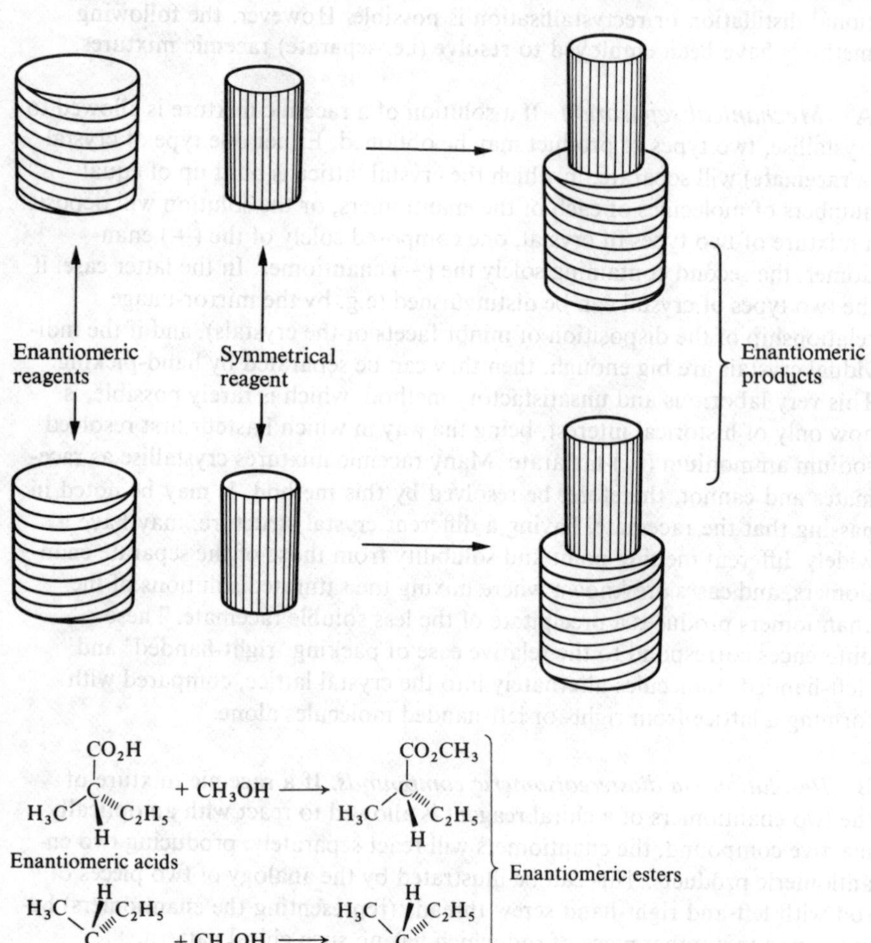

$$\begin{array}{c}CO_2H\\ |\\ H_3C\text{—}C\text{—}C_2H_5\\ H\end{array} + CH_3OH \longrightarrow \begin{array}{c}CO_2CH_3\\ |\\ H_3C\text{—}C\text{—}C_2H_5\\ H\end{array}$$

Enantiomeric acids

$$\begin{array}{c}H_3C\quad H\quad C_2H_5\\ C\\ |\\ CO_2H\end{array} + CH_3OH \longrightarrow \begin{array}{c}H_3C\quad H\quad C_2H_5\\ C\\ |\\ CO_2CH_3\end{array}$$

Enantiomeric esters

hand glove, resulting in a situation in which (left-hand + right-hand glove) is not the mirror image of (right hand + right-hand glove). It should also be

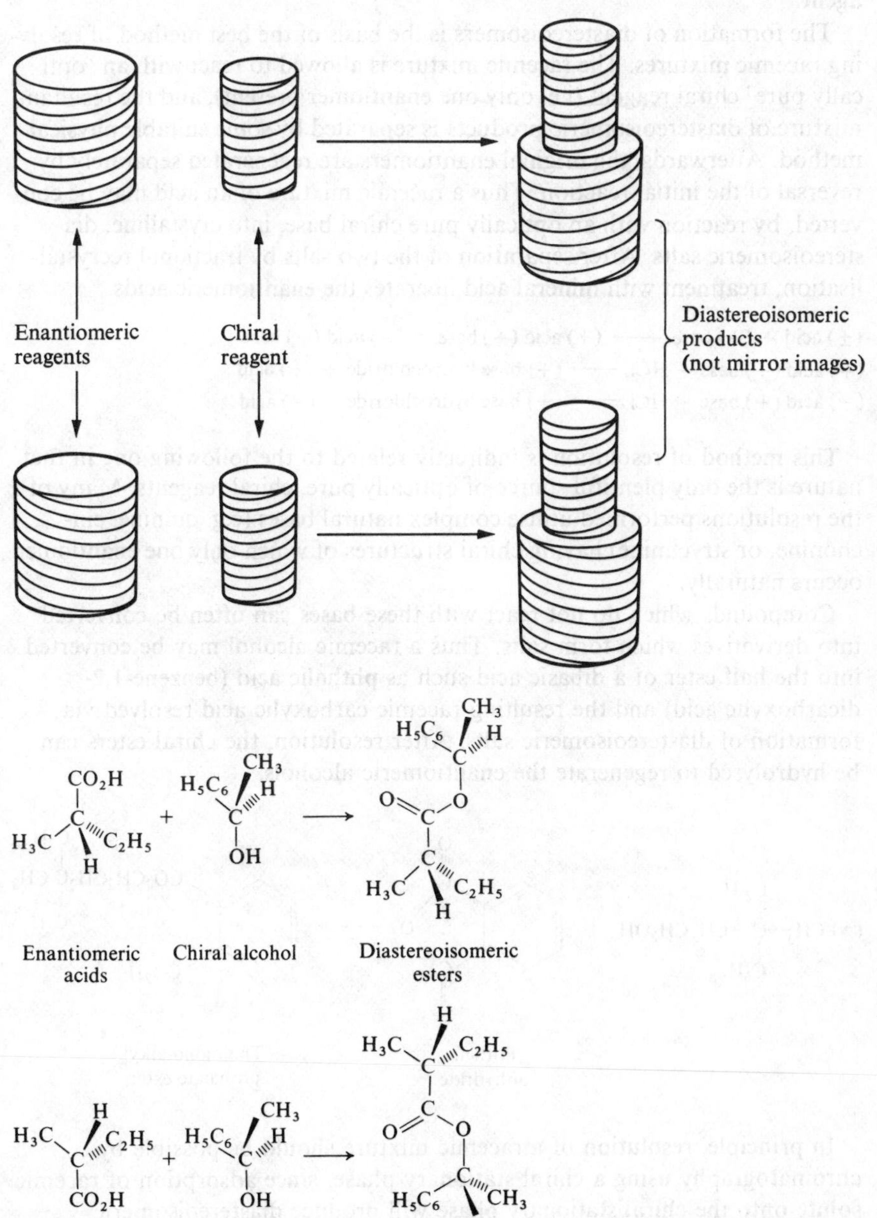

noted that the latter process is much easier than the former, corresponding to the different rates at which enantiomeric compounds react with a chiral reagent.

The formation of diastereoisomers is the basis of the best method of resolving racemic mixtures. The racemic mixture is allowed to react with an 'optically pure' chiral reagent (i.e. only one enantiomer present), and the resultant mixture of diastereoisomeric products is separated by some suitable physical method. Afterwards, the original enantiomers are regenerated separately by reversal of the initial reaction. Thus a racemic mixture of an acid may be converted, by reaction with an optically pure chiral base, into crystalline, diastereoisomeric salts. After separation of the two salts by fractional recrystallisation, treatment with mineral acid liberates the enantiomeric acids.

(±) acid + (+) base ⟶ (+) acid (+) base + (−) acid (+) base
(+) acid (+) base + HCl ⟶ (+) base hydrochloride + (+) acid
(−) acid (+) base + HCl ⟶ (+) base hydrochloride + (−) acid

This method of resolution is indirectly related to the following one in that nature is the only plentiful source of optically pure, chiral reagents. Many of the resolutions performed utilise complex natural bases (e.g. quinine, cinchonine, or strychnine) having chiral structures of which only one enantiomer occurs naturally.

Compounds which do not react with these bases can often be converted into derivatives which form salts. Thus a racemic alcohol may be converted into the half ester of a dibasic acid such as phthalic acid (benzene-1,2-dicarboxylic acid) and the resulting racemic carboxylic acid resolved via formation of diastereoisomeric salts. After resolution, the chiral esters can be hydrolysed to regenerate the enantiomeric alcohols.

(±) $CH_3-C(C_2H_5)(C_6H_5)-CH_2CH_2OH$ + Phthalic anhydride ⟶ The mono-alkyl phthalate ester ($C_6H_4(CO_2CH_2CH_2C(C_2H_5)(C_6H_5)CH_3)(CO_2H)$)

Phthalic anhydride

The mono-alkyl phthalate ester

In principle, resolution of a racemic mixture should be possible by chromatography using a chiral stationary phase, since adsorption of racemic solute onto the chiral stationary phase will produce diastereoisomeric adsorbent–solute combinations. One enantiomer of the racemic solute (the less tightly bound) should be eluted more rapidly than the other. In practice this method has achieved varying degrees of success, lactose, starch, cellulose, and optically active quartz being examples of the chiral adsorbents employed. Paper chromatography has also been used to achieve some resolutions either using the cellulose of the paper as the chiral adsorbent or by impregnation of the paper with some other chiral reagent. Thus 2-aminophenylethanoic acid, $C_6H_5CHNH_2CO_2H$, has been resolved by chromatography on paper impregnated with (+)-camphor-10-sulphonic acid.

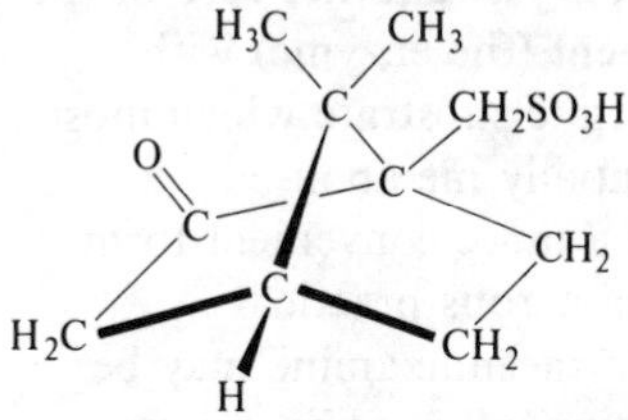

(+) Camphor-10-sulphonic acid

If a racemic mixture of one chiral reagent (A) reacts with a racemic mixture of a second chiral reagent (B), then four products are obtained:

$$(\pm)\text{A} + (\pm)\text{B} \longrightarrow \begin{cases} (+)\text{A}(+)\text{B} \quad (-)\text{A}(-)\text{B} \\ (+)\text{A}(-)\text{B} \quad (-)\text{A}(+)\text{B} \end{cases}$$

As the products are set out above, horizontal pairs are enantiomeric, vertical or diagonal pairs are diastereoisomeric.

C *Biological resolution of racemic mixtures.* If a racemic mixture can be fed to a living organism, then it is often found that one enantiomer is preferentially metabolised. If this is so, then the unwanted isomer can sometimes be recovered. When a racemic mixture of mevalonic acid (3,5-dihydroxy-3-methylpentanoic acid) is fed to rats, one optical isomer is totally absorbed,

Mevalonic acid

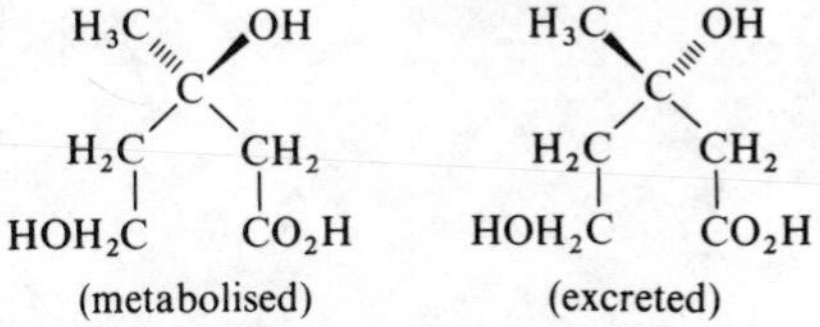

(metabolised) (excreted)

and almost all the other is excreted in the urine, from which it can be recovered. Similarly moulds or other micro-organisms will utilise one enantiomer of a racemic mixture in the culture medium. However, this method suffers from several disadvantages, compounds may be poisonous or not assimilated at all; and even if the method works, one of the enantiomers is always lost.

This method of resolution is really another example of resolution via formation of diastereoisomers. Reactions occurring in living systems are controlled by protein catalysts (enzymes) which are themselves chiral. The ability of an organism to metabolise a substance depends upon the presence of enzymes which will adsorb the molecules prior to catalysing their chemical transformation (p. 313) as part of the process of digestion. The initial formation of the enzyme-substrate complex is just another case of the interaction of a single enantiomer of a chiral reagent (the enzyme) with a racemic compound, and that enantiomer of a racemic substrate which most readily combines with the enzyme will be preferentially metabolised.

This method of resolution has been adapted to a more convenient form by the use of purified enzymes to overcome the numerous practical disadvantages of employing living systems. Thus a racemic amine may be resolved by convertion into its *N*-ethanoyl derivative followed by enzymic hydrolysis of the racemic amide. By appropriate choice of enzyme and conditions, one of the enantiomers of the amide can be hydrolysed selectively, leaving a mixture of amine and amide which is easily separated by standard chemical means e.g.:

$$\underset{\text{D, L }(R, S)}{(\pm)CH_3\underset{\large NH_2}{\underset{|}{C}}HCO_2H} \xrightarrow[Na_2CO_3]{(CH_3CO)_2O} \underset{\text{D, L }(R, S)}{(\pm)CH_3\underset{\large NHCOCH_3}{\underset{|}{C}}HCO_2H}$$

$$\downarrow \text{Hog acylase, } p\text{H8}$$

$$\underset{\text{D }(R)}{(+)CH_3\underset{\large NHCOCH_3}{\underset{|}{C}}HCO_2H} + \underset{\text{L }(S)}{(+)CH_3\underset{\large NH_2}{\underset{|}{C}}HCO_2H}$$

Racemisation. It is sometimes found that, on keeping, optically pure compounds slowly lose their optical activity, being converted reversibly into the enantiomers. This process, which eventually converts either enantiomer into a racemic mixture, is known as racemisation. Racemisation occurs when enantiomers can be converted into some symmetrical compound, which can then revert to either of the alternative mirror-image structures. Enolisation of carbonyl compounds is one such process, and results in very rapid racemisation of simple ketones of the type R′RCHCOR″.

Symmetrical intermediate

The designation of chirality by D and L, and the representation of absolute configuration*

We have seen that the most important observable difference between enantiomers is the action on polarised light, and for over a century following the discovery of optical isomerism the only certain method of distinguishing between enantiomers was by reference to the direction of rotation, hence the use of (+) and (−) in nomenclature. Although it was realised that the rotation of polarised light corresponded to alternative configurations of the molecular structure, there was no way of determining the absolute configuration (i.e. the true spatial arrangement of the groups in the molecule), and it was very soon discovered that *there is no simple relationship between the direction of rotation of polarised light and the molecular configuration*. Thus a dextrorotatory alcohol might form a laevorotatory ethanoate (acetate) and a dextrorotatory benzoate, or a laevorotatory amine might form a dextrorotatory protonated cation, and there are many such examples, in which reactions, which cannot involve a change in configuration at the asymmetric centre, give products with optical activity widely different from that of the starting material.

In spite of the ignorance of the true configuration, it was nevertheless possible to relate the configuration of compounds on the basis of chemical interconversions, which did not interfere with the configuration of the chiral centre. Thus, although (+)-lactic acid, $CH_3CH(OH)CO_2H$ ($\alpha = +3{\cdot}82°$), on esterification forms (−)-methyl lactate, $CH_3CH(OH)CO_2CH_3$ ($\alpha = -8{\cdot}25°$),

* Stereoisomeric structures, interconversion of which requires the breaking and re-formation of covalent bonds, are said to have different molecular **configurations**. (Cf. molecular conformations (p. 188), which are interconvertible by rotation about bonds.)

it is not expected that the configuration at position 2 has changed. If we arbitrarily assign one of the two possible configurations to (+)-lactic acid, we must automatically assign the same configuration to (−)-methyl lactate and all other esters formed by (+)-lactic acid, irrespective of their rotatory powers. For reasons which will become apparent later (p. 237), the dextrorotatory isomer of glyceraldehyde (p. 68) was taken as the standard compound, arbitrarily assigned the configuration shown below, and named D-glyceraldehyde. Its enantiomer was named L-glyceraldehyde. In these

CHO | C; H, CH_2OH, OH

D-(+)-Glyceraldehyde

CHO | C; HOH_2C, H, HO

L-(−)-Glyceraldehyde

names D and L refer to configuration only, and are not related to optical rotation. (They should not be confused with *d* and *l*, see footnote, p. 172.) Since the configurations of (+)-glyceraldehyde and (−)-lactic acid can be demonstrated to be the same by a sequence of chemical transformations which effectively interconvert these two compounds, then we can say that (−)-lactic acid and its (+)-methyl ester (see above) both have the D-configuration and are represented by:

CO_2H | C; H, CH_3, OH

D-(−)-Lactic acid

CO_2CH_3 | C; H, CH_3, OH

(+)-Methyl D-lactate

When recently it became possible to determine the absolute configuration of enantiomers,* it was found that, purely by chance, the correct assignment of configuration had been made. The projection formulae above, therefore, represent the true absolute configurations of the enantiomers.

The Fischer projection. Whilst it is possible to draw two-dimensional diagrams, which are reasonably clear perspective diagrams of three-dimensional structures having one or two chiral centres, this is not so for more complex molecules, and a stylised representation known as the Fischer projection is used for this purpose. The two enantiomers of the species C ① ② ③ ④ are represented by:

* By a method employing diffraction of X-rays by crystals in a way which permitted the absolute spatial disposition of atoms to be determined.

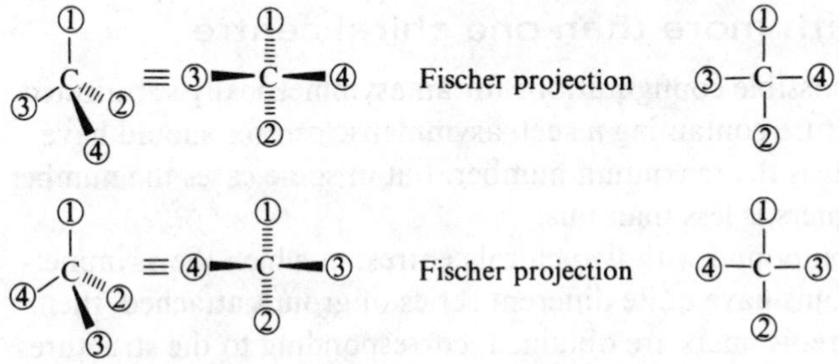

①
|
③–C–②, ④ (3D) ≡ ③–C–④ (① above, ② below) Fischer projection

$$\begin{array}{c}①\\|\\③—C—④\\|\\②\end{array}$$

①
|
④–C–②, ③ (3D) ≡ ④–C–③ (① above, ② below) Fischer projection

$$\begin{array}{c}①\\|\\④—C—③\\|\\②\end{array}$$

the convention being that the horizontal bonds in the Fischer projection, project forward from the plane of the paper, whilst the vertical bonds project backwards into the paper. Thus in the Fischer projection of D-glyceraldehyde, the hydroxyl group is to the right of the carbon chain, when this is represented with the aldehyde group at the top, and vice versa for L-glyceraldehyde.

CHO, H, CH_2OH, OH (3D)

$$\begin{array}{c}CHO\\|\\H—C—OH\\|\\CH_2OH\end{array}$$

D-(+)-Glyceraldehyde

CHO, HO, CH_2OH, H (3D)

$$\begin{array}{c}CHO\\|\\HO—C—H\\|\\CH_2OH\end{array}$$

L-(−)-Glyceraldehyde

When using the Fischer projection, care must be taken about rotation of the projection formulae, e.g.

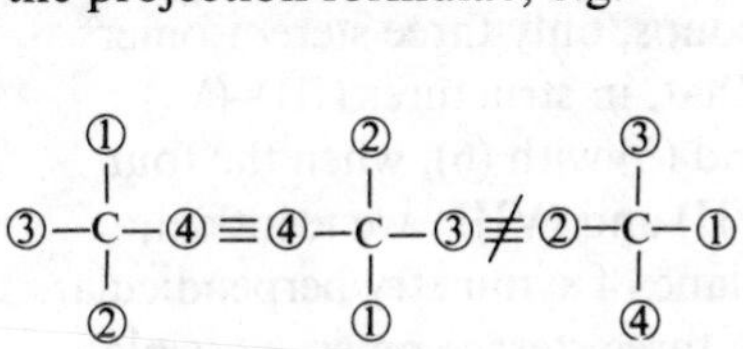

$$\begin{array}{c}①\\|\\③—C—④\\|\\②\end{array} \equiv \begin{array}{c}②\\|\\④—C—③\\|\\①\end{array} \not\equiv \begin{array}{c}③\\|\\②—C—①\\|\\④\end{array}$$

i.e. rotation of the projection formulae through 180° produces the projection of the same configuration, rotation through 90° produces the projection of the enantiomer, as exchange of the vertical and horizontal bonds in the projection is equivalent to a mirror-image inversion of the three-dimensional structure.

Compounds with more than one chiral centre

Since there are two possible configurations for an asymmetrically substituted carbon atom, a structure containing *n* such asymmetric centres should have 2^n stereoisomers. This is the maximum number, but in some cases the number of possible stereoisomers is less than this.

If we consider a compound with two chiral centres, in which the asymmetrically substituted atoms have quite different series of groups attached, then the four expected stereoisomers are obtained, corresponding to the structures (III)–(VI):

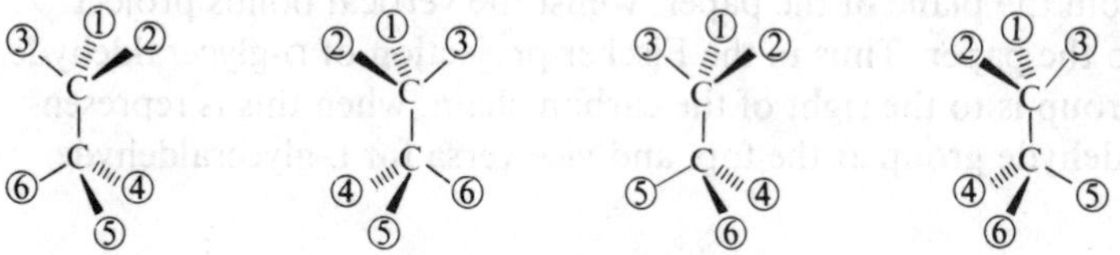

(III) (IV) (V) (VI)

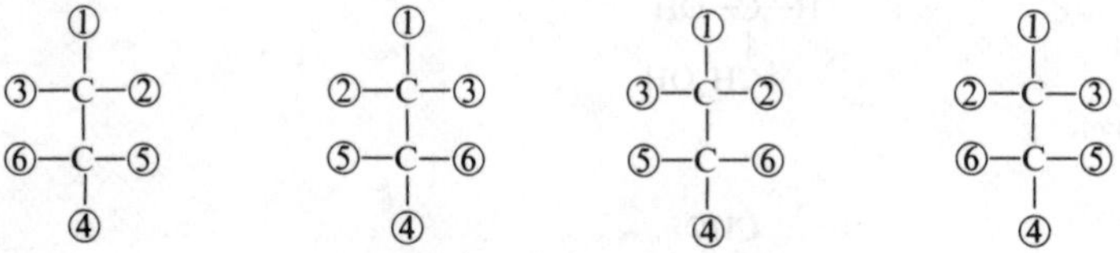

Fischer projections

In these four structures, (III) and (IV) are enantiomers, as are (V) and (VI), and in general, with *n* chiral centres there will be 2^{n-1} pairs of enantiomers.

If, however, we consider the case of a compound with two chiral centres, each bearing identical substituent groups, only three stereoisomers are possible. This can be seen by supposing that, in structures (III)–(VI) group (1) is identical with (4), (2) with (5), and (3) with (6), when the four stereoisomers become (VII)–(X). Of these (VII) and (VIII) are identical, superposable structures* since they have a plane of symmetry perpendicular to the C—C bond. There are, therefore, only three stereoisomers possible in this case, of which two, (IX) and (X), are enantiomeric and therefore optically active, and one, represented by (VII) or (VIII), is optically inactive since it is superposable upon its mirror image. Such an optically inactive stereoisomer is designated by the prefix '*meso*'.

* The use of molecular models makes this much easier to see.

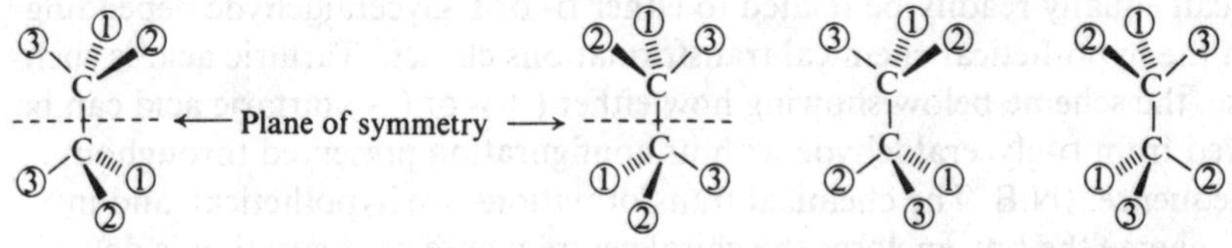

← Plane of symmetry →

(VII) (VIII) (IX) (X)

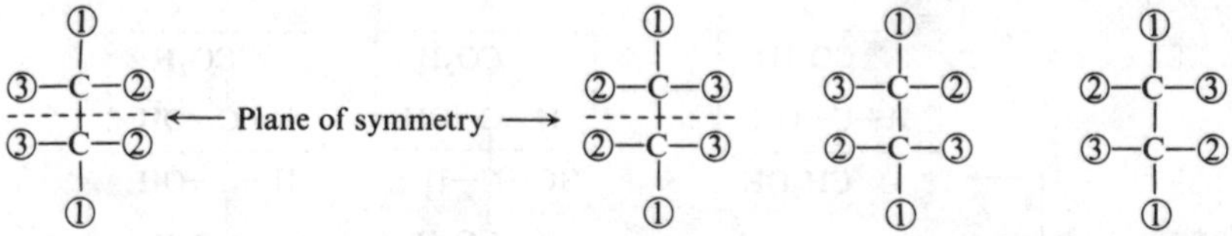

← Plane of symmetry →

Tartaric acid, $HO_2CCH(OH)CH(OH)CO_2H$, is a well-known example of this phenomenon. Three stereoisomers are known, whose structures and Fischer projections are shown below. Whilst (+)- and (−)-tartaric acid, being enantiomers, will have identical physical properties, this is not so for the

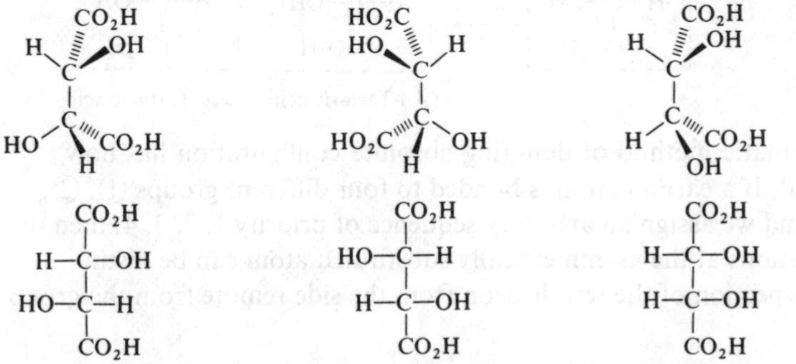

(+)-Tartaric acid (−)-Tartaric acid *meso*-Tartaric acid

meso-stereoisomer, which will have different physical properties (p. 216).

If, in a chemical reaction, the functional groups in a *meso*-compound react in such a way as to destroy the molecular symmetry, then two enantiomeric products will be obtained. Thus *meso*-tartaric acid forms two optically active, enantiomeric, monomethyl esters, $CH_3O_2CCH(OH)CH(OH)CO_2H$, both of which, on further esterification, form the same optically inactive dimethyl ester, $CH_3O_2CCH(OH)CH(OH)CO_2CH_3$.

Designation of chirality by *R* and *S*. (Cahn–Ingold–Prelog rules)

The use of D, L, and the Fischer projection to describe absolute configuration has a number of disadvantages arising from the necessity of relating compounds structurally to glyceraldehyde. In some cases the same chiral structure can equally readily be related to either D- or L-glyceraldehyde depending upon the hypothetical chemical transformations chosen. Tartaric acid is such a case, the scheme below showing how either (+)- or (−)-tartaric acid can be derived from D-glyceraldehyde with its configuration preserved throughout the sequence. (N.B. The chemical transformations are hypothetical, and in each scheme the box encloses the chiral centre whose configuration is derived from D-glyceraldehyde.)

A more systematic method of denoting absolute configuration has now been proposed. If a carbon atom is bonded to four different groups ①, ②, ③, and ④, and we assign an arbitrary sequence of priority 1, 2, 3, 4, then the two configurations at the asymmetrically substituted atom can be distinguished by inspection of the tetrahedron from the side remote from the group

of lowest priority (i.e. ④), when the other three groups will be seen in a clockwise (XI) or anticlockwise (XII) sequence of decreasing priority. These configurations are described by *R* (from *rectus*) and *S* (from *sinister*) respectively.

The arbitrary priority rules adopted are that the highest priority goes to the group joined to the chiral centre by the atom of highest atomic number (e.g. I, Br, Cl, SH, F, OH, NH_2, CH_3, H in this order of decreasing priority). Hydrogen, therefore, always has the lowest priority. If the chiral centre is attached to two isotopes of the same atom then the higher priority goes to the heavier isotope (e.g. T > D > H, and $^{14}C > ^{13}C > ^{12}C$). If two groups have identical atoms joined to the chiral centre then a distinction between these groups is sought by comparison of the atomic numbers of the atoms one stage further removed (e.g. $CH_2Cl > CH_2OH > CH_2CH_3 > CH_3$) When, at this point, more than one atom of highest atomic number has to be considered, then higher priority goes to the group with most such atoms in second place (e.g. $C(CH_3)_3 > CH(CH_3)_2 > CH_2CH_3$), and additionally, doubly bonded atoms count twice and triply bonded atoms count three times (e.g. $—C{\equiv}CH > —CH{=}CH_2 > —CH_2CH_3$; $CO_2H > CHO > CH_2OH$). If it is still not possible to assign a priority sequence, then the atomic numbers of the next set of atoms are taken into account. It is important to realise that the process of priority assignment stops immediately a distinction can be made, irrespective of the atomic numbers and frequency of atoms which might be taken into account at the next step. Thus $CH(CH_3)_2 > CH_2CBr_3$ since the former has two carbon atoms in second place as opposed to one in the latter group.

Although this system of nomenclature is almost invariably employed by chemists, it has not yet entirely displaced the older D/L designation for simple biological molecules. Throughout the remainder of this book absolute configurations will be described in terms of the Fischer projection using both the D/L and *R*/*S* nomenclatures where this is possible, since the D/L system is likely to persist in biochemical usage, notwithstanding its disadvantages.

Asymmetric substitution of atoms other than carbon

Although optical isomerism has, so far, been described solely in terms of carbon compounds, any tetrahedrally substituted atom is potentially a source of chirality.

Ammonia and amines have a distorted tetrahedral structure (p. 7) with a lone pair of electrons acting as the fourth substituent. However, ammonia and amines invert extremely rapidly, so that although chirality may be present in the structures, enantiomers can never be separated. Quaternary ammonium

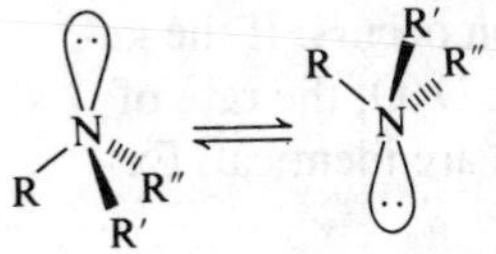

cations and tertiary amine *N*-oxides, on the contrary, have stable configurations, and enantiomers can be resolved.

$CH(CH_3)_2$ / ^{+}N / H_5C_6 CH_3 / C_2H_5 $\qquad$ $CH(CH_3)_2$ / N^{+} / H_3C C_6H_5 / C_2H_5

O^- / N^+ / H_3C C_2H_5 / C_6H_5 $\qquad$ O^- / N^+ / H_5C_2 CH_3 / C_6H_5

Sulphonium cations (p. 75) have tetrahedral configurations, which, unlike amines, do not readily invert, and sulphoxides (p. 76) also have stable tetrahedral configurations. Examples of both types of compound have been resolved into enantiomers.

S^+ / H_5C_6 CH_3 / C_2H_5 $\qquad$ S^+ / H_3C C_6H_5 / C_2H_5

S^+ / H_5C_6 O^- / CH_3 $\qquad$ S^+ / ^-O C_6H_5 / CH_3

Problems

1. Draw 'three-dimensional' diagrams showing the different configurations of the molecules below, and designate the chirality in each case according to the Cahn–Ingold–Prelog rules.

   $CH_3CHBrCN \qquad CH_3CH_2CH(OH)NH_2 \qquad CH_3CH(SCH_3)CO_2CH_3$

2. If one enantiomer of the chiral ketone $C_6H_5CH(CH_3)COC_6H_5$ is shaken with aqueous sodium carbonate solution, racemisation occurs. If the ketone is shaken with a solution of sodium carbonate in D_2O, the rate of racemisation and the rate of deuterium incorporation are identical. Explain this observation.

3. Draw 'three-dimensional' diagrams showing the configurations at the chiral centres of the molecules, whose Fischer projections are given below.

$$
\begin{array}{c}
CO_2H \\
| \\
C_2H_5\!-\!C\!-\!NH_2 \\
| \\
C_6H_5
\end{array}
\qquad
\begin{array}{c}
CHO \\
| \\
CH_3\!-\!C\!-\!H \\
| \\
H\!-\!C\!-\!OH \\
| \\
CH_2OH
\end{array}
\qquad
\begin{array}{c}
CO_2H \\
| \\
HO\!-\!C\!-\!H \\
| \\
C_6H_5\!-\!C\!-\!H \\
| \\
H\!-\!C\!-\!OCH_3 \\
| \\
H\!-\!C\!-\!NH_2 \\
| \\
C_3H_7
\end{array}
$$

For the last molecule, a projection formula based on a zig-zag carbon chain will probably be most convenient:

e.g. C C C C C C

4. Draw the Fischer projections for all the stereoisomeric forms of $CH_2Br(CHBr)_2CH_2Br$ and $CH_2Br(CHBr)_3CH_2Br$. Label all the optically inactive '*meso*' stereoisomers, all pairs of stereoisomers which are related as enantiomers, and one pair related as diastereoisomers.

13

Stereochemistry part II. Conformational and *cis-trans* (geometrical) isomerism

Conformational isomerism in open-chain molecules

There is much evidence to suggest that in simple, open-chain molecules rotation about single bonds can and does occur readily. Thus in ethane the two methyl groups rotate independently about the central σ bond, whose electron cloud is axially symmetrical about the line joining the carbon nuclei (p. 7), leading to an infinite number of **conformations*** for the ethane molecule. However, although this rotation occurs very easily, it is not wholly unrestricted since the different conformers of the ethane molecule vary slightly in energy on account of the differing separation between hydrogen nuclei attached to adjacent carbon atoms and the consequent variation in interaction between the electron clouds of the C—H bonds.

Newman projections of ethane

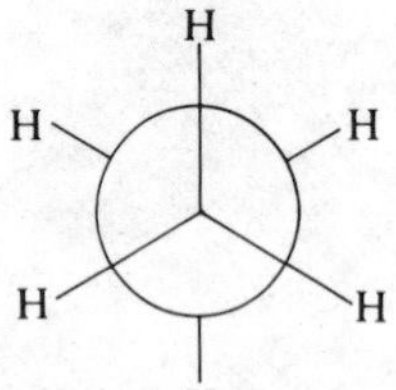

Staggered conformation
maximum H - - - H separation
lowest energy

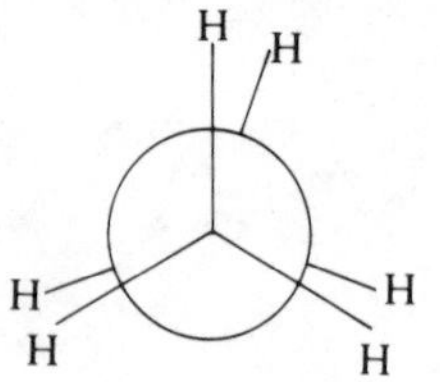

Eclipsed conformation
minimum H - - - H separation
highest energy

* Different conformations of a molecular structure can be interconverted by rotation about bonds, i.e. without the breaking and re-formation of chemical bonds (cf. 'configuration', p. 179).

These variations, and in particular the conformations of maximum and minimum energy, are best illustrated by the **Newman projections**, which are visualisations of the molecule viewed along the bond about which rotation is being considered.

In the foreground of these projections are three bonds linked to the near carbon atom, behind which the circle represents the electron cloud of the σ bond. The three bonds to the remote carbon atom project from the rear of this electron cloud. The relative orientations of C—H bonds about the C—C bond are easily seen from this projection and the highest energy 'eclipsed' and lowest energy 'staggered' conformations are readily discerned and easily portrayed.

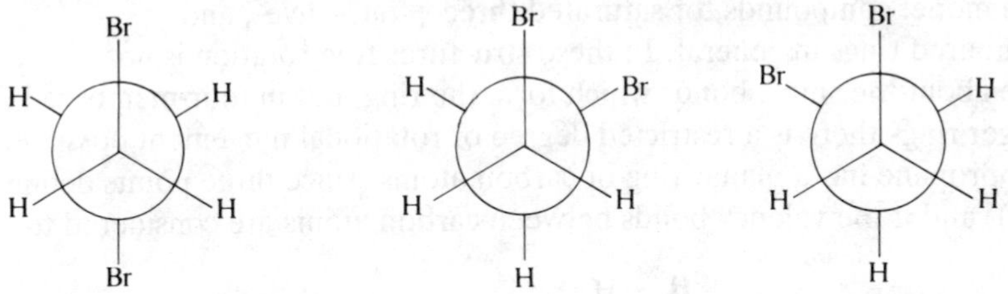

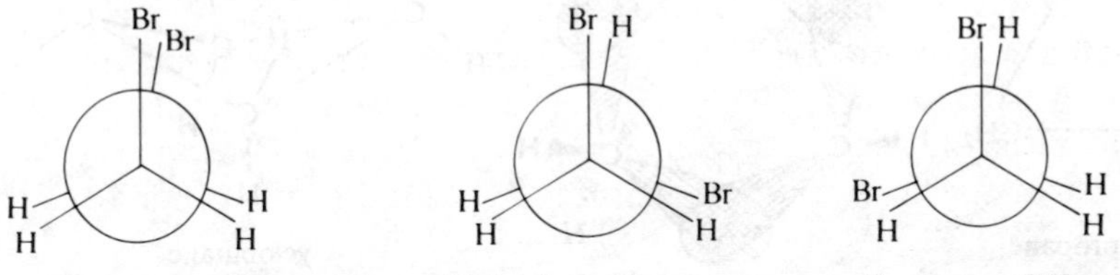

The type of interaction which gives rise to very small energy differences between the conformational extremes for ethane causes much greater energy differences where there are larger substituents on the carbon atoms. Thus 1,2-dibromoethane, CH_2BrCH_2Br, has three staggered conformations, of which the one with the maximum separation between the bulky bromine atoms has the lowest energy, and three eclipsed conformations of which that with the two C—Br bonds eclipsed giving the closest approach of the large bromine atoms has the highest energy. At any moment in a sample of this compound a substantial proportion of the molecules will be in conformations close to that of lowest energy and only a minute fraction will be in the highest energy eclipsed conformation. Exactly similar considerations apply to all singly bonded systems where free rotation can occur, and the conformational preferences described for dibromoethane apply equally well to conformers arising from rotation about the central bond of butane. Where the chemical properties of these or other compounds depend upon the precise shape of the molecule, conformational preferences of these types may be very important in determining the course or rate of reactions.

Conformational isomerism in cyclic systems

The conformational features of open-chain molecules described above have significant effects upon the shapes of small, cyclic molecules and their derivatives, which will be considered briefly here. The lower cycloalkanes serve as model compounds for saturated three-, four-, five-, and six-membered rings in general. In these structures free rotation is not possible about the single bonds which form the ring, but in four-membered and larger rings there is a restricted degree of rotational movement possible.

Cyclopropane has a planar ring of carbon atoms (since three points define a plane), and if the valency bonds between carbon atoms are considered to

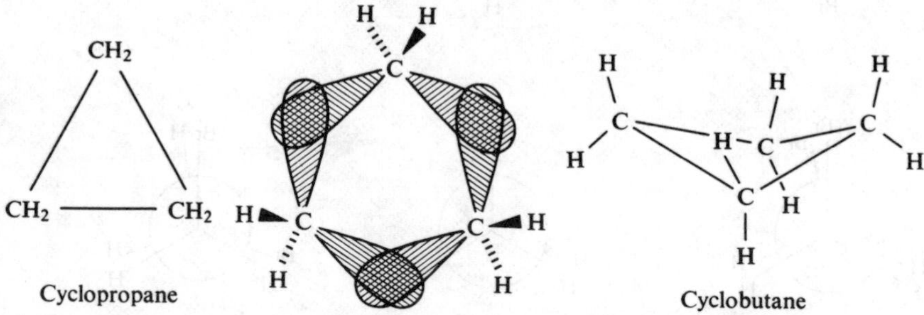

run linearly between the carbon nuclei then the C—C—C bond angle = 60°, which is very considerably less than the angle of 109·5° found in tetrahedrally substituted carbon. Alternatively the small ring can be considered to be formed by less than optimum overlap of orbitals, as shown above. It is found that cyclopropane is a very reactive hydrocarbon (catalytic hydrogenation converts it into propane) and is said to have a 'strained' ring. Around this planar ring adjacent C—H bonds are inevitably eclipsed. Cyclobutane, which if planar would have ∠C—C—C = 90°, is slightly buckled, with a dihedral angle of approximately 170°. Although this buckling reduces the angle ∠C—C—C to slightly less than 90°, increasing the ring strain with respect to the planar structure, it reduces the unfavourable eclipsing interactions between adjacent C—H bonds in the planar conformation. Small rotations about the ring bonds will interconvert the two possible buckled structures. Cyclobutane is less strained than cyclopropane and is therefore less reactive, being inert to hydrogenation like all the larger cycloalkanes.

A planar ring of five carbon atoms would have ∠C—C—C = 108°, which is very close to the tetrahedral angle, but to relieve the associated eclipsing C—H interactions cyclopentane adopts a slightly buckled structure rather like an open envelope with the flap raised. Here too, limited rotation about

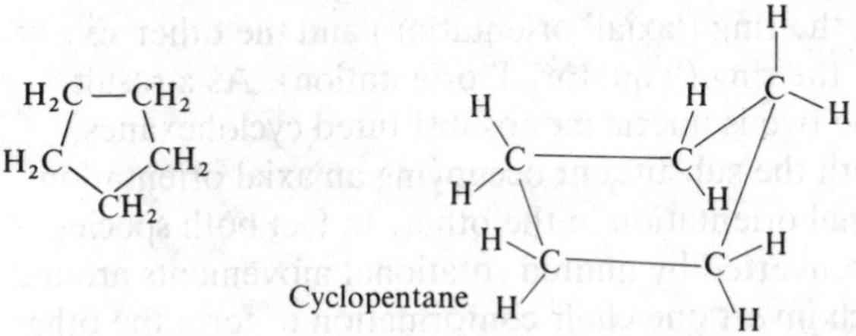

Cyclopentane

rings bonds interconverts a series of geometrically identical conformations in which different carbon atoms occupy the raised position.

If cyclohexane had a regular planar ring of carbon atoms the ∠C—C—C would be 120°, which is considerably greater than the preferred tetrahedral angle. Since puckering of a planar ring results in a decrease of the angles, cyclohexane adopts buckled structures in which ∠C—C—C = 109·5°. Two such structures are possible, known as the 'boat' and 'chair' conformations. The boat form is not important in practice since in this structure two of the hydrogen atoms (marked with asterisks) approach so closely that their electron clouds overlap, causing strong repulsion, and neighbouring C—H

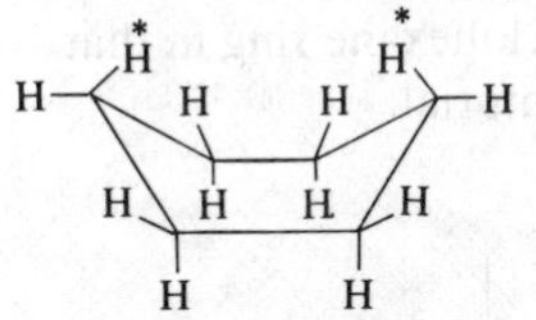

Cyclohexane (boat conformation)

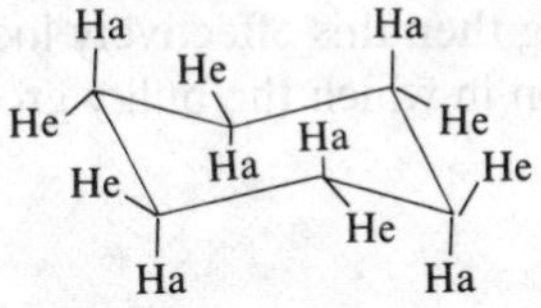

Cyclohexane (chair conformation)

C—Ha . . . axial bond
C—He . . . equatorial bond

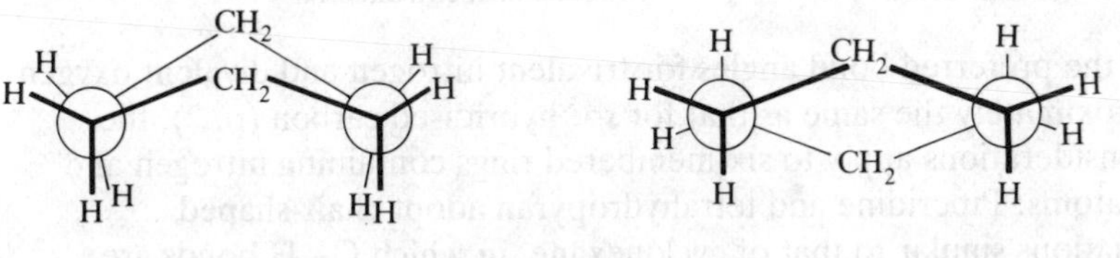

Newman projections

bonds along the sides of the boat are eclipsed. Together these give the boat conformation of cyclohexane a higher energy than the alternative chair conformation such that at room temperature only one molecule of cyclohexane in a thousand is in the boat conformation. In the chair conformation these 'non-bonded interactions' are minimised, with adjacent CH_2 groups being staggered. However, in the chair form of cyclohexane two types of C—H bond orientation can be discerned, with six C—H bonds lying parallel to the axis of the ring ('axial' orientation) and the other six directed out to the sides of the ring ('equatorial' orientation). As a result there could, in principle, be two isomeric monosubstituted cyclohexanes, e.g. chlorocyclohexane, with the substituent occupying an axial orientation in one case and an equatorial orientation in the other. In fact both species exist and are rapidly interconverted by limited rotational movements around the bonds of the ring, which invert one chair conformation to form the other via the intermediate formation of a boat conformation. Usually, the conformation in which a substituent group is in an axial position is energetically disfavoured with respect to the equatorial conformer. In the axial orientation close approach to the axial hydrogen atom (or other groups) on the same side of the ring causes repulsive overlap of the electron clouds. In the equatorial orientation the corresponding interactions with neighbouring but more distant equatorial hydrogen atoms is less significant. In general therefore, substituted cyclohexane molecules tend to adopt the chair conformation in which a substituent (or, for multiply substituted derivatives, the most substituents) lies in an equatorial orientation. If a very large substituent group such as *t*-butyl, $C(CH_3)_3$, is attached to a six-membered ring then this effectively locks the cyclohexane ring in that chair conformation in which the bulky group is equatorial.

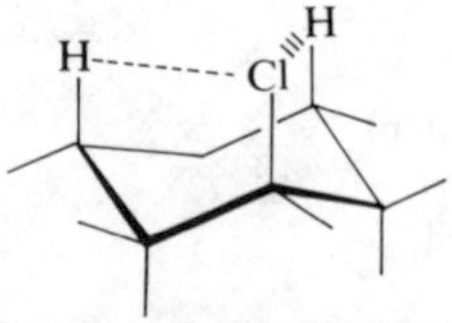

Strong non-bonded interaction

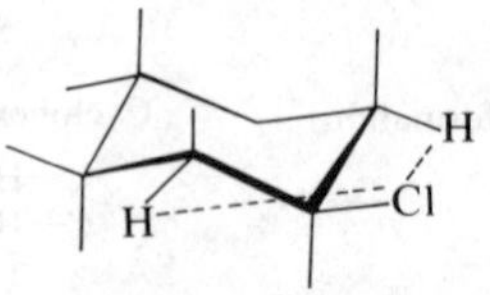

Weak non-bonded interaction

Since the preferred bond angles for trivalent nitrogen and divalent oxygen are approximately the same as that for sp^3 hybridised carbon (p. 7), the same considerations apply to six-membered rings containing nitrogen and oxygen atoms. Piperidine and tetrahydropyran adopt chair-shaped conformations similar to that of cyclohexane, in which C—H bonds are replaced by lone pairs at the heteroatom sites.

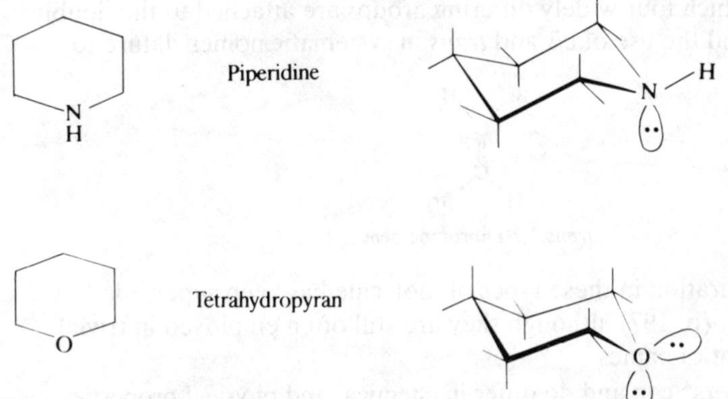

Larger saturated rings containing seven or more carbon atoms are known. All have buckled structures in which all ∠C—C—C = 109·5°, but the flexibility of the rings makes their stereochemistry much more difficult to describe in general terms.

Cis-trans isomerism in molecules with double bonds

The free rotation about σ bonds described above, which is a characteristic feature of open-chain, singly bonded molecules, does not readily occur about double bonds. In a molecule where two groups are joined by a double bond the σ bond has the customary axial symmetry, but the π bond requires that *p* orbitals on adjacent atoms should be aligned parallel in space to achieve maximum overlap (p. 10). Any independent rotation of these groups will shift the *p* orbitals from this optimum alignment, thereby reducing overlap, and for rotation to occur freely the π bond must be broken. Since the breaking of a chemical bond requires the provision of a great deal of energy, molecules of the type (A) (B)C═C(A) (B) can exist in two different configurations:

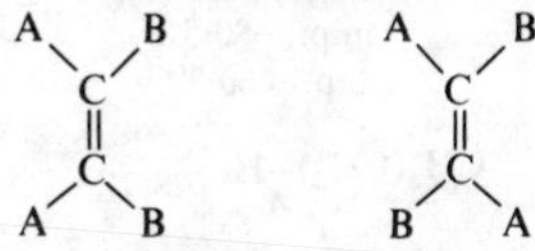

which can often be isolated or separated as distinct compounds with different chemical and physical properties. These stereoisomers are often distinguished by use of the prefixes *cis*- and *trans*-, *cis*- being employed to denote the compound in which similar groups are on the same side of the plane of the π bond, whilst *trans*- describes the isomer in which similar groups are on opposite sides. This convention is unsuitable for describing compounds in which four widely differing groups are attached to the doubly bonded atoms and the use of *cis* and *trans* in systematic nomenclature to

cis-But-2-ene

trans-1,2-Dibromoethene

designate configuration in these types of molecule has been superseded by the E/Z notation (p. 197) although they are still often employed in trivial (i.e. non-systematic) names

Cis-trans isomers* can and do differ in chemical and physical properties. Since in the solid state the crystal lattices of the isomers are built of quite different-shaped units, it is not surprising that differences occur in solubility and melting point, which reflect the differing stabilities of the crystal lattices. In simple cases it is frequently the isomer with similar groups *trans* which has the higher melting point and lower solubility since the more symmetrical *trans*-molecule packs better into a three-dimensional structure than the less symmetrical molecule with similar groups *cis*. In view of these differences, separation is usually possible by physical methods such as fractional distillation, fractional recrystallisation, and chromatography. Simple *cis-trans* isomers have no effect on polarised light, though in complex molecules *cis-trans* and optical isomerism may occur simultaneously.

The physical properties of some cis–trans *isomers*

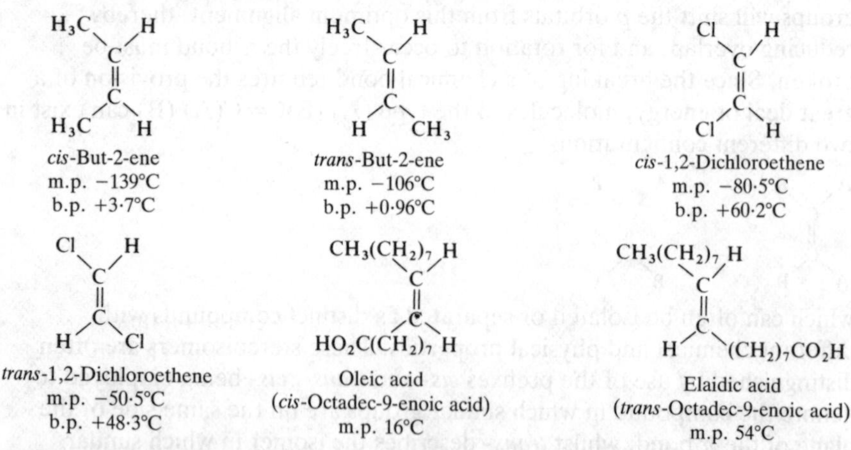

cis-But-2-ene m.p. −139°C b.p. +3·7°C

trans-But-2-ene m.p. −106°C b.p. +0·96°C

cis-1,2-Dichloroethene m.p. −80·5°C b.p. +60·2°C

trans-1,2-Dichloroethene m.p. −50·5°C b.p. +48·3°C

Oleic acid (*cis*-Octadec-9-enoic acid) m.p. 16°C

Elaidic acid (*trans*-Octadec-9-enoic acid) m.p. 54°C

* These are sometimes referred to as *cis-trans* diastereoisomers, since they are stereoisomers which are not enantiomers (p. 175).

The heats of formation of *cis-trans* isomers from their elements are not necessarily the same. The reason for this can be seen if the two isomers of a compound (A) (B)C=C(A) (B) are considered, in which A is a bulky group compared with B (e.g. $CH_3CH{=}CHCH_3$, ICH=CHI). In the *cis* isomer the two large A groups are close neighbours, and repulsion due to overlap of the

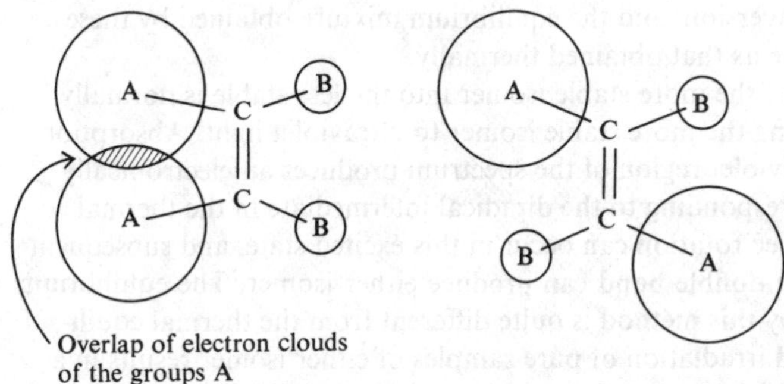

Overlap of electron clouds of the groups A

electron clouds of the two groups A will make this isomer less stable than the *trans* isomer in which large and small groups are adjacent. Similar considerations will apply if A or B is associated with a large dipole moment. It is usually found that, in simple cases, the *trans* isomer is more stable than the *cis*.

The interconversion of *cis-trans* isomers requires the breaking and re-formation of a chemical bond. In cases where the isomerism is caused by double bonds, it is the π bond which is broken in the interconversion, and the method employed depends upon whether the less stable isomer is to be converted into the more stable or vice versa. Usually heating the less stable isomer will convert it more or less readily into the more stable isomer by a process thought to involve a diradical intermediate produced by homolytic fission of the π bond. In this intermediate, free rotation of the previously doubly

bonded groups can occur, since these are now joined by a single (σ) bond only. Re-formation of the second (π) bond may result in formation of either isomer, but the equilibrium will favour the more stable one (i.e. that with the lower energy of formation). If the heats of formation differ by 11 kJ/mole the equilibrium mixture will contain only 1 per cent of the less stable isomer. A similar conversion occurs in the presence of traces of bromine or nitric oxide. It is thought that nitric oxide (a stable molecule with one unpaired electron) adds to the double bond to give a species in which free rotation can occur. Subsequent re-formation of the double bond by extrusion of nitric

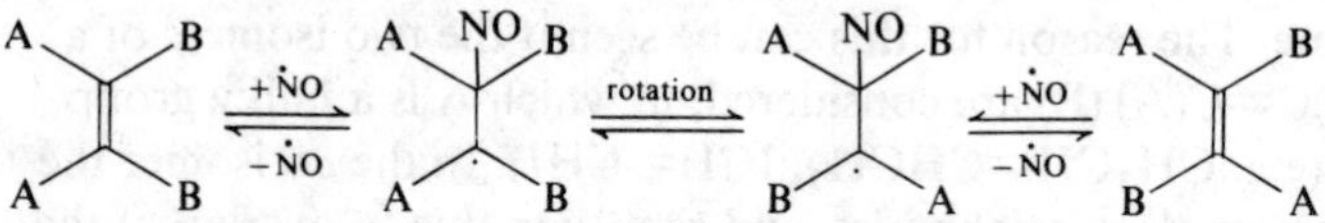

$$\text{AB C=C AB} \underset{-\dot{N}O}{\overset{+\dot{N}O}{\rightleftharpoons}} \text{(A)(NO)(B)C–}\dot{C}\text{(A)(B)} \underset{}{\overset{\text{rotation}}{\rightleftharpoons}} \text{(A)(NO)(B)C–}\dot{C}\text{(B)(A)} \underset{-\dot{N}O}{\overset{+\dot{N}O}{\rightleftharpoons}} \text{AB C=C BA}$$

oxide can result in the formation of either isomer. Bromine atoms can bring about a similar conversion, and the equilibrium mixture obtained by these methods is the same as that obtained thermally.

The conversion of the more stable isomer into the less stable is normally achieved by exposing the more stable isomer to ultraviolet light. Absorption of light in the ultraviolet region of the spectrum produces an electronically excited species corresponding to the diradical intermediate in the thermal interconversion. Free rotation can occur in this excited state, and subsequent re-formation of the double bond can produce either isomer. The equilibrium mixture obtained by this method is quite different from the thermal equilibrium mixture, and irradiation of pure samples of either isomer results in a mixture of both, which can be separated by the usual methods.

Although the chemistry of *cis-trans* isomers may differ appreciably, no general description of the differences is possible, since these depend entirely on the nature of the functional groups in the molecule. One common distinction between *cis-* and *trans*-isomers is the formation of cyclic derivatives when suitable groups are present on the same side of the plane of the π bond, e.g.:

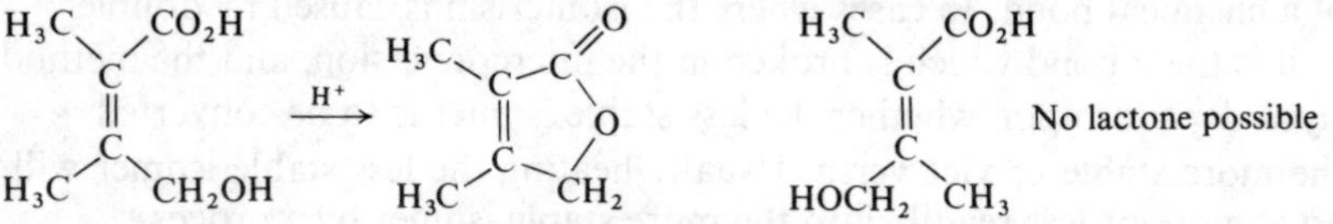

$$(H_3C)(CO_2H)C{=}C(H_3C)(CH_2OH) \xrightarrow{H^+} \text{lactone: } H_3C\text{–C=C(H}_3\text{C)–CH}_2\text{–O–C(=O)–}$$

$$(H_3C)(CO_2H)C{=}C(HOCH_2)(CH_3) \quad \text{No lactone possible}$$

When the functional groups are on opposite sides of the double bond they are too far apart to interact. Reactions such as this have frequently been used to assign configurations to geometrical isomers.

Cis-trans isomerism is not confined to compounds having doubly bonded carbon atoms, being found also in compounds with C=N and N=N links. Examples of isomerism in these compounds are given below, but since the chemistry of these functional groups has not been described in detail, differences between the isomers will not be considered further.

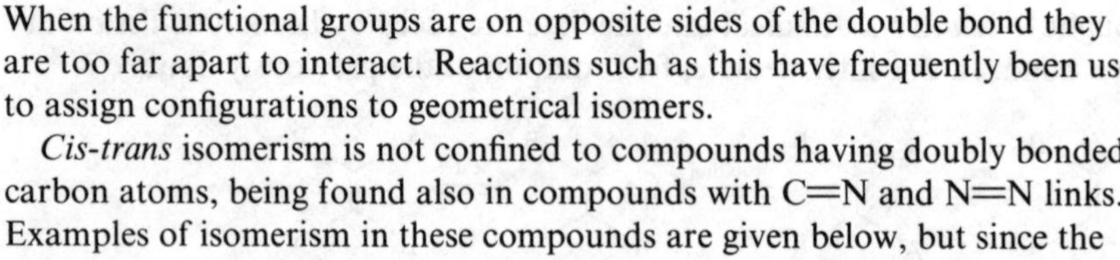

$H_5C_6(H)C{=}\ddot{N}\text{–}OH$	$H_5C_6(H)C{=}\ddot{N}\text{–}$ (HO on other side)	$H_5C_6\text{–}\ddot{N}{=}\ddot{N}\text{–}C_6H_5$	$H_5C_6\text{–}\ddot{N}{=}\ddot{N}\text{–}C_6H_5$ (same side)
m.p. 34°C	m.p. 127°C	m.p. 68°C	m.p. 71°C
Benzaldoxime		*Azobenzene*	

Designation of configuration by *E* and *Z*

The confusion about the precise meaning of the prefixes *cis* and *trans* and the difficulty associated with their usage when the four substituents on a doubly bonded system have no obvious relationship (e.g. $CH_3CH{=}CClBr$) have led to this method for designating configuration being replaced by one based on the Cahn–Ingold–Prelog rules for designating chirality (p.183). Using these rules, a priority sequence can be assigned to the substituents at each end of the doubly bonded system, leading to two possibilities:

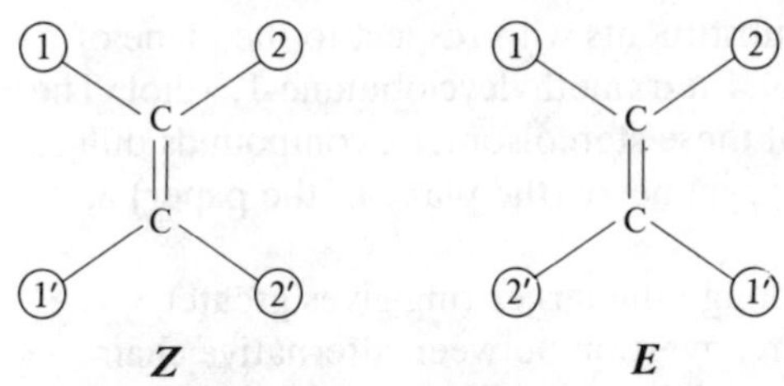

The configuration in which the high priority substituents are on the same side is described as *Z* (German *zusammen*—together) and the alternative configuration is described as *E* (German *entgegen*—opposed). This system can be used for *cis-trans* isomers involving other than C=C bonds, and in these cases, where one of the doubly bonded atoms bears a lone pair of electrons, the lone pair has lower priority even than H.

E.g.

E-1-Chloro-2-methoxypropene

Z-Benzaldoxime

Cis-trans isomerism in cyclic compounds

The phenomenon of restricted rotation is not confined to molecules containing double bonds. In saturated cyclic molecules the difficulty of distortion of bond lengths and bond angles restricts the extent to which rotation may occur about the σ bonds comprising the ring, and *cis-trans* isomers can be formed

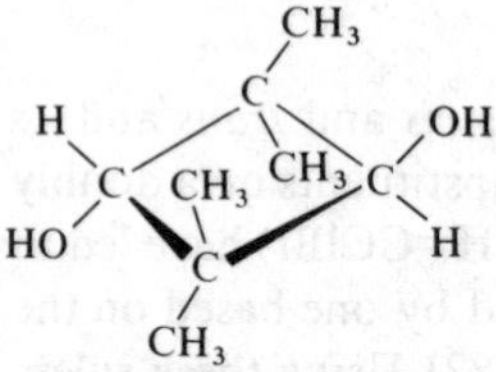

trans-2,2,4,4-Tetramethyl-
cyclobutane-1,3-diol
m.p. 148°C

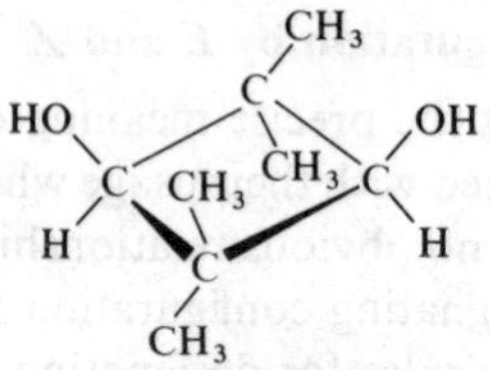

cis-2,2,4,4-Tetramethyl-
cyclobutane-1,3-diol
m.p. 163°C

which differ in the orientation of substituents with respect to the plane of the ring. This is illustrated for 2,2,4,4-tetramethylcyclobutane-1,3-diol. The physical and chemical properties of these stereoisomeric compounds differ, but both molecules have a plane of symmetry (the plane of the paper) and are therefore optically inactive.

Similarly, with cyclohexane, although the larger ring gives greater flexibility to the molecule and interconvertion between alternative chair conformations of each isomer is possible, a *cis*-disubstituted cyclohexane cannot be converted into its *trans* isomer by rotational processes alone, e.g. for 1,4-dibromocyclohexanes:

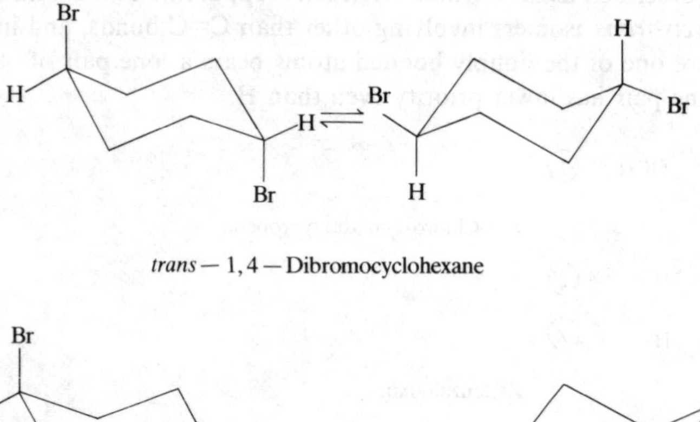

trans — 1,4 — Dibromocyclohexane

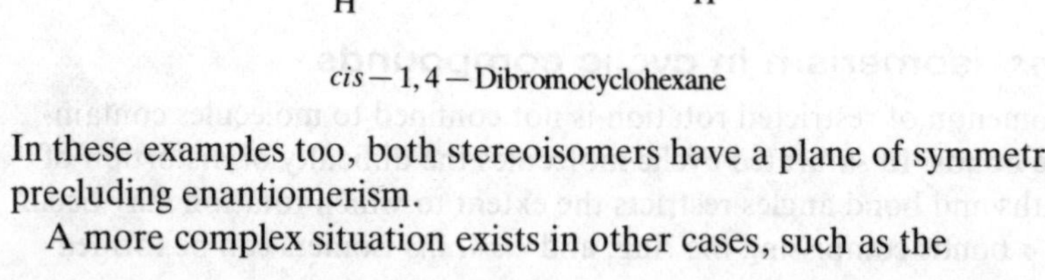

cis — 1,4 — Dibromocyclohexane

In these examples too, both stereoisomers have a plane of symmetry, precluding enantiomerism.

A more complex situation exists in other cases, such as the

(Cyclopropane-1(R), 2(S)-dicarboxylic acid)

cis-Cyclopropane-1,2-dicarboxylic acid, m.p. 139°; pK_1 3·40

(Cyclopropane-1(R), 2(R)-dicarboxylic acid)

(Cyclopropane-1(S), 2(S)-dicarboxylic acid)

Enantiomeric *trans*-Cyclopropane-1,2-dicarboxylic acids, m.p. 175°; pK_1 3·68

cyclopropane-1,2-dicarboxylic acids. The *cis* isomer has a plane of symmetry (perpendicular to the ring, passing through C(3) and bisecting the bond between C(1) and C(2)) and is therefore optically inactive. The *trans* isomer has no plane of symmetry and therefore exists in enantiomeric forms. Whilst the enantiomeric *trans*-dicarboxylic acids will have identical chemical and physical properties (except for the rotation of polarised light) the *cis*-isomer has properties quite different from either of the *trans*-isomers. Although the structures of the compounds here are somewhat more complicated, the cyclopropane-1,2-dicarboxylic acids provide another example of the general case of a molecule with two identical chiral centres previously illustrated by the tartaric acids (p. 183). Similar considerations apply to *cis* and *trans*-1,2- or 1,3-disubstituted cyclohexanes.

There are no general methods of interconverting *cis-trans* isomers arising in cyclic systems. The action of heat, light, or nitric oxide, which affect the π bond in doubly bonded compounds, is usually ineffective with saturated molecules. Interconversion by chemical means is sometimes possible depending on the functional groups present in the molecule. Thus dimethyl *cis*-cyclobutane-1,3-dicarboxylate may be converted into the *trans*-isomer in the presence of strong base via formation of the enolate anion, but *cis*-1,3-dimethylcyclobutane is not affected by these conditions as the corresponding anion is not formed.

The chemical differences between *cis-trans* isomers depend on the functional groups present in the molecule. As with *cis-trans* isomers in unsaturated molecules, the formation of cyclic derivatives is commonly employed to distinguish between alternative configurations.

Problems

1. Both *cis*- and *trans*-but-2-ene are known, but cyclohexene is known only in the *cis* form. Explain.

2. How might the two compounds below be distinguished by chemical means?

3. If you had two isomeric compounds, C_6H_{12}, which you believed were *cis-trans* isomers of $CH_3CH{=}CHCH(CH_3)_2$, how would you attempt to prove this by chemical means?

4. Draw both chair conformations for *cis*-1,3-dimethylcyclohexane and *trans*-1,2-dimethylcyclohexane. Indicate in each case which conformation has the lower energy.

14

Mechanistic and stereochemical aspects of some reactions

Hitherto, many reactions have been described without reference to stereochemical details. In this chapter the mechanisms and three-dimensional aspects of selected reactions will be described more fully. All of the conclusions are supported by extensive study of numerous examples, but space does not permit more than the most concise description here. More advanced texts should be consulted if greater detail is required.

Kinetic study of reactions. The majority of the information about the mechanism of chemical reactions comes from a study of the kinetics, i.e. the effect of factors such as reagent concentration, temperature, variation in solvent, etc., on the rate of the reaction. As a result of this study a rate equation can be written which shows how the rate is related to reagent concentration. In organic reactions, where covalent bonds are formed and broken, a reaction may proceed by way of several stages, e.g. the reaction of nitrous acid with primary amines (p. 86) or acid-catalysed esterification (p. 137). In such cases, measurements of the rate of disappearance of reagents or formation of products will measure the slowest of the various stages, called the 'rate-determining' step, and the effect of reagent concentration on the overall rate of reaction will be the effect on the rate-determining step. Paradoxically, this can lead to a situation in which, although a particular reagent is essential for the occurrence of a reaction, the rate of the reaction is independent of the concentration of this reagent, since it is not involved in the reaction process of the rate-determining step (see S_N1 reactions, p. 202). A drastic change in reagent concentration, temperature, or solvent type may result in a change in the rate-determining step of a multi-stage process.

Nucleophilic substitution at saturated carbon atoms

The essential details of nucleophilic displacement at saturated carbon atoms have been described in the chapter on alkyl halides (p. 51). Kinetic study of reactions of the type:

$$R{-}X + Y^- \longrightarrow R{-}Y + X^-$$

e.g. $CH_3{-}I + CN^- \longrightarrow CH_3{-}CN + I^-$

shows that the rate equation takes one of two forms. Where the group R can form a stable carbonium ion i.e. tertiary alkyl groups like $(CH_3)_3C$, or mixed alkyl aryl groups $C_6H_5CH_2$, $(C_6H_5)_2CH$, $C_6H_5(CH_3)_2$, etc., the rate equation is of the type:

Rate $= k\,[R{-}X]$ 'Unimolecular' nucleophilic substitution . . . S_N1.

where k is known as the rate constant. If R is a primary alkyl group (e.g. C_2H_5) then the rate equation is always of the form:

Rate $= k\,[R{-}X]\,[Y^-]$ 'Bimolecular' nucleophilic substitution . . . S_N2.

If R is a secondary alkyl group the rate equation may be of either type depending on the structure of the alkyl group and the polarity of the solvent. In very polar solvents (water, aqueous ethanol, aqueous propanone, liquid sulphur dioxide) the rate equation will be like that for tertiary halides. In solvents of low polarity (dry ethanol, propanone, ether, benzene, petrol) which do not stabilise carbonium ions by solvation, secondary alkyl halides react by a bimolecular process with S_N2 kinetics.

The rate equation for reactions of tertiary halides is taken to imply that the rate-determining step, which involves only the alkyl halide molecule, is the slow ionisation of the R—X bond followed by very much more rapid reaction of the resultant carbonium ion with the nucleophile:

(i) $R{-}X \longrightarrow R^+ + X^-$ (slow)
(ii) $R^+ + Y^- \longrightarrow R{-}Y$ (fast)

The alternative data for primary halides (S_N2 reaction) suggest that both nucleophile and alkyl halide participate in the rate-determining process, and this is interpreted as meaning that departure of the leaving group X^- is assisted by the approach of the nucleophile, reaction occurring via an 'activated complex' in which the R—X bond is partly broken and the R—Y bond is already partly formed

$$Y^- + R{-}X \longrightarrow (Y\text{---}R\text{---}X)^- \longrightarrow Y{-}R + X^-$$

Activated complex

The stereochemistry of nucleophilic substitution. It is possible to examine the stereochemical result of nucleophilic displacement reactions by using optically active alkyl halides having the halogen atom attached at the chiral centre, e.g. $C_2H_5CHClCH_3$. The observations obtained in this way are fully consistent with the mechanistic interpretation of the kinetic results described above. If an optically active chiral halide undergoes an S_N1 reaction then the product of reaction is racemic. If the nucleophilic substitution is an S_N2 reaction (ascertained by kinetic study) then the products are optically active, and if an optically pure chiral halide is employed, the products are also optically pure (i.e. no racemisation has occurred).

Racemisation during an S_N1 reaction of an optically active halide is explicable in terms of the symmetry of the intermediate carbonium ion. Carbonium ions are planar about the trivalent carbon atom, the substituents being attached by sp^2 orbitals (p. 9), and the central carbon atom having a vacant *p* orbital. The incoming nucleophile can attack the planar carbonium ion equally well from either side, resulting in equal probability of formation of the enantiomeric products.

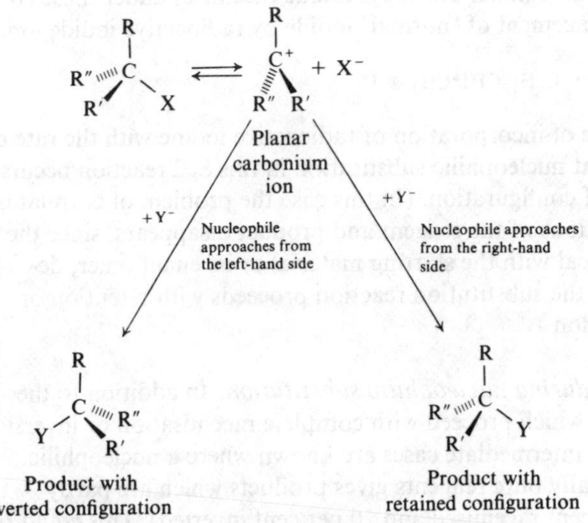

The retention of optical purity in S_N2 reactions means that displacement occurs either wholly with retention, or wholly with inversion of configuration. The three-dimensional relationship of the two possible processes may be represented by:

$+Y^-$ Approach from the same side as X → Product with retained configuration $+ X^-$

$+Y^-$ Approach from the opposite side to X → Product with inverted configuration $+ X^-$

Since there is no simple relationship between the direction of rotation of light and absolute configuration, it was difficult to distinguish which of these alternatives was correct. It is now known that all nucleophilic substitutions, following S_N2 kinetics, proceed with inversion of configuration at the centre from which the displacement occurs. This conclusion, derived initially by means of roundabout reaction sequences, was ingeniously confirmed by studying the reaction between isotopically labelled (radioactive) iodide ion and optically pure 2-iodo-octane. The alkyl iodide racemises under these conditions by an S_N2 displacement of 'normal' iodide by radioactive iodide ion.

$$C_6H_{13}CHICH_3 + I^{*-} \longrightarrow C_6H_{13}CHI^*CH_3 + I^-$$

Comparison of the rate of incorporation of radioactive iodine with the rate of racemisation shows that nucleophilic substitution in this S_N2 reaction occurs solely with inversion of configuration. (In this case the problem of correlating the absolute configurations of the reagent and product disappears, since the product is either identical with the starting material or its enantiomer, depending upon whether the substitution reaction proceeds with retention or inversion of configuration.)

Partial racemisation during nucleophilic substitution. In addition to the cases described above, which proceed with complete racemisation or inversion of configuration, some intermediate cases are known where a nucleophilic substitution with optically pure reagents gives products which are partly racemised (e.g. 80 per cent racemised and 20 per cent inverted). This could be caused by S_N1 and S_N2 processes occurring simultaneously, but such cases are extremely rare. Usually the reaction is S_N1 (shown by kinetic study) proceeding through a very reactive carbonium ion of short lifetime. Such a carbonium ion will react before the leaving group has had time to diffuse into the surrounding medium, and the bulk of the departing leaving group will tend to shield one face of the carbonium ion. The incoming nucleophile, therefore, has a greater chance of reacting on the opposite face, giving a higher proportion of product with inverted configuration.

Bulky X^- shielding the right-hand face of the carbonium ion

$+Y^-$ Approach from L.H.S. (more probable)

$+Y^-$ Approach from R.H.S. (less probable)

Major product (inverted configuration)

Minor product (retained configuration)

The mechanism and stereochemistry of alkene-forming eliminations

The general pattern of elimination reactions may be represented by:

$$-\underset{X}{\overset{|}{C}}-\underset{Y}{\overset{|}{C}}- \xrightarrow{-XY} \;\rangle C{=}C\langle$$

We will confine our attention to eliminations in which Y is hydrogen (although the same stereochemical principles apply in many other cases). In principle the elimination of HX could occur by three pathways:

(i) $-\underset{H}{\overset{|}{C}}-\underset{X}{\overset{|}{C}}- \xrightarrow{-\overset{+}{B}H} -\overset{|}{\underset{\cdot\cdot}{C}}{}^{-}-\underset{X}{\overset{|}{C}}- \xrightarrow{-X^-} \;\rangle C{=}C\langle$ (B = base)

(ii) $-\underset{H}{\overset{|}{C}}-\underset{X}{\overset{|}{C}}- \xrightarrow{-X^-} -\underset{H}{\overset{|}{C}}-\overset{|}{C}{}^{+} \xrightarrow{-\overset{+}{B}H} \;\rangle C{=}C\langle$ (E1)

(iii) $-\underset{H}{\overset{|}{C}}-\underset{X}{\overset{|}{C}}- \xrightarrow[\overset{+}{B}H \text{ and } X^-]{\text{Concerted loss of}} \;\rangle C{=}C\langle$ (E2)

In the first two processes, elimination occurs in a stepwise fashion, either H^+ or X^- being lost initially. The first type of elimination, via carbanion formation, is comparatively rare and will not be considered further. (An example of this type of reaction is the base-catalysed dehydration of aldols (p. 110).) In general this type of reaction requires that the initial carbanion be stabilised in some way, e.g. by conjugation with an adjacent carbonyl group. The second type of elimination starts by initial formation of a carbonium ion followed by loss of a proton. This is likely to occur only if the intermediate carbonium ion is tertiary and it should be noted that this ionisation is identical with the first stage of an S_N1 reaction. It is observed, in practice, that alkene formation is an important side reaction during nucleophilic substitution of tertiary alkyl halides, though insignificant in S_N2 reactions. The third pathway for elimination involves a concerted attack of base on the proton and loss of X^-. This is the most important elimination pathway for primary and secondary alkyl halides. (Several examples of eliminations will be found on p. 29.)

If the ionic intermediates in the schemes (i) and (ii) have lifetimes long enough to permit rotation about single bonds before the second stage of elimination ensues, then no special steric requirement will be observed for the relative orientation of the C—H and C—X bonds in the starting material. However, it is found that in the concerted (E2) elimination (iii), the reaction proceeds most readily if the C—H and C—X bonds are 'antiperiplanar', i.e. lie in the same plane and on opposite sides of the common C—C bond. Elimination is not impossible if the relevant bonds are otherwise oriented,

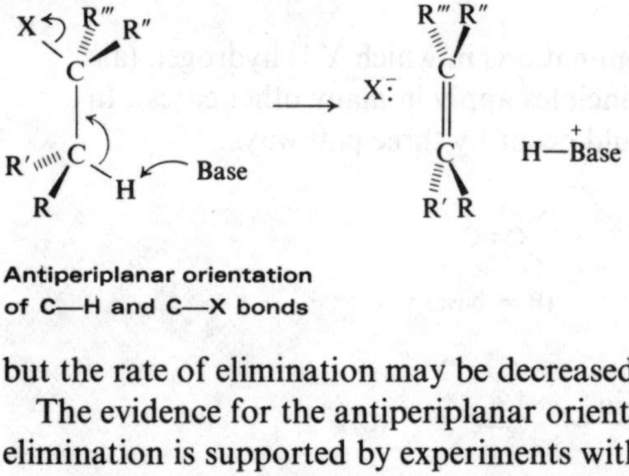

Antiperiplanar orientation of C—H and C—X bonds

but the rate of elimination may be decreased by a factor of several thousand.

The evidence for the antiperiplanar orientation required for a concerted elimination is supported by experiments with rigid systems such as steroids (p. 326) in which the bonds are held in fixed relative positions. Similar results are observed in simpler systems. Thus menthyl chloride reacts with sodium ethoxide to give one substituted cyclohexene exclusively, since in neither of the chair conformations is it possible to get the C—H and C—Cl bonds in the antiperiplanar arrangement required for elimination in the other direction.

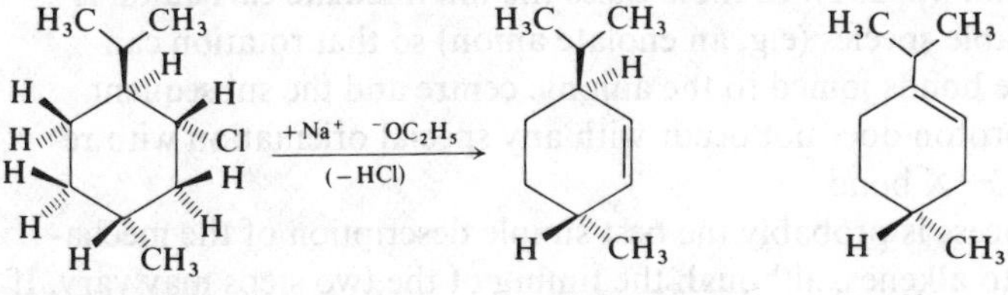

Menthyl chloride $\xrightarrow[(-\mathrm{HCl})]{+\mathrm{Na}^+\ ^-\mathrm{OC_2H_5}}$ 100% + 0%

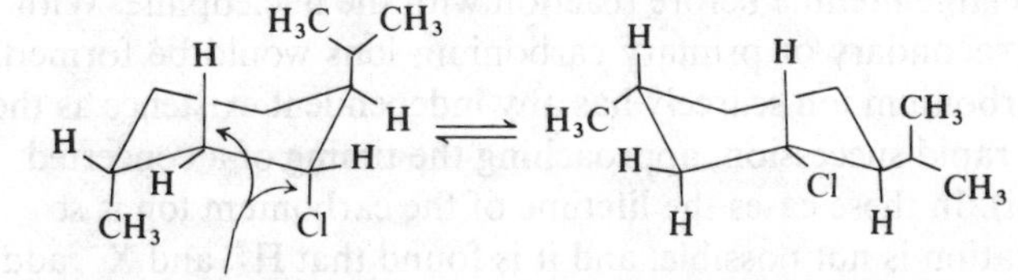

Axial C—H and C—Cl bonds in antiperiplanar orientation

The equatorial C—Cl bond has no antiperiplanar C—H bond in this conformation

Note that in cyclic systems such as those shown above, the elimination requires not only that the leaving groups be formally *trans* about a common C—C bond in a planar diagram, but to achieve the necessary coplanarity both bonds must be axial in the conformation from which elimination occurs.

The mechanism and stereochemistry of addition to alkenes

The reverse of an elimination reaction is an addition reaction, and just as three types of elimination process can be distinguished, so the reverse processes may, in principle, occur during addition reactions. Attention will be directed here to addition of hydrogen halides, HX, although the same principles apply in many other cases (p. 31).

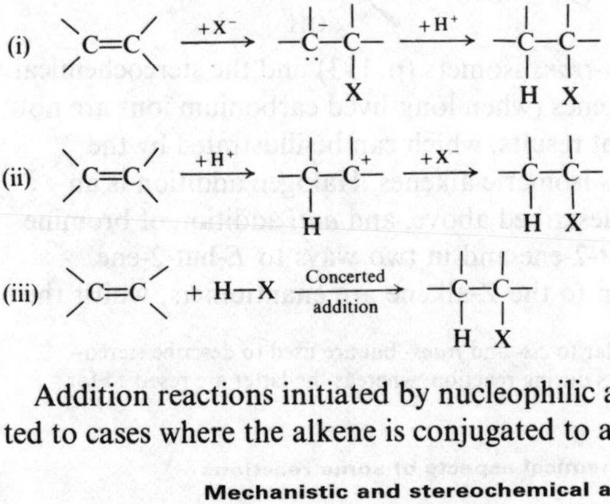

(i) $\mathrm{>C{=}C<} \xrightarrow{+\mathrm{X}^-} \mathrm{-\overset{-}{C}-C(X)-} \xrightarrow{+\mathrm{H}^+} \mathrm{-C(H)-C(X)-}$

(ii) $\mathrm{>C{=}C<} \xrightarrow{+\mathrm{H}^+} \mathrm{-C(H)-\overset{+}{C}-} \xrightarrow{+\mathrm{X}^-} \mathrm{-C(H)-C(X)-}$

(iii) $\mathrm{>C{=}C<} + \mathrm{H{-}X} \xrightarrow[\text{addition}]{\text{Concerted}} \mathrm{-C(H)-C(X)-}$

Addition reactions initiated by nucleophilic attack on alkenes (i) are restricted to cases where the alkene is conjugated to an electronegative group, i.e.

the Michael addition (p. 231). In these cases the intermediate carbanion is usually a fairly stable species (e.g. an enolate anion) so that rotation can occur about single bonds joined to the anionic centre and the subsequent attachment of a proton does not occur with any special orientation with respect to the new C—X bond.

The second process is probably the best simple description of the mechanism of addition to alkenes, although the timing of the two steps may vary. If a very stable carbonium ion (i.e. tertiary) is formed in the first step then this may have an appreciable lifetime before reaction with the nucleophile. With reactions in which secondary or primary carbonium ions would be formed, the intermediate carbonium ion scarcely has any independent existence as the two steps follow in rapid succession, approaching the timing of a concerted addition process (iii). In these cases the lifetime of the carbonium ion is so short that bond rotation is not possible, and it is found that H^+ and X^- add to the double bond in a *anti* orientation.* *Syn*-additions to alkenes are known

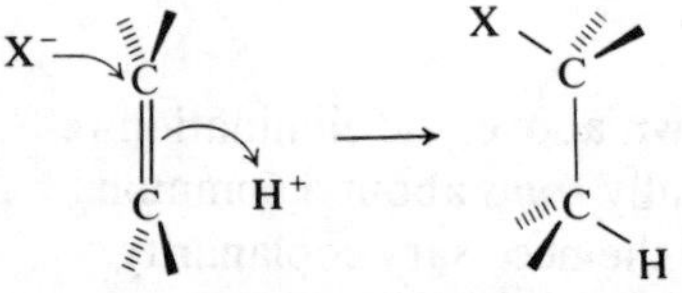

in two very similar cases. Alkenes are converted into *vic*-diols by reaction with alkaline solutions of permanganates or osmium tetroxide (p. 33). Both MnO_4^- and OsO_4 form cyclic intermediates whose decomposition results in the formation of diols, the overall reaction being a *syn*-addition of two hydroxyl groups.

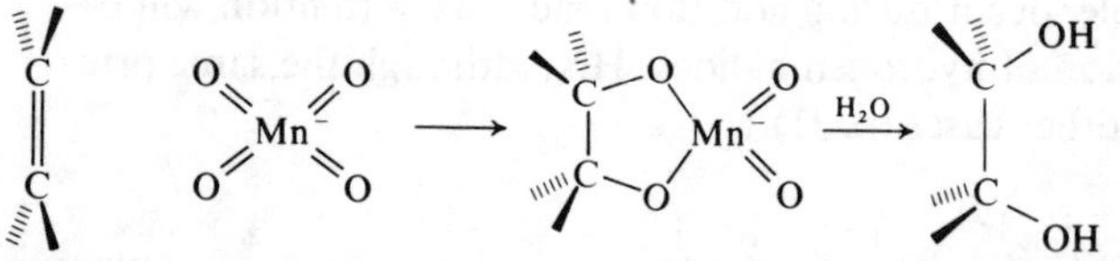

Many alkenes can exist as *cis-trans* isomers (p. 193) and the stereochemical requirements of addition to alkenes (when long-lived carbonium ions are not intermediates) lead to important results, which can be illustrated by the addition of bromine to *cis-trans* isomeric alkenes. Halogen addition is an *anti* addition, similar to those described above, and *anti* addition of bromine can occur in two ways to *Z*-but-2-ene and in two ways to *E*-but-2-ene. The products from the addition to the *Z*-alkene are enantiomers, whilst the

* *Syn-* and *anti-* have meanings similar to *cis-* and *trans-* but are used to describe stereochemical relationships between reagents during reaction, whereas the latter are reserved for description of molecular structure.

alternative additions to the *E*-alkene give identical *meso* products. Analogous results arise from stereospecific *syn* addition (e.g. hydroxylation by osmium tetroxide).

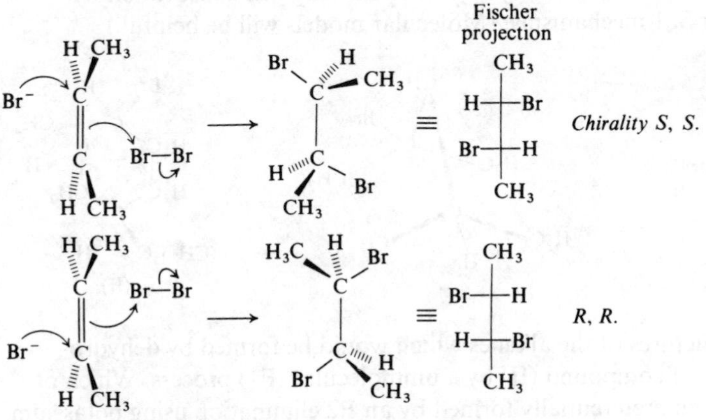

Fischer projection

Chirality *S*, *S*.

R, *R*.

Anti-additions to *Z*-but-2-ene

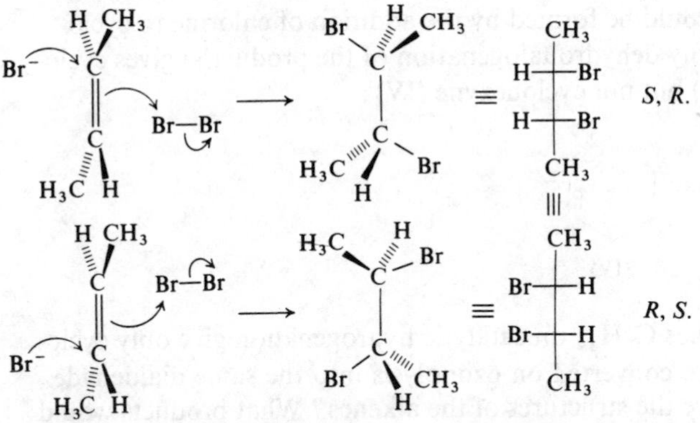

S, *R*.

R, *S*.

Anti-additions to *E*-but-2-ene

Note that if an unsymmetrical alkene (e.g. pent-2-ene) were to be used, then *anti*-addition of halogen to the *Z*-stereoisomer would give one pair of enantiomeric products and *anti*-addition to the *E*-isomer would give another pair of enantiomers.

Problems

1. What are the products of reaction between 2(*R*)-bromobutane and sodium thiophenate ($NaSC_6H_5$) if (a) the reaction follows S_N2 kinetics; (b) the

reaction follows S_N1 kinetics? Show the configurations of the products by means of suitable 'three-dimensional' diagrams.

2. Explain why the compound (I) is inert to nucleophilic substitution by either S_N2 or S_N1 mechanisms. (Molecular models will be helpful.)

(I) (II)

3. Give the structures of the alkenes which would be formed by dehydrobromination of compound (II) by a unimolecular (E1) process. Which of these would be preferentially formed by an E2 elimination using potassium *t*-butoxide? (Consider the chair conformations of (II).)

4. What product(s) would be formed by the addition of chlorine to cyclohexene? Explain why dehydrohalogenation of the product(s) gives cyclohexa-1,3-diene (III) but not cyclohexyne (IV).

(III) (IV)

5. Two isomeric alkenes C_8H_{14} on catalytic hydrogenation give only cyclooctane and both are converted on ozonolysis into the same dialdehyde $C_8H_{14}O_2$. What are the structures of the alkenes? What products would be obtained by hydroxylation with dilute potassium permanganate solution?

6. Dehydrobromination of $(CH_3)_3CCH_2CHBrCH_3$ with base by the E2 mechanism gives $(CH_3)_3CH{=}CHCH_3$. Using Newman projections (p. 189) to portray conformations about the CH_2—CHBr bond, explain why the product is predominantly the *E*-stereoisomer.

15

Hydroxy-acids and keto-acids

Numerous, biochemically important carboxylic acids are known, bearing hydroxyl or carbonyl substituents. We shall consider the chemistry of representative examples of these classes of compound.

Hydroxy-acids

Several types of hydroxy-acid are of biological significance, distinguished by the position of hydroxyl substitution relative to the carboxylic acid group. These are usually differentiated by use of the trivial nomenclature employing Greek letters α, β, γ, δ, etc., to indicate substitution at positions 2, 3, 4, 5 relative to the carboxyl carbon atom.

α-Hydroxy-acids

Preparation of α-hydroxy-acids. α-Hydroxy-acids can be obtained by the hydrolysis of the corresponding halogen substituted acid, or from the amino acid by reaction with nitrous acid, or by reduction of the corresponding keto-acid.

$$RCHClCO_2H \xrightarrow{^-OH/H_2O} RCH(OH)CO_2^- \xrightarrow{H^+} RCH(OH)CO_2H \xleftarrow{HNO_2} RCH(NH_2)CO_2H$$

$$RCOCO_2H \xrightarrow{Ni/H_2} RCH(OH)CO_2H$$

A more important general method of synthesis is from aldehydes or ketones via the cyanohydrins

$$RR'C{=}O \xrightarrow[+CN^-]{+HCN} RR'C(OH)CN \xrightarrow{H^+/H_2O} RR'C(OH)CO_2H$$

Properties and reactions of α-hydroxy-acids. α-Hydroxy-acids are solids or high-boiling liquids, usually very soluble in water (note the enhanced possibilities for hydrogen bond formation). Many, but not all, exhibit optical isomerism. Most of the biologically important α-hydroxy-acids have a secondary alcohol function, and the convention for designating the configuration of these compounds in terms of D and L is that in the Fischer projection with the carbon chain vertical and the carboxyl group at the top, the D enantiomer has the hydroxyl group on the right-hand side.

$$\begin{array}{c} CO_2H \\ | \\ H\text{—}C\text{—}OH \\ | \\ R \end{array} \qquad \begin{array}{c} CO_2H \\ | \\ C \\ H \;\; R \\ OH \end{array} \qquad D$$

$$\begin{array}{c} CO_2H \\ | \\ HO\text{—}C\text{—}H \\ | \\ R \end{array} \qquad \begin{array}{c} CO_2H \\ | \\ C \\ HO \;\; R \\ H \end{array} \qquad L$$

If the priority sequence is $OH > CO_2H > R$, then D≡*R* and L≡*S*.

Numerous reactions of α-hydroxy-acids are characteristic of the carboxylic acid and hydroxyl groups reacting independently. Thus two series of esters can be formed:

$$RCH(OH)CO_2H \xrightarrow{CH_3OH/H^+} RCH(OH)CO_2CH_3$$

$$RCH(OH)CO_2H \xrightarrow{(CH_3CO)_2O} RCH(OCOCH_3)CO_2H$$

and in the presence of bases of suitable strength either a monoanion or dianion can be obtained (cf carboxylic acids and alcohols)

$$RCH(OH)CO_2H \xrightarrow{NaHCO_3} RCH(OH)CO_2^- \xrightarrow{K^+OC(CH_3)_3} RCH(O^-)CO_2^-$$

The carboxylic acid group can be converted into a nitrile or amide, but no α-hydroxyacyl chloride can be formed as the acyl chloride function would immediately react with the alcoholic hydroxyl group. Reaction of α-hydroxy-acids with phosphorus halides forms the α-haloacyl halide.

$$RCH(OH)CO_2H \xrightarrow{PCl_5} RCHClCOCl$$

Oxidation of the α-hydroxy-acids gives the corresponding α-keto-acid (2-oxo-acid) when the hydroxyl group is part of a primary or secondary alcohol function. The keto-acids obtained are readily oxidised further (p. 218).

$$HOCH_2CO_2H \xrightarrow{CrO_3} OHCCO_2H$$

$$RCH(OH)CO_2H \xrightarrow{CrO_3} RCOCO_2H$$

If the hydroxyl group is tertiary, oxidation degrades the α-hydroxy-acid to the lower ketone

$$R_2C(OH)CO_2H \xrightarrow{CrO_3} R_2CO + CO_2 + H_2O$$

Apart from these reactions, most of which are shown by the majority of hydroxy-acids, irrespective of the relative positions of the two functional groups, there are a few reactions which are characteristic of α-hydroxy-acids, since they depend on the two groups being closely adjacent. If α-hydroxy-acids are heated with catalytic quantities of mineral acids, dimeric esters, known as lactides are formed, in which the carboxylic acid group of one molecule esterifies the hydroxyl group of the second and vice versa.

$$2\,RCH(OH)CO_2H \xrightarrow{+H^+} \text{Lactide (cyclic: RCH–O–C(=O)–CHR–O–C(=O))}$$

Lactide

Cyclic complex ions ('**chelates**') can be formed with metal ions, iron(III) chloride giving intensely yellow complex ions with α-hydroxy-acids:

$$RCH(OH)CO_2H + Fe^{+3} \longrightarrow \left[\text{RCH–C(=O)–O–Fe–O (ring)}\right]^+$$

(both of these reactions depend upon the ease of formation of five-and six-membered rings).

If α-hydroxy-acids are heated with sulphuric acid, decomposition into the lower aldehyde or ketone occurs, with formation of carbon monoxide.

$$RCH(OH)CO_2H \xrightarrow{H_2SO_4} RCHO + CO + H_2O$$

Since the α-hydroxy-acid has hydroxyl groups on adjacent carbon atoms, oxidation by periodic acid or lead tetra-acetate is possible:

$$R_2C(OH)CO_2H \xrightarrow{Pb(OCOCH_3)_4} R_2CO + CO_2 + H_2O$$

β-Hydroxy-acids can be prepared by reduction of the esters of the corresponding keto-acids, followed by hydrolysis:

$$RCOCH_2CO_2C_2H_5 \xrightarrow{Ni/H_2} RCH(OH)CH_2CO_2C_2H_5 \xrightarrow[\text{(ii) } H^+]{\text{(i) } ^-OH/H_2O} RCH(OH)CH_2CO_2H$$

or by the hydrolysis of the corresponding nitrile, which can be obtained from alkenes via the bromohydrin:

$$R{-}CH{=}CH_2 \xrightarrow{Br_2/H_2O} R{-}\underset{OH}{\underset{|}{C}H}{-}\underset{Br}{\underset{|}{C}H_2} \xrightarrow{CN^-} R{-}\underset{OH}{\underset{|}{C}H}{-}CH_2{-}CN \xrightarrow{H^+/H_2O} RCH(OH)CH_2CO_2H$$

Simple β-hydroxy-acids having hydrogen atoms in the α position are dehydrated to the corresponding unsaturated carboxylic acids so readily that their preparation is very difficult. Attempts to perform reactions with simple β-hydroxy-acids normally result in extensive dehydration and little of the desired product.

$$RCH(OH)CH_2CO_2H \xrightarrow{heat} RCH{=}CHCO_2H + H_2O$$

γ-Hydroxy-acids and δ-hydroxy-acids are also known. The free acids rapidly form cyclic internal esters known as 'lactones' in which the hydroxyl function is esterified by the carboxylic acid group of the same molecule. Note

$$RCH(OH)CH_2CH_2CO_2H \longrightarrow \text{R–CH}\underset{\text{O}}{\overset{\text{H}_2\text{C–CH}_2}{}}\text{C=O} + H_2O$$

γ-Lactone

$$RCH(OH)CH_2CH_2CH_2CO_2H \longrightarrow \text{R–CH}\underset{\text{O–C(=O)}}{\overset{\text{H}_2\text{C–CH}_2}{}}\text{CH}_2 + H_2O$$

δ-Lactone

that, here again, it is the cyclic compounds with five- and six-membered rings which are easily formed; α- and β-hydroxy-acids cannot be converted directly into the corresponding α- and β-lactones (three- and four-membered rings).

Where hydroxy-acids have the two functional groups separated by more than three carbon atoms, the hydroxyl and carboxylic acid groups react independently like simple alcohols and carboxylic acids.

Some naturally occurring hydroxy-acids

Glycollic acid (Hydroxyethanoic acid), $CH_2(OH)CO_2H$, is an important intermediate in some metabolic pathways, and occurs in the juice of beet and unripe grapes. Its chemistry is that of a typical α-hydroxy-acid, and since there is no chiral centre in the molecule glycollic acid does not form enantiomers.

Lactic acid (2-Hydroxypropanoic acid), $CH_3CH(OH)CO_2H$, is widely distributed in nature, the trivial name arising from its occurrence in sour milk. It possesses an asymmetrically substituted carbon atom and can exist in two enantiometric forms, m.p. 53°C, and a racemate, m.p. 18°C.

$$\begin{array}{c} CO_2H \\ | \\ H—C—OH \\ | \\ CH_3 \end{array} \qquad \begin{array}{c} CO_2H \\ | \\ HO—C—H \\ | \\ CH_3 \end{array}$$

D-(−)-Lactic acid — 2(*R*)-Hydroxypropanoic acid; L-(+)-Lactic acid — 2(*S*)-Hydroxypropanoic acid

The contraction of muscle tissue during work is accompanied by the formation of L-lactic acid from glycogen if only a limited supply of oxygen is available; and fermentation of carbohydrates by micro-organisms frequently produces L- or DL-lactic acid.

Malic acid (Hydroxysuccinic acid), can exist as enantiomers, which are crystalline solids. The L(−) enantiomer is an intermediate in the tricarboxylic

$$\begin{array}{c} CO_2H \\ | \\ H—C—OH \\ | \\ CH_2CO_2H \end{array} \qquad \begin{array}{c} CO_2H \\ | \\ HO—C—H \\ | \\ CH_2CO_2H \end{array}$$

D-(+)-Malic acid — 2(*R*)-Hydroxybutanedioic acid; L-(−)-Malic acid — 2(*S*)-Hydroxybutanedioic acid

acid cycle (p. 234) and also occurs widely in the juice of many green fruits, being readily isolated from unripe apples. It forms salts and esters normally, and exhibits reactions typical of both α- and β-hydroxy-acids. Thus malic acid gives the yellow complex ion with ferric salts, is oxidised by periodic acid, and eliminates carbon monoxide on treatment with warm sulphuric acid. Heating malic acid dehydrates it to the corresponding unsaturated acids (p. 232), a reaction characteristic of a β-hydroxy-acid.

$$\begin{array}{l} CH(OH)CO_2H \\ | \\ CH_2CO_2H \end{array} \xrightarrow{\text{heat}} \begin{array}{l} CHCO_2H \\ \| \\ CHCO_2H \end{array}$$

Oxidation gives the corresponding keto-acid (oxaloacetic acid) which on further oxidation is degraded to malonic acid.

$$\begin{array}{l} CH(OH)CO_2H \\ | \\ CH_2CO_2H \end{array} \xrightarrow{\text{Oxidation}} \begin{array}{l} COCO_2H \\ | \\ CH_2CO_2H \end{array} \longrightarrow \begin{array}{l} CO_2H \\ | \\ CH_2CO_2H \end{array} + CO_2$$

Oxaloacetic acid
2-Oxobutanedioic acid

Tartaric acid (2,3-Dihydroxysuccinic acid), $HO_2CCH(OH)CH(OH)CO_2H$, has two identical chiral centres giving rise to two enantiomeric forms, m.p. 167–170°C, and an optically inactive stereoisomer (*meso*-tartaric acid), m.p. 140°C (for Fischer projections see p. 183). The racemate DL-tartaric acid ('racemic acid'), m.p. 206°C, is a substance of some historical significance, since it was Pasteur's investigation of the relationship between the three stereoisomers and the racemate which first led to the appreciation of the significance of the three-dimensional aspects of organic chemistry and the tetrahedral array of covalent bonds around carbon atoms.

Dextrorotatory tartaric acid occurs in the juice of many plants, particularly grape juice, and is deposited as a scaly precipitate of the monopotassium salt ('argol') during wine-making. It is an important metabolic intermediate in some biochemical processes.

The hydroxyl and carboxylic acid groups of tartaric acid exhibit the usual reactions. Oxidation with periodic acid gives methanoic acid and carbon dioxide, whilst cautious oxidation forms the hydrated form of dioxosuccinic acid which on further oxidation gives oxalic acid (ethanedioic acid).

$$\begin{array}{c}CO_2H\\|\\CHOH\\|\\CHOH\\|\\CO_2H\end{array}\xrightarrow{\text{Oxidation}}\begin{array}{c}CO_2H\\|\\C(OH)_2\\|\\C(OH)_2\\|\\CO_2H\end{array}\longrightarrow\begin{array}{c}CO_2H\\|\\CO_2H\end{array}$$

'Dihydroxytartaric acid'

Tartaric acid shows reactions characteristic of both α- and β-hydroxy-acids. Thus complex ions are formed with ferric salts (yellow) and also with cupric salts (e.g. Fehling's solution, p. 113). Heating the acid gives pyruvic acid by a route commencing with a dehydration typical of a β-hydroxy-acid. The compound formed initially is the enol tautomer (p. 105) of oxaloacetic acid. This rearranges to the keto-form, a β-keto-acid, which rapidly decarboxylates (p. 218) forming pyruvic acid.

$$\begin{array}{c}CO_2H\\|\\CHOH\\|\\CHOH\\|\\CO_2H\end{array}\xrightarrow[-H_2O]{\text{heat}}\begin{array}{c}CO_2H\\|\\C{-}OH\\\|\\CH\\|\\CO_2H\end{array}\longrightarrow\begin{array}{c}CO_2H\\|\\C{=}O\\|\\CH_2\\|\\CO_2H\end{array}\xrightarrow{-CO_2}\begin{array}{c}CO_2H\\|\\C{=}O\\|\\CH_3\end{array}$$

Pyruvic acid

Citric acid occurs widely in nature, particularly in fruit juices, lemon juice being a source from which extraction was once carried out on a commercial scale.

Several series of salts and esters can be formed, varying in the number or position of the reacting carboxylic acid groups. Citric acid shows the characteristic reactions of an α-hydroxy-acid, giving complexes with iron(III) ion and copper(II) ion (Benedict's reagent, p. 113) and being converted to the corresponding carbonyl compound by strong acids. Since the alcoholic hydroxyl group is tertiary, oxidation is accompanied by breakdown of the

$$\underset{\text{Acetonedicarboxylic acid}}{O{=}C(CH_2CO_2H)_2} \xleftarrow[-CO\ -H_2O]{+H_2SO_4} \underset{\text{Citric acid}}{HO{-}C(CO_2H)(CH_2CO_2H)_2} \xrightarrow{\text{heat}} \underset{\text{Aconitic acid}}{CHCO_2H{=}C(CO_2H){-}CH_2CO_2H}$$

carbon skeleton with the formation of ethanoic and oxalic acids. On heating, citric acid behaves as a β-hydroxy-acid, losing water to form the unsaturated acid (aconitic acid).

Isocitric acid, is simultaneously an α-, β-, and γ- hydroxy-acid. In addition to the normal properties of α- and β-hydroxy-acids it is readily converted into the γ-lactone.

$$\underset{\text{Isocitric acid}}{HO_2C{-}CH(OH){-}CH(CO_2H){-}CH_2CO_2H} \xrightarrow{-H_2O} HO_2C{-}CH{-}O{-}C(=O){-}CH_2{-}CH(CO_2H)\ \text{(ring)}$$

Keto-acids (oxo-acids)

Carboxylic acids containing other carbonyl groups are frequently found in living tissue, and the chemistry of α- and β-keto-acids will be considered here. Compounds, in which the carbonyl and carboxylic acid groups are more widely separated behave simply as normal ketones and carboxylic acids.

α-Keto-acids

Pyruvic acid (2-Oxopropanoic acid), CH_3COCO_2H, is a typical α-keto-acid. It may be prepared by the hydrolysis of the corresponding nitrile or dihalo-substituted acid, or by oxidation of the corresponding hydroxy-acid:

$$CH_3COCl \xrightarrow{CuCN} CH_3COCN \xrightarrow{H^+/H_2O} CH_3COCO_2H$$

$$CH_3CBr_2CO_2H \xrightarrow{H_2O} CH_3COCO_2H$$

$$CH_3CH(OH)CO_2H \xrightarrow{CrO_3} CH_3COCO_2H$$

Pyruvic acid is, however, most conveniently obtained by pyrolysis of tartaric acid, a preparation not available for other α-keto-acids.

Pyruvic acid is a viscous liquid, b.p. 165°C, miscible with water. It is an important intermediate in many metabolic processes. It has most of the normal properties of a carboxylic acid, forming salts, esters, amides, etc., and also exhibits many normal ketonic properties forming an oxime, phenylhydrazone, and being reduced to the alcohol (*RS*-lactic acid).

$$CH_3COCO_2H \xrightarrow{CH_3OH/H^+} CH_3COCO_2CH_3$$
$$CH_3COCO_2H \xrightarrow{CrO_3} CH_3CO_2H + CO_2$$
$$CH_3COCO_2H \xrightarrow{Cl_2} CH_2ClCOCO_2H$$
$$CH_3COCO_2H \xrightarrow{H_2NOH} CH_3C(=NOH)CO_2H$$
$$CH_3COCO_2H \xrightarrow{C_6H_5NHNH_2} CH_3C(=NNHC_6H_5)CO_2H$$
$$CH_3COCO_2H \xrightarrow{Ni/H_2} CH_3CH(OH)CO_2H$$

Oxidation readily degrades α-keto-acids forming carbon dioxide and the lower carboxylic acid, even mild oxidants like Fehling's solution being effective (p. 113).

Pyruvic acid reacts with halogens readily, substitution occurring via formation of the enol, $CH_2{=}C(OH)CO_2H$. (p. 105). The enol tautomer is also of biological significance, occurring as its phosphate ester 'phospho-enolpyruvic acid', which, in biological systems, can act both as a phosphorylating agent or an alkylating agent (p. 161).

$$CH_2{=}C(CO_2H)OPO_3H_2 + ROH \xrightarrow{\text{Phosphorylation}} ROPO_3H_2 + CH_3COCO_2H$$
$$CH_2{=}C(CO_2H)OPO_3H_2 + ROH \xrightarrow{\text{Alkylation}} RO{-}C(=CH_2)CO_2H + H_3PO_4$$

'Phospho-enolpyruvic acid'

β-Keto-acids

Acetoacetic acid (3-Oxobutanoic acid), $CH_3COCH_2CO_2H$, is a typical β-keto-acid. It can be prepared by the careful hydrolysis of its esters at low temperature, but is very unstable. β-Keto-acids decompose very readily even at room temperature, losing carbon dioxide and forming the corresponding ketone.

$$RCOCH_2CO_2H \longrightarrow RCOCH_3 + CO_2$$

Acetoacetic acid, or its thio-ester with coenzyme A is an important metabolic intermediate, being formed during the oxidation of fats (p. 234). In some pathological conditions (e.g. diabetes) it is excreted in the urine along with its decarboxylation product acetone (propanone).

Although acetoacetic acid and most free β-keto-acids are of little chemical interest on account of their instability, the esters are of considerable significance. Ethyl acetoacetate can be prepared by the condensation of two molecules of ethyl acetate in the presence of strong bases:

$$2\,CH_3CO_2C_2H_5 \xrightarrow{Na^+\bar{O}C_2H_5} CH_3COCH_2CO_2C_2H_5 + C_2H_5OH$$

This reaction is initiated by formation of the resonance-stabilised enolate anion of ethyl acetate by proton abstraction by the base.

$$C_2H_5O^- + CH_3{-}\overset{\overset{:O:}{\|}}{C}{-}OC_2H_5 \rightleftharpoons C_2H_5OH + :\bar{C}H_2{-}\overset{\overset{:O:}{\|}}{C}{-}OC_2H_5 \longleftrightarrow CH_2{=}\overset{\overset{:\ddot{O}:^-}{|}}{C}{-}OC_2H_5$$

$$CH_3{-}\overset{\overset{:O:}{\|}}{C}{-}OC_2H_5 + :\bar{C}H_2{-}\underset{\underset{:O:}{\|}}{C}{-}OC_2H_5 \longrightarrow CH_3{-}\overset{\overset{:\ddot{O}:^-}{|}}{\underset{\underset{CH_2-\underset{\underset{O}{\|}}{C}-OC_2H_5}{|}}{C}}{-}OC_2H_5 \xrightarrow{-\bar{O}C_2H_5} CH_3{-}\overset{\overset{O}{\|}}{C}{-}CH_2{-}CO_2C_2H_5$$

The nucleophilic enolate anion attacks a molecule of ethyl acetate forming ethyl acetoacetate (by a mechanism similar to type III on p. 142). A similar condensation ('the Claisen condensation') can be performed with any ester

$$R{-}\overset{\overset{R}{|}}{\underset{\underset{H}{|}}{C}}{-}CO_2C_2H_5 \xrightarrow{\text{Strong base}} R{-}\overset{\overset{R'}{|}}{\underset{\underset{H}{|}}{C}}{-}\overset{\overset{O}{\|}}{C}{-}\overset{\overset{R'}{|}}{\underset{\underset{R}{|}}{C}}{-}CO_2C_2H_5 + C_2H_5OH$$

possessing an α-H atom or a reaction can be procured between different esters, though this normally leads to a mixture of all possible products.

Ethyl acetoacetate is a compound having a methylene group adjacent to two carbonyl groups. As in the case of diethyl malonate (p. 165), the methylene group can readily lose a proton to form a resonance-stabilised anion. The *pK* for this ionisation is 10·7, so that ethyl acetoacetate dissolves in 1M sodium hydroxide solution with complete conversion into the anion.

$$CH_3{-}\overset{\overset{O}{\|}}{C}{-}CH_2{-}\overset{\overset{O}{\|}}{C}{-}OC_2H_5 \xrightarrow{-H^+} CH_3{-}\overset{\overset{:O:}{\|}}{C}{-}\ddot{\bar{C}}H{-}\overset{\overset{:O:}{\|}}{C}{-}OC_2H_5$$

$$\updownarrow$$

$$CH_3{-}\overset{\overset{:O:}{\|}}{C}{-}CH{=}\overset{\overset{:\ddot{O}:^-}{|}}{C}{-}OC_2H_5 \longleftrightarrow CH_3{-}\overset{\overset{:\ddot{O}:^-}{|}}{C}{=}CH{-}\overset{\overset{:O:}{\|}}{C}{-}OC_2H_5$$

The resonance-stabilised anion can react with alkyl halides, to give alkyl-substitution on the methylene group, and by successive reactions both of the hydrogen atoms may be replaced.

$$CH_3COCH_2CO_2C_2H_5 \xrightarrow{Na^+\ ^-OC_2H_5} CH_3CO-\ddot{C}H^--CO_2C_2H_5 \xrightarrow{+RI} CH_3CO-CH(R)-CO_2C_2H_5$$

$$CH_3CO-CH(R)-CO_2C_2H_5 \xrightarrow{Na^+\ ^-OC_2H_5} CH_3CO-\ddot{C}^-(R)-CO_2C_2H_5 \xrightarrow{R'I} CH_3CO-C(R)(R')-CO_2C_2H_5$$

The mono- or dialkylated β-keto-esters prepared in this way can be converted into the corresponding ketones by hydrolysis with hot dilute acid, the initially formed β-keto-acid decarboxylating rapidly.

$$CH_3CO-C(R)(R')-CO_2C_2H_5 \xrightarrow{H^+/H_2O} CH_3CO-C(R)(R')-CO_2H \xrightarrow{-CO_2} CH_3COCHRR'$$

Alternatively, heating the β-keto-esters with alcoholic potassium hydroxide cleaves the molecule into the salts of two carboxylic acids by a reversal of the mechanism of the Claisen condensation.

$$CH_3-C(=\ddot{O}:)-C(R)(R')-CO_2C_2H_5 + {}^-:\ddot{O}H \longrightarrow CH_3-C(-\ddot{O}:^-)(OH)-C(R)(R')-CO_2C_2H_5$$

$$\longrightarrow CH_3-C(=\ddot{O})-OH + :\bar{C}(R)(R')-CO_2C_2H_5$$

$$CH_3CO_2^- \xleftarrow{\bar{O}H} CH_3-C(=O)OH$$

$$:\bar{C}(R)(R')-CO_2C_2H_5 \xrightarrow{+H_2O} H-C(R)(R')-CO_2C_2H_5 \xrightarrow{H\bar{O}/H_2O} RR'CHCO_2^- + C_2H_5OH$$

Ethyl acetoacetate is therefore a valuable intermediate in the synthesis of both ketones and carboxylic acids.

Apart from the hydrolysis to form the acid, ethyl acetoacetate shows few of the normal reactions of esters, nucleophiles reacting preferentially with the ketonic carbonyl group. The ketonic carbonyl group can be reduced to the alcohol and forms a normal derivative with 2,4-dinitrophenylhydrazine, but most of the reagents which give condensation compounds with aldehydes and

$$CH_3COCH_2CO_2C_2H_5 \xrightarrow{Ni/H_2} CH_3CH(OH)CH_2CO_2C_2H_5$$

$$\downarrow + H_2NNH-C_6H_3(NO_2)_2$$

$$CH_3C(=N-NH-C_6H_3(NO_2)_2)CH_2CO_2C_2H_5$$

ketones form cyclic derivatives with ethyl acetoacetate by loss of ethanol from the initially formed condensation compound.

$$CH_3COCH_2CO_2C_2H_5$$

$$\xrightarrow{+C_6H_5NHNH_2} CH_3C(=N-NH-C_6H_5)CH_2CO_2C_2H_5 \xrightarrow{-C_2H_5OH} \text{(ring: } CH_3-C=N-N(C_6H_5)-C(=O)-CH_2\text{)}$$

$$\xrightarrow{+H_2NOH} CH_3C(=N-OH)CH_2CO_2C_2H_5 \xrightarrow{-C_2H_5OH} \text{(ring: } CH_3-C=N-O-C(=O)-CH_2\text{)}$$

The tautomerism of ethyl acetoacetate. The correct structure of ethyl acetoacetate was the subject of a prolonged controversy before the phenomenon of tautomerism (p. 105) was understood. It is now appreciated that normal samples of the ester contain both tautomers in an equilibrium mixture composed of 93 per cent keto and 7 per cent enol tautomer.* By special techniques it is possible to separate pure samples of the tautomers, which, in the

* Amongst the arguments in favour of the enol structure was quoted the reaction with alkali metals to form salts, as at one time it was thought that this was characteristic of hydroxyl groups. The formation of the enolate anion, however, does not require the prior existence of the enol, as the keto-isomer can be deprotonated directly, and acidic hydrocarbons are by no means unknown, e.g. ethyne (p. 35) and cyclopentadiene (p. 276).

absence of basic or acidic catalysts, are only slowly interconverted. Since even the alkaline surface of soda glass is an effective catalyst, equilibration under normal laboratory conditions is a very rapid process. One method of

$$\underset{\text{'keto'}}{CH_3{-}\overset{\displaystyle O}{\overset{\|}{C}}{-}CH_2{-}CO_2C_2H_5} \rightleftharpoons \underset{\text{'enol'}}{CH_3{-}\overset{\displaystyle OH}{\overset{|}{C}}{=}CH{-}CO_2C_2H_5}$$

separation consists of distilling the equilibrium mixture in a silica (quartz) apparatus (silica does not catalyse the tautomerisation). The enol tautomer, which is the more volatile on account of internal hydrogen bonding, distils over leaving the residue enriched in the keto tautomer.

H- bonded enol tautomer

Compounds which exist to an appreciable extent as enols give purple colours with iron(III) salts (e.g. phenols). Ethyl acetoacetate forms a dark red-purple chelate complex with iron(III) chloride, and similar complexes are obtained with many other metal ions. Complexes of this type are not observed with simple ketones, such as propanone, on account of the low equilibrium concentration of enol and the instability of the metal complexes.

Decarboxylation of carboxylic acids

The decarboxylation of a carboxylic acid $R{-}CO_2H \longrightarrow R{-}H + CO_2$ is a reaction occurring very readily in some cases and with the utmost difficulty in others. Simple aliphatic and aromatic carboxylic acids are usually stable up to high temperatures and can be decarboxylated, usually in very poor yield, only by heating their sodium salts to high temperatures with sodium hydroxide (soda-lime).

$$CH_3CO_2^- Na^+ + NaOH \longrightarrow CH_4 + Na_2CO_3$$

$$C_6H_5CO_2^- Na^+ + NaOH \longrightarrow C_6H_6 + Na_2CO_3$$

The mechanism of this type of reaction, which takes place in the solid phase, is obscure, and the reaction is of no synthetic importance. At the other extreme, β-keto-acids decarboxylate readily even at 0°C, and in between these limits are examples of carboxylic acids of a range of stabilities.

The decarboxylation of a carboxylic acid involves the breaking of a C—C bond between the R and carbonyl groups. Neglecting the possibility of radical pathways, two heterolytic mechanisms can be considered for this step:

$$\mathrm{R{-}C({=}O){-}OH} \longrightarrow \mathrm{R{:}^{-}} + \mathrm{{}^{+}C({=}O){-}OH}$$

$$\mathrm{R{-}C({=}O){-}OH} \longrightarrow \mathrm{R^{+}} + \mathrm{{:}\bar{C}({=}O){-}OH}$$

The former alternative could be assisted by prior ionisation of the carboxylic acid group to give the anion:

$$\mathrm{R{-}C({=}\ddot{O}{:}){-}\ddot{O}{-}H} \xrightarrow{-\,H^{+}} \mathrm{R{-}C({=}\ddot{O}{:}){-}\ddot{O}{:}^{-}} \longrightarrow \mathrm{R{:}^{-}} + \mathrm{CO_2}$$

whereas no such help is available for the latter process.

It appears that decarboxylation of carboxylic acids in the normal range of laboratory conditions (i.e. up to *ca.* 250°C) proceeds via formation of the anion $R{:}^{-}$, and the ease of decarboxylation is closely related to the stability (i.e. ease of formation) of this carbanion. Where R is a simple alkyl or aryl group the carbanion is of high energy and therefore difficult to form. Consequently these acids are stable (e.g. benzoic acid, $C_6H_5CO_2H$, boils without decomposition at 250°C and dodecanoic acid (lauric acid), $CH_3(CH_2)_{10}CO_2H$, boils at 298°C). The carboxylic acids which decarboxylate readily are those which give rise to carbanions which are stabilised (i.e. have a lower energy of formation) by substituents. Stabilisation of the carbanion can be achieved in a number of ways.

Inductive stabilisation of $R{:}^{-}$ occurs where the carbon atom bearing the carboxylic acid group also bears highly electronegative substituents which, by inductive electron displacement, withdraw electron density and electrical charge from the negatively charged atom of the carbanion. Examples of this type are trichloroethanoic acid and 2,4,6-trinitrobenzoic acid, both of which decompose in boiling water.

$$Cl_3C—C(=O)—O—H \quad OH_2 \longrightarrow Cl_3C{:}^- + CO_2 + H_3O^+$$

$$Cl_3C{:}^- + H_3O^+ \longrightarrow Cl_3CH + H_2O$$

$$\text{2,4,6-}(NO_2)_3C_6H_2CO_2H \xrightarrow[\text{water}]{\text{Boiling}} \text{2,4,6-}(NO_2)_3C_6H_2{:}^- + CO_2 + H_3\overset{+}{O}$$

$$\text{2,4,6-}(NO_2)_3C_6H_2{:}^- + H_3O^+ \longrightarrow \text{1,3,5-}(NO_2)_3C_6H_3$$

Conjugative stabilisation of $R{:}^-$ is the more common cause of easy decarboxylation. In these cases the atomic orbital containing the lone pair in the carbanion overlaps with other orbitals which delocalise the electron density and negative charge. The simplest example of this case is where the β-atom of the carboxylic acid has a vacant atomic orbital:

$$\overset{+}{C}—CH_2—C(=O)—O—H \longrightarrow C{=}CH_2 + CO_2 + H^+$$

and although this situation is not attainable in a stable molecule, such species can be obtained by protonation of an unsaturated acid. Thus cinnamic acid decarboxylates readily in strongly acidic aqueous solution at 100° C.

$$\underset{\text{Cinnamic acid}}{C_6H_5\,CH{=}CH\,CO_2H} \underset{-\,H^+}{\overset{+\,H^+}{\rightleftharpoons}} C_6H_5\overset{+}{C}H—CH_2CO_2H \longrightarrow C_6H_5\,CH{=}CH_2 + CO_2 + H^+$$

Alternatively, if there is no vacant atomic orbital on the β-carbon atom, a leaving group attached there can be displaced by the electron density accumulating on the α-carbon atom as the σ-bond to the carbonyl group is cleaved.

$$-\overset{|}{\underset{\underset{X}{|}}{C}}-\overset{|}{\underset{|}{C}}-\overset{O}{\overset{\|}{C}}-O-H \quad \text{Base} \longrightarrow \;>C=C< \; + X^- + CO_2 + \text{H-}\overset{+}{\text{Base}}$$

E.g.

$$Cl-CH_2-\overset{CH_2Cl}{\overset{|}{\underset{\underset{CH_2Cl}{|}}{C}}}-CO_2H \xrightarrow[+\text{ base}]{\text{Heat}} CH_2=C\begin{matrix} CH_2Cl \\ CH_2Cl \end{matrix}$$

This mechanism is a slightly complicated type of elimination reaction (p 29). A biologically important example of a decarboxylation of this type is found in the convertion of phosphorylated mevalonic acid (p. 177) into 'isopentenyl pyrophosphate' (3-methylbut-3-enyl pyrophosphate):

$$\text{Base} \quad H-O-\underset{\underset{O}{\|}}{C}-CH_2-\overset{CH_3 \quad OPO_3H_2}{C}-CH_2CH_2O\overset{O}{\overset{\|}{\underset{\underset{OH}{|}}{P}}}O\overset{O}{\overset{\|}{\underset{\underset{OH}{|}}{P}}}-OH$$

$$\downarrow$$

$$\overset{+}{\text{Base}}-H + CO_2 + \underset{CH_2}{\overset{CH_3}{C}}-CH_2\,CH_2\,O\overset{O}{\overset{\|}{\underset{\underset{OH}{|}}{P}}}O\overset{O}{\overset{\|}{\underset{\underset{OH}{|}}{P}}}-OH$$

Isopentenyl pyrophosphate

which is one step in the conversion of ethanoic acid into steroids (p. 326) via mevalonic acid and squalene (p. 34).

The last type of conjugative interaction which facilitates decarboxylation involves stabilisation of the developing carbanion by conjugation with an unsaturated system in which negative charge can be concentrated onto a highly electronegative atom. The β-keto-acids are good examples of this, the carbanion developing on the α-carbon atom being stabilised as an enolate anion.

$$R-\overset{\overset{O}{\|}}{C}-CH_2-\overset{\overset{O}{\|}}{C}-O-H \quad Base \longrightarrow \left\{ \begin{array}{c} R-\overset{\overset{O^-}{|}}{C}=CH_2 \\ \updownarrow \\ R-\overset{\overset{O}{\|}}{C}-\bar{C}H_2 \end{array} \right\} + CO_2 + H\text{-}\overset{+}{Base}$$

$$\xrightarrow{+H^+} R-\overset{\overset{O}{\|}}{C}-CH_3$$

Subsequently the enolate anion, being a strong base, will abstract a proton from the medium or from the protonated base. The decarboxylation of β-keto-acids such as acetoacetic acid (p. 218) and oxalacetic acid (p. 216) proceeds by this mechanism under basic conditions.

The carbonyl group is not the only example of a group which can assist decarboxylation in this way. Any group which can stabilise the electron density and charge of the intermediate carbanion (i.e. can act as an 'electron sink') will be effective. Both α-cyano and α-nitro acids are readily decarboxylated on heating on account of the mesomeric stabilisation of the anion.

$$N\equiv C-CH_2-\overset{\overset{O}{\|}}{C}-OH \xrightarrow[-CO_2]{-H^+} :N\equiv C-\ddot{C}H_2^- \longleftrightarrow \ddot{N}^-=C=CH_2 \xrightarrow{+H^+} N\equiv C-CH_3$$

$$O_2N-CH_2-CO_2H \xrightarrow[-CO_2]{-H^+} \begin{array}{c} :\ddot{O} \\ \| \\ \overset{+}{N}-CH_2^- \\ | \\ :\ddot{O}^- \end{array} \longleftrightarrow \begin{array}{c} :\ddot{O}^- \\ | \\ \overset{+}{N}=CH_2 \\ | \\ :\ddot{O}^- \end{array} \xrightarrow{+H^+} O_2N-CH_3$$

However, unsaturated hydrocarbon groups are ineffective, since although they can provide the π-system adjacent to the developing carbanion they contain no highly electronegative atom onto which the electron density and charge can be shunted. Thus pent-3-enoic acid, $CH_3CH{=}CHCH_2CO_2H$, is quite stable to heat, boiling without decomposition at 193°C.

All the mechanisms so far described involve the initial abstraction of the proton from the carboxylic acid group by an external base (not necessarily a strong base, water will suffice in many cases). β-Keto-acids can undergo a slightly different method of decarboxylation involving an un-ionised molecule, in which the ketonic carbonyl group acts as a base in a synchronised process of proton transfer and loss of carbon dioxide.

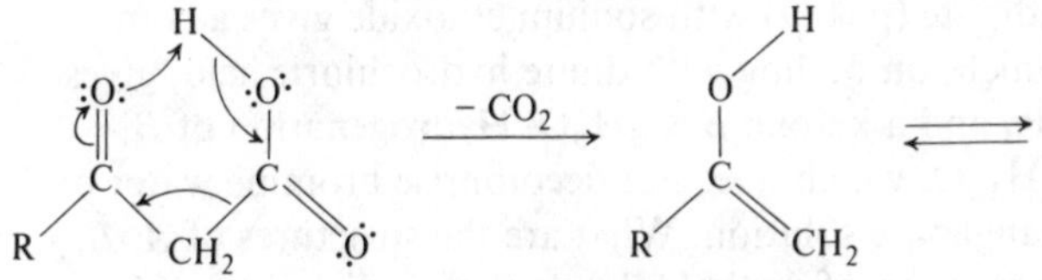

$$\xrightarrow{-CO_2} \quad \rightleftharpoons \quad RCOCH_3$$

Note that it is the enol that is formed initially by this pathway, although a subsequent tautomeric change may give the keto form as the major end-product of reaction. Cases are known of β-keto-acids in which the molecular geometry prevents the formation of the enol even though the keto form is quite stable. In such cases the β-keto-acid does not decarboxylate.

Ketopinic acid is an example of this. At first sight it might appear that this should decarboxylate readily to form the known ketone (I) but in fact it is quite stable to heat (m.p. 234°C) since the cage structure of the molecule

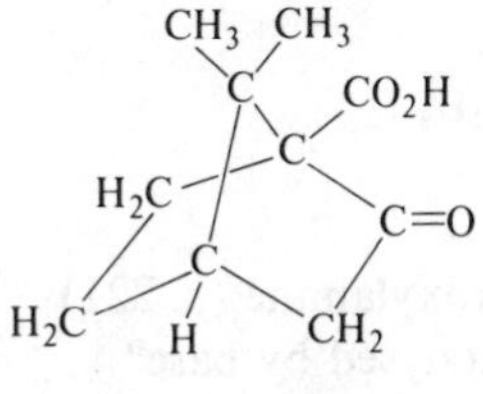

Ketopinic acid

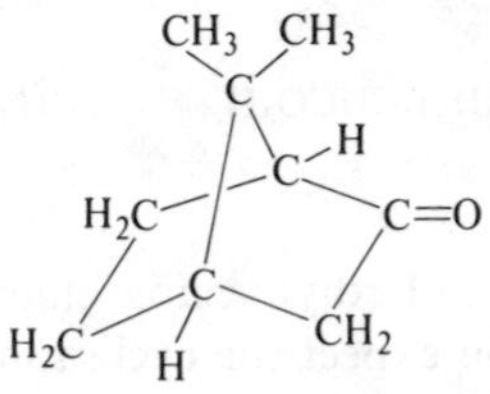

(I)

(II)

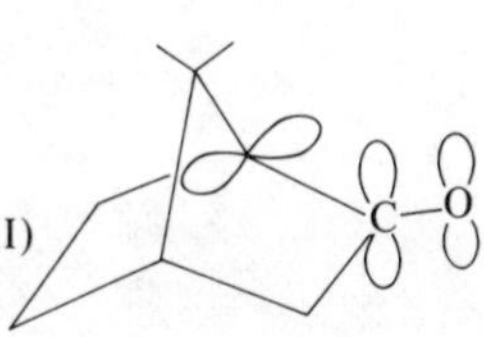

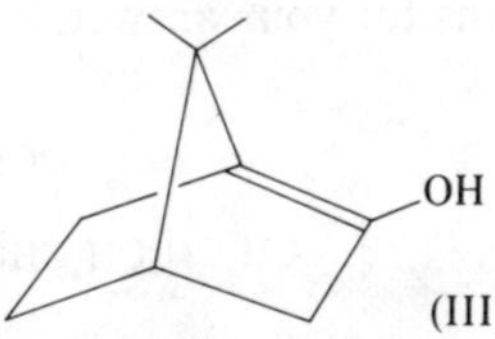

(III)

holds the π-orbitals of the ketonic carbonyl group almost perpendicular to the σ-bond which would be broken during decarboxylation (II). Since the orbitals cannot achieve a parallel alignment no overlap and no stabilisation of the carbanion can occur, nor can the enol (III) be formed from (I).

Problems

1. By what mechanism is 2-hydroxybutyric acid converted into its lactide on heating with acid? How many stereoisomeric lactides are formed in this

reaction from racemic 2-hydroxybutyric acid? Draw suitable projection diagrams to illustrate the configurations of the products.

2. Treatment of diethyl adipate (p. 169) with sodium ethoxide gives a compound *A*, $C_8H_{12}O_3$, which, on boiling with dilute hydrochloric acid, gives ethanol, carbon dioxide, and a ketone *B*, C_5H_8O. Hydrogenation of *B* gives an alcohol *C*, $C_5H_{10}O$, which does not decolourise bromine water or dilute potassium permanganate solution. What are the structures of *A*, *B*, and *C*? Formulate the reaction of diethyl adipate with sodium ethoxide mechanistically.

3. Which of the following compounds would you expect to give a purple colour with iron(III) chloride solution?

$CH_3COCH(CH_3)CO_2C_2H_5$ $\quad$ $CH_3COC(CH_3)_2CO_2C_2H_5$

$(CH_3)_3CCOCH_2CO_2C_2H_5$ $\quad$ $CH_3COCH_2CH_2CO_2C_2H_5$

4. Starting from ethyl acetoacetate, outline schemes for synthesis of the following compounds.

$CH_3COCH(CH_3)(CH_2C_6H_5)$ $\quad$ $(CH_3)_2CHCO_2H$ $\quad$ $CH_3-C(CH_3)(CH{=}CH_2)-CO_2C_2H_5$

5. Rationalise the reaction of ethyl acetoacetate with hydroxylamine (p. 221) mechanistically. Do you expect the cyclisation to be catalysed by base?

6. Which of the following compounds are expected to decarboxylate readily? Explain the reasons for your answer.

$(CH_3)_2CHCH(CHO)CO_2H$ $\quad$ $CH_3CH_2CH_2CH(OCH_3)CO_2H$

$C_6H_5CH(\overset{+}{N}(CH_3)_3\,I^-)CH_2CO_2H$ $\quad$ $CH_3CH(CN)CH_2CH_2CO_2H$

$CH_3COCH{=}CHCH_2CO_2H$ $\quad$ $O_2N-C_6H_4-CH_2CO_2H$

16

αβ-Unsaturated aliphatic carboxylic acids

Numerous classes of aliphatic unsaturated carboxylic acid are known, differing in the relative positions of the alkene and carboxylic acid groups. Where these are isolated, the chemistry of the molecule is a simple combination of the chemistry of an alkene and a carboxylic acid, but where the groups are adjacent, resulting in conjugation between the alkene and the carbonyl groups, some marked changes in behaviour are apparent. We shall be concerned here solely with the conjugated unsaturated acids, which are usually known as *αβ*-unsaturated acids following an archaic system of nomenclature (p. 125).

The preparation of these *αβ*-unsaturated acids may be achieved by dehydration of hydroxy-acids or dehydrohalogenation of halogen substituted acids.

$$CH_3CH(OH)CH_2CO_2H \xrightarrow[-H_2O]{\text{heat}} CH_3CH{=}CHCO_2H$$

$$\left.\begin{array}{l} CH_3CHClCH_2CO_2H \\ CH_3CH_2CHClCO_2H \end{array}\right\} \xrightarrow[-HCl]{\text{strong base}} CH_3CH{=}CHCO_2H$$

Where the substitution of the molecule is suitable, mixtures of *cis-trans* isomers may be obtained.

The reactions of the carboxylic acid group are not greatly affected by the adjacent double bond in *αβ*-unsaturated carboxylic acids. Conversion of the carboxyl group into esters, acyl chloride, anhydride, amides, etc., is possible by the usual methods. However, the chemistry of the alkene group is very greatly modified by conjugation with the carbonyl group of the carboxylic acid, and, as indicated by the canonical structures, the electronegative carbonyl group depletes the electron density of the π bond (p. 17) resulting in a decrease of the reactivity of the C=C group towards electrophiles. The

$$R_2C{=}C(R){-}C({=}\ddot{O})OH \longleftrightarrow R_2\overset{+}{C}{-}C(R){=}C(OH){-}\ddot{O}{:}^-$$

characteristically rapid reaction of simple alkenes with electrophilic reagents, such as bromine, proceeds very much more slowly when the alkene is conjugated to a carbonyl function, and furthermore, addition always occurs with the more electronegative part of the addendum being attached to the β carbon atom (irrespective of Markownikoff's rule).

$$RCH{=}C(CH_3){-}CO_2H \xrightarrow{+HBr} RCHBr{-}CH(CH_3){-}CO_2H$$

This reaction is probably initiated by the electrophile attacking the carbonyl oxygen atom to give a resonance-stabilised cation, which subsequently reacts with an anion followed by rearrangement of the product.

$$RCH{=}C(R){-}C({=}\ddot{O})\ddot{O}H \xrightarrow{+H^+} RCH{=}C(R){-}C({=}\overset{+}{O}H)\ddot{O}H \longleftrightarrow RCH{=}C(R){-}C(\ddot{O}H){=}\overset{+}{O}H$$

$$\updownarrow \qquad\qquad \updownarrow$$

$$RCH(Br){-}C(R){=}C(\ddot{O}H)\ddot{O}H \xleftarrow{+Br^-} R\overset{+}{C}H{-}C(R){=}C(\ddot{O}H)\ddot{O}H \longleftrightarrow RCH{=}C(R){-}\overset{+}{C}(\ddot{O}H)\ddot{O}H$$

$$\downarrow \text{keto–enol tautomerism}$$

$$RCH(Br){-}CH(R){-}C({=}O)OH$$

Acid-catalysed hydration follows a similar pathway.

$$RCH{=}CHCO_2H \xrightarrow{H^+/H_2O} RCH(OH)CH_2CO_2H$$

The electron displacement in the conjugated system, which decreases the reactivity towards electrophiles, produces a corresponding increase in reactivity towards nucleophiles, which is a characteristic feature of $\alpha\beta$-unsaturated carbonyl compounds. The general mechanism of nucleophilic addition of

HX to an $\alpha\beta$-unsaturated carbonyl compound proceeds by the mechanism:

$$\mathrm{RCH{=}CH{-}C(=\ddot{O})Y} + \mathrm{X{:}^-} \longrightarrow \mathrm{RCH(X){-}CH{=}C(\ddot{O}{:}^-)Y} \xrightarrow{\mathrm{H{-}X}} \mathrm{RCH(X){-}CH(H){-}C(=O)Y}$$

Where HX does not dissociate spontaneously to form the nucleophilic anion X^-, the reaction requires a catalytic amount of anion to be added (e.g. hydroxide ion for the addition of water, ethoxide ion for the addition of ethanol, etc.). Alternatively if the undissociated molecule HX is itself a nucleophile (e.g. NH_3 and amines), addition may occur via the intermediate

$$\mathrm{RCH(\overset{+}{X}{-}H){-}CH{=}C(\ddot{O}{:}^-)Y}$$

This nucleophilic addition to $\alpha\beta$-unsaturated carbonyl compounds (known as the 'Michael addition') is not restricted to acids but occurs generally with $\alpha\beta$-unsaturated esters, ketones, aldehydes, and also nitriles. $\alpha\beta$-Unsaturated carboxylic acids, in fact, react less readily than their esters or nitriles since, under the reaction conditions, the carboxylic acid group is usually converted into the anion (most good nucleophiles are also bases), which, being negatively charged, is less susceptible to attack by nucleophiles than an uncharged species. However the derivatives react readily, e.g.

$$\mathrm{RCH{=}CHCO_2C_2H_5} \xrightarrow{\mathrm{HCN/CN^-}} \mathrm{RCH(CN)CH_2CO_2C_2H_5}$$

$$\mathrm{RCH{=}CHCO_2C_2H_5} \xrightarrow{\mathrm{CH_2(CO_2C_2H_5)_2}\ +\ {}^-\mathrm{CH(CO_2C_2H_5)_2}} \mathrm{RCH(CH(CO_2C_2H_5)_2)CH_2CO_2C_2H_5}$$

$$\mathrm{RCH{=}CHCN} \xrightarrow{\mathrm{NH_3}} \mathrm{RCH(NH_2)CH_2CN}$$

Reduction of the C=C group of $\alpha\beta$-unsaturated carboxylic acids is possible with many reagents which do not reduce isolated double bonds. Even dissolving metal reducing systems (see p. 61) will reduce the conjugated alkene group in these molecules.

Crotonic acid (*E-but-2-enoic acid*). $CH_3—CH=CH—CO_2H$, is one of the simplest $\alpha\beta$-unsaturated carboxylic acids. Its chemistry is typical of this group of compounds. The metabolism of long-chain fatty acids with even numbers of carbon atoms produces the coenzyme A thioester of crotonic acid in the penultimate stage of degradation (p. 234).

The geometrical isomer of crotonic acid *Z*-but-2-enoic acid is also known, but is very rapidly transformed into the more stable *E* isomer.

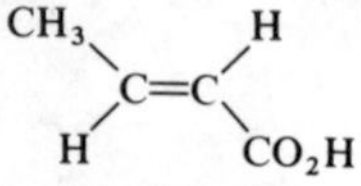

trans-But-2-enoic acid
E-But-2-enoic acid
Crotonic acid

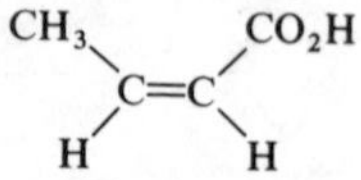

cis-But-2-enoic acid
Z-But-2-enoic acid
Isocrotonic acid

Maleic and fumaric acids are the best-known examples of *cis-trans* isomers. Heating malic acid (p. 215) results in the elimination of water and formation of the two isomeric acids. Separation of these products is achieved by distillation, when maleic acid—the *cis*-isomer—forms a cyclic anhydride, which is

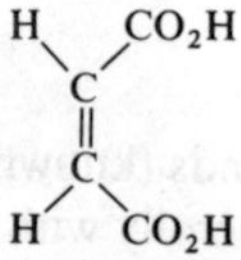

Maleic acid
cis-Butenedioic acid
Z-Butenedioic acid

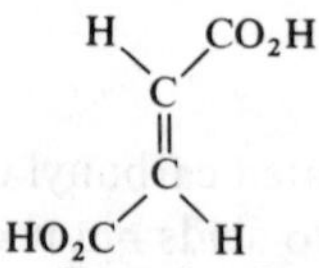

Fumaric acid
trans-Butenedioic acid
E-Butenedioic acid

volatile (m.p. 60°C, b.p. 196°C), whilst fumaric acid is comparatively involatile and cannot form a simple anhydride.

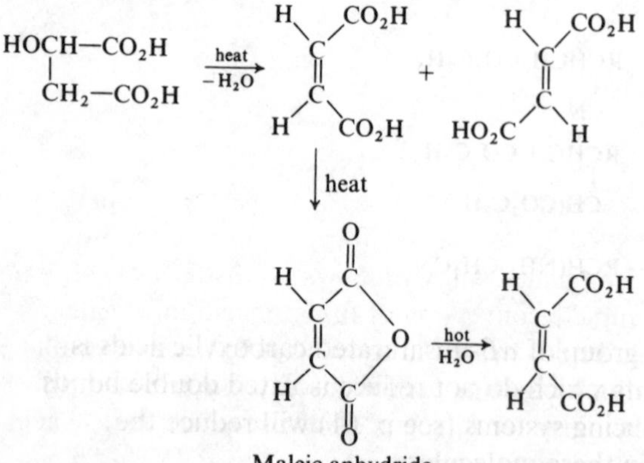

Maleic anhydride

The wide differences, which may occur between the physical properties of geometrical isomers, are conveniently illustrated by reference to these compounds and their derivatives.

PROPERTY	MALEIC ACID	FUMARIC ACID
M.p. of acid	130°C	300°C
M.p. of amide	266°C	267°C
M.p. of methyl ester	−8·4°C	102°C
B.p. of methyl ester	205°C	192°C
pK_1 of acid	2·00	3·03
pK_2 of acid	6·26	4·47
Solubility g/100 g H_2O (20°C)	41·1	0·5

The large difference between the pK values of maleic acid compared with fumaric acid is especially noteworthy, and is attributable to the greater interaction of the functional groups in the *cis*-isomer. Internal hydrogen-bonding may occur in the mono-anion of maleic acid.

The reactions of these acids are normal. Both give two series of esters and salts. Electrophilic addition to the double bond occurs normally, but is slow (p. 229). Ozonolysis gives glyoxylic acid ($CHO.CO_2H$). Nucleophilic substitution of the double bond also occurs, since the alkene is conjugated to two carbonyl groups. Heating either isomer with aqueous sodium hydroxide forms racemic malic acid, and the reaction with alkoxides or amines gives the corresponding substituted succinic acids.

$$CHCO_2H{=}CHCO_2H \xrightarrow{^-OH/H_2O} CH(OH)CO_2H{-}CH_2CO_2H$$

$$CHCO_2H{=}CHCO_2H \xrightarrow{ROH/RO^-} CH(OR)CO_2H{-}CH_2CO_2H$$

$$CHCO_2H{=}CHCO_2H \xrightarrow{R_2NH} CH(NR_2)CO_2H{-}CH_2CO_2H$$

$$CHCO_2H{=}CHCO_2H \xrightarrow[\text{slowly}]{+Br_2} CHBrCO_2H{-}CHBrCO_2H$$

The interconversion of maleic and fumaric acids is achieved by normal methods. If fumaric acid is heated, maleic acid is formed in low concentration and this dehydrates forming maleic anhydride, which distils away, ultimately resulting in the complete conversion of fumaric acid into maleic anhydride. However, if maleic acid is heated in a sealed tube the true equilibrium mixture is produced, which contains predominantly fumaric acid. In the presence of light, traces of iodine, or nitrogen oxides, esters of maleic acid are converted into the corresponding fumaric esters.

Fumaric acid is an intermediate in the tricarboxylic acid cycle and as such is widely distributed in living systems.

Biological occurrence of $\alpha\beta$-unsaturated carboxylic acids. Although very simple $\alpha\beta$-unsaturated acids are not of any great significance in biochemistry, their derivatives occur as intermediates in the synthesis and breakdown of long-chain fatty acids, which are an important constituent of animal fats (p. 150). The enzymic process of degradation is summarised in the scheme below. Desaturation (oxidation) of the saturated acyl coenzyme A occurs to give the ester of a conjugated unsaturated acid. This adds the elements of water, probably via nucleophilic attack (p. 230), and the resulting β-hydroxyacyl coenzyme A is oxidised to the corresponding β-oxo-acyl coenzyme A. This intermediate is then cleaved by interaction with the thiol group of another molecule of coenzyme A in a process analogous to the cleavage of β-keto-esters by alcoholic potassium hydroxide (p. 220). Repetition of this sequence results in the degradation of a long chain fatty acid into C_2 units, which can then be used for further cell reactions.

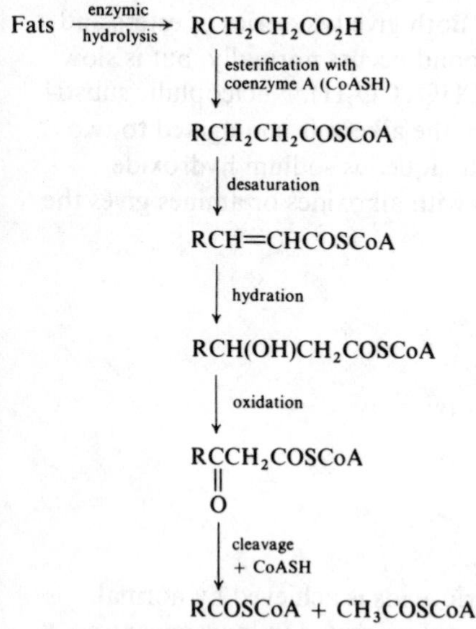

The tricarboxylic acid cycle

The tricarboxylic acid cycle is one of the best known biochemical processes, and is typical of many such cell-reaction sequences, by which relatively large quantities of substrates may be transformed by a cyclic series of reactions involving very small quantities of intermediates. In the tricarboxylic acid cycle (also known as the Krebs' cycle, or the citric acid cycle) the overall reaction is the oxidation of acetic acid to carbon dioxide and water, but the process may serve either as an energy source, or as a source of the intermediate compounds for use in biological synthesis.

Acetic acid is introduced into the cycle as acetylcoenzyme A ($CH_3COSCoA$) and the subsequent transformations are shown in the scheme below. A comparison of the reactions with those of 'normal' chemistry is instructive, even though many important biochemical aspects of the cycle will not be considered.

$CH_3COSCoA$
Acetylcoenzyme A

Oxaloacetic acid $COCO_2H$ / CH_2CO_2H —A→ CH_2CO_2H / $C(OH)CO_2H$ / CH_2CO_2H Citric acid

—B→ $(H)(CO_2H)C{=}C(CO_2H)(CH_2CO_2H)$ *cis*-Aconitic acid

—C→ $CH(OH)CO_2H$ / $CHCO_2H$ / CH_2CO_2H Isocitric acid (2*R*, 3*S*)

—D→ $COCO_2H$ / $CHCO_2H$ / CH_2CO_2H

—E→ (CO_2) $COCO_2H$ / CH_2 / CH_2CO_2H 2-Oxoglutaric acid

—F→ (CO_2) CH_2CO_2H / CH_2CO_2H Succinic acid

—G→ $(H)(CO_2H)C{=}C(HO_2C)(H)$ Fumaric acid

—H→ $CH(OH)CO_2H$ / CH_2CO_2H *S*-Malic acid

—I→ Oxaloacetic acid

A The combination of acetylcoenzyme A and oxaloacetic acid is an example of the aldol reaction (p. 108) and could reasonably occur by nucleophilic attack of the enolate anion of acetylcoenzyme A ($:\bar{C}H_2COSCoA$) on the ketonic carbonyl group of oxaloacetic acid.

B The dehydration of citric acid to the so-called '*cis*-aconitic acid' is a typical dehydration of a β-hydroxy-acid (p. 214) (see p. 318 for a discussion of this reaction).

C The conversion of '*cis*-aconitic acid' into isocitric acid is a hydration of an $\alpha\beta$-unsaturated acid, probably initiated by nucleophilic attack on the conjugated system (cf. the conversion of maleic and fumaric acids into malic acid).

D This step is simply the oxidation of a secondary alcohol to a ketone.

E This step is the normal decarboxylation of a β-keto-acid (p. 218).

F This step is the oxidation of an α-keto-acid to the lower carboxylic acid (p. 218).

G The dehydrogenation of succinic acid to fumaric acid is the only reaction of this cycle which is not readily simulated in the laboratory. It is very similar in type to the enzymic desaturation of fatty acid esters of coenzyme A in the degradation of fats (p. 234). Note that, in the tricarboxylic acid cycle, this dehydrogenation is stereospecific, giving only the *trans* compound and none of the *cis* isomer.

H This step is the hydration of an $\alpha\beta$-unsaturated acid, like step C.

I This step is the oxidation of a secondary alcohol to a ketone.

It can be seen that, with the exception of step G, all the reactions are in accord with the chemistry of the compounds involved. The principal difference between cell reactions and 'laboratory chemistry' is the rapidity and efficiency with which cell reactions occur, on account of the presence of highly specific protein catalysts (enzymes), which have evolved during the millions of years of this planet's biological history. Although the chemist is not yet able to simulate all enzymic reactions, there is no reason to suppose that this is inherently impossible.

Problems

1. Starting from diethyl malonate and ethyl crotonate how would you attempt to synthesise the cyclic anhydride

```
                    O
                  //
         CH2—C
        /          \
CH3—CH            O
        \          /
         CH2—C
                  \\
                    O
```

2. The Michael addition of ammonia to maleic acid gives a mixture of the ammonium salts of racemic 2-amino-succinic acid. Explain why this addition can give rise to enantiomeric products.

17

Carbohydrates

The carbohydrates are a class of compounds, examples of which occur in all living systems. The name was derived from the early observation that many members have molecular formulae of the type $C_m(H_2O)_n$, but is no longer applied in this restricted sense, and is used loosely to describe many aliphatic polyhydroxy compounds and their derivatives.

The simple carbohydrates, or sugars, have formulae $C_nH_{2n}O_n$, ($n \geqslant 3$),* are known as monosaccharides, and are polyhydroxyaldehydes or polyhydroxyketones. The groups of isomeric compounds $C_3H_6O_3$, $C_4H_8O_4$, $C_5H_{10}O_5$, $C_6H_{12}O_6$, are known as trioses, tetroses, pentoses, and hexoses respectively, and these groups comprise the majority of naturally occurring sugars. Examples of heptoses, octoses, etc., are known in nature but occur much less frequently.

The simplest molecular formula, $C_3H_6O_3$, corresponds to three isomeric trioses, two of which are enantiomeric:

```
  CHO               CHO             CH2OH
   |                 |                |
H—C—OH           HO—C—H             CO
   |                 |                |
  CH2OH             CH2OH           CH2OH
```

D-Glyceraldehyde	L-Glyceraldehyde	Dihydroxyacetone
R-Glyceraldehyde	*S*-Glyceraldehyde	Dihydroxypropanone

D- and L-glyceraldehyde may be regarded as the parents of two series of enantiomeric sugars—the aldoses—which can schematically be derived by insertion of hydroxymethylene groups ($>$CHOH) between the aldehyde function and the adjacent chiral centre. Similarly dihydroxyacetone is the parent of a series of sugars known as ketoses, distinguished by the presence of a

* The chemistry of methanal (formaldehyde), CH_2O, and glycolaldehyde, CH_2OHCHO differs from that of sugars, and they are not normally regarded as monosaccharides even though their molecular formulae fit the general case.

ketonic carbonyl group. The general structures of these series of compounds are given by the Fischer projections:

```
   CHO            CHO                CH2OH          CH2OH
    |              |                  |              |
 (CHOH)n        (CHOH)n               CO             CO
    |              |                  |              |
H—C—OH        HO—C—H              (CHOH)m        (CHOH)m
    |              |                  |              |
   CH2OH          CH2OH           H—C—OH        HO—C—H
                                      |              |
                                     CH2OH          CH2OH

D-Aldoses      L-Aldoses          D-Ketoses      L-Ketoses
```

It can be seen from the charts of D-aldoses and D-ketoses (p. 240) that any sugar can give rise to two epimeric* sugars containing one more hydroxymethylene group, since this extra group introduces another chiral centre. The common feature of all D-sugars is the configuration at the penultimate carbon atom of the chain. When the Fischer projection of a D-sugar is drawn with the carbon chain vertical and the aldehyde or ketone function at the top, D-sugars always (by definition) have the hydroxyl group on the penultimate carbon atom on the right-hand side of the chain. Likewise L-sugars, so represented, have this hydroxyl group on the left-hand side. When this convention was originally proposed, it was not possible to discover which of two alternative structures was the correct absolute configuration of D-glyceraldehyde, and hence of all the sugars. Fortunately the arbitrary choice has since been shown to be correct, so that the Fischer projections in the charts of D-aldoses and D-ketoses represent the true three-dimensional structures.

It should be noted that the specific names given to the stereoisomeric sugars imply a sequence of chiral centres of fixed *relative* configuration.† Thus D-ribose and L-ribose do not differ merely in the configuration of a single chiral centre, but have mirror-image related structures, in which the configurations of all the chiral centres are changed. The epimer of D-ribose in which only the

```
   CHO            CHO              CHO
    |              |                |
H—C—OH        HO—C—H          H—C—OH
    |              |                |
H—C—OH        HO—C—H          H—C—OH
    |              |                |
H—C—OH        HO—C—H         HO—C—H
    |              |                |
   CH2OH          CH2OH            CH2OH

D-Ribose       L-Ribose        L-Lyxose (not L-Ribose)
```

* Diastereoisomeric structures, which differ in the configuration of only one of several chiral centres, are known as '**epimers**'.

† In the nomenclature of compounds with multiple chiral centres, but otherwise unrelated to carbohydrates, prefixes derived from the names of sugars are frequently used to designate relative or absolute configuration (see p. 243, e.g. D-*gluco*-saccharic acid).

D-*Aldoses*

D-Glyceraldehyde
CHO
H—C—OH
CH_2OH

D-Erythrose	D-Threose
CHO	CHO
H—C—OH	HO—C—H
H—C—OH	H—C—OH
CH_2OH	CH_2OH

D-Ribose	D-Arabinose	D-Xylose	D-Lyxose
CHO	CHO	CHO	CHO
H—C—OH	HO—C—H	H—C—OH	HO—C—H
H—C—OH	H—C—OH	HO—C—H	HO—C—H
H—C—OH	H—C—OH	H—C—OH	H—C—OH
CH_2OH	CH_2OH	CH_2OH	CH_2OH

D-Allose	D-Altrose	D-Glucose	D-Mannose	D-Gulose	D-Idose	D-Galactose	D-Talose
CHO	CHO	CHO	CHO	CHO	CHO	CHO	CHO
H—C—OH	HO—C—H	H—C—OH	HO—C—H	H—C—OH	HO—C—H	H—C—OH	HO—C—H
H—C—OH	H—C—OH	HO—C—H	HO—C—H	H—C—OH	H—C—OH	HO—C—H	HO—C—H
H—C—OH	H—C—OH	H—C—OH	H—C—OH	HO—C—H	HO—C—H	HO—C—H	HO—C—H
H—C—OH	H—C—OH	H—C—OH	H—C—OH	H—C—OH	H—C—OH	H—C—OH	H—C—OH
CH_2OH	CH_2OH	CH_2OH	CH_2OH	CH_2OH	CH_2OH	CH_2OH	CH_2OH

D-*Ketoses*

$$\begin{array}{c} CH_2OH \\ | \\ CO \\ | \\ CH_2OH \end{array}$$

Dihydroxyacetone

$$\begin{array}{rcl} & CH_2OH & \\ & | & \\ & CO & \\ & | & \\ H— & C & —OH \\ & | & \\ & CH_2OH & \end{array}$$

D-Erythrulose

$$\begin{array}{rcl} & CH_2OH & \\ & | & \\ & CO & \\ & | & \\ H— & C & —OH \\ & | & \\ H— & C & —OH \\ & | & \\ & CH_2OH & \end{array}$$

D-Ribulose

$$\begin{array}{rcl} & CH_2OH & \\ & | & \\ & CO & \\ & | & \\ HO— & C & —H \\ & | & \\ H— & C & —OH \\ & | & \\ & CH_2OH & \end{array}$$

D-Xylulose

$$\begin{array}{rcl} & CH_2OH & \\ & | & \\ & CO & \\ & | & \\ H— & C & —OH \\ & | & \\ H— & C & —OH \\ & | & \\ H— & C & —OH \\ & | & \\ & CH_2OH & \end{array}$$

D-Allulose
(D-Psicose)

$$\begin{array}{rcl} & CH_2OH & \\ & | & \\ & CO & \\ & | & \\ HO— & C & —H \\ & | & \\ H— & C & —OH \\ & | & \\ H— & C & —OH \\ & | & \\ & CH_2OH & \end{array}$$

D-Fructose

$$\begin{array}{rcl} & CH_2OH & \\ & | & \\ & CO & \\ & | & \\ H— & C & —OH \\ & | & \\ HO— & C & —H \\ & | & \\ H— & C & —OH \\ & | & \\ & CH_2OH & \end{array}$$

D-Sorbose

$$\begin{array}{rcl} & CH_2OH & \\ & | & \\ & CO & \\ & | & \\ HO— & C & —H \\ & | & \\ HO— & C & —H \\ & | & \\ H— & C & —OH \\ & | & \\ & CH_2OH & \end{array}$$

D-Tagatose

configuration of the penultimate carbon atom is changed is L-lyxose (cf. the Fischer projection of D-lyxose).

The physical properties of carbohydrates. The simple carbohydrates are usually crystalline solids, but some are known only as viscous syrups and great difficulty is frequently encountered in attempts to crystallise sugars (cf. the very slow crystallisation of honey or 'golden syrup', which are supersaturated solutions of glucose and sucrose). Because of the possibilities of hydrogen bonding arising from multiple hydroxyl groups, sugars usually have rather harder crystals than is normal with organic compounds, and they are usually very soluble in water, only very sparingly soluble in ethanol, and quite insoluble in non-hydroxylic solvents such as ether, chloroform or benzene.

D-Glucose

D-Glucose* is one of the most abundant sugars in nature, and a study of its chemistry will serve to illustrate the chemistry of all the simple aldoses.

The ring structure of glucose. Although it is often convenient to represent the simple carbohydrates by the open-chain structures, the correct structures of pentoses and hexoses are cyclic ones, in which the carbonyl function is converted into a hemiacetal (p. 100) by combination with one of the hydroxyl groups in the same molecule. Normally only five- or six-membered rings are produced in this way, and are known as the furanose and pyranose structures respectively, after the parent heterocyclic compounds furan and pyran. In this

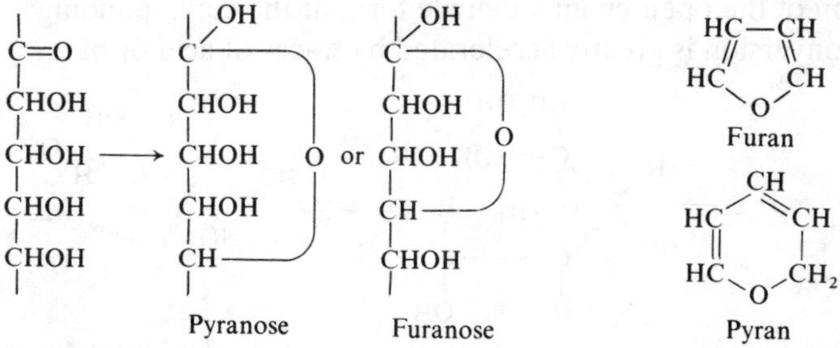

```
 |           | OH            | OH
C=O          C               C
 |           |               |
CHOH        CHOH            CHOH
 |           |               |          O
CHOH  --->  CHOH      O  or CHOH
 |           |               |
CHOH        CHOH            CH
 |           |               |
CHOH        CH              CHOH
 |           |               |
          Pyranose        Furanose
```

Furan: HC—CH, HC, CH, O

Pyran: CH, HC, CH, HC, CH_2, O

reaction, a new chiral centre is produced at the carbonyl carbon atom, so two epimeric hemiacetals could be formed, and in many cases both are known.

Glucose, in the free state, exists entirely as the pyranose form, and many of its derivatives retain this ring structure. This six-membered ring will have two alternative chair conformations (see p. 191). The preferred conformation (shown below) will be that which has more of the substituent groups in equatorial orientations. In the text following, both 'flat' and 'chair' diagrams will be used to show ring structures since the former

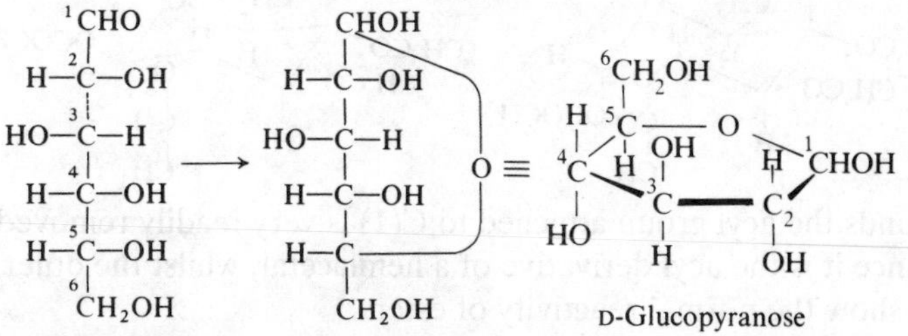

```
     1CHO              CHOH
     2|                 |
  H—C—OH            H—C—OH
     3|                 |
 HO—C—H     --->   HO—C—H        O  ≡
     4|                 |
  H—C—OH            H—C—OH
     5|                 |
  H—C—OH            H—C
     6|                 |
    CH2OH              CH2OH
```

$^{6}CH_2OH$; H ^{5}C — O; ^{4}C H OH; H ^{1}CHOH; ^{3}C — C_2; HO; H; OH

D-Glucopyranose

* The name 'glucose', unqualified by the prefix D- or L- could mean either of the two enantiomeric molecules or the racemate. However, throughout the rest of this text, unless otherwise specified, the name will be used with reference to D-glucose only.

portrays more clearly the relative configurations of adjacent chiral centres, whereas the latter gives a more realistic indication of molecular shape.

By recrystallisation of glucose from various solvents, two different forms of D-glucose can be obtained (named 'α' and 'β'), which differ in melting point and also in the specific rotation of their freshly prepared solutions (showing that the difference is one of molecular structure and not of crystal architecture as in polymorphism). However, if freshly prepared solutions of α-D-glucose (specific rotation = +110°) and β-D-glucose (specific rotation = +19·7°) are allowed to stand, the rotatory powers of the solutions change slowly and eventually achieve a common value equivalent to a specific rotation of +52·5°. This phenomenon, known as '**mutarotation**', is caused by the slow equilibration of the C(1) epimers*, probably via a very low concentration of the open-chain aldehyde form or the corresponding diol. This interconversion is greatly accelerated by traces of acid or base.

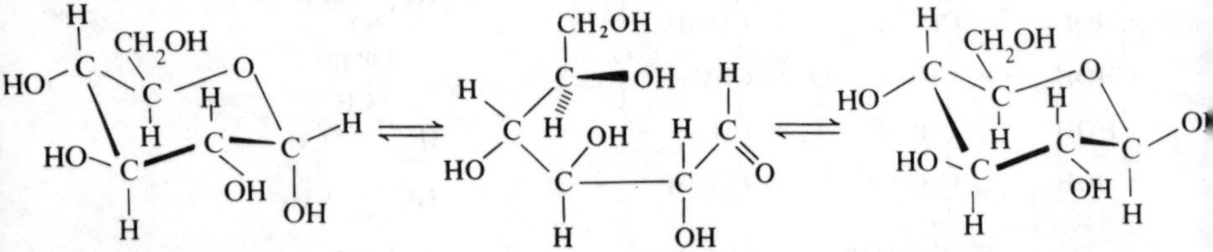

α-D-Glucopyranose ⇌ (open-chain form) ⇌ β-D-Glucopyranose

Reactions of glucose. The hydroxyl groups of glucose can be acylated by acid anhydrides in the normal way and, depending on the conditions, two anomeric penta-acyl derivatives can be obtained.

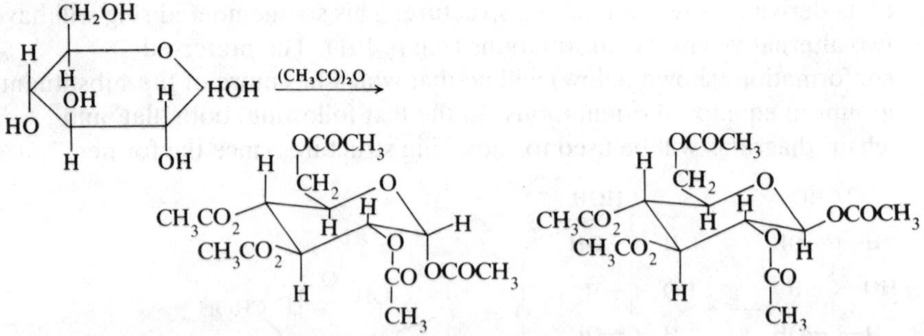

$\xrightarrow{(CH_3CO)_2O}$

In these compounds the acyl group attached to C(1) is very readily removed by hydrolysis, since it is the acyl derivative of a hemiacetal, whilst the other acylated groups show the normal reactivity of esters.

* Carbohydrate stereoisomers which differ only in the configuration of the carbon atom of the hemiacetal function, C(1), are called '**anomers**'. The 'α'-anomer is defined as that which has the C(1)—O and C(5)—C(6) bonds *trans* with respect to the ring.

Warming glucose with concentrated sulphuric acid produces a voluminous deposit of carbon, a reaction typical of many aliphatic polyhydroxy compounds, and used to prepare pure carbon from sucrose.

Glucose shows some, but not all, of the reactions characteristic of simple aldehydes. The very low concentration of the true aldehyde in solutions of glucose is responsible for the failure to give Schiff's test (p. 117), except when performed with very concentrated solutions. Reduction of glucose (e.g. with sodium amalgam and water) gives the corresponding alcohol sorbitol (or

```
   CH2OH                 CO2H
    |                     |
 H—C—OH               H—C—OH
    |                     |
HO—C—H               HO—C—H                      CH2OH
    |                     |                     H  |
 H—C—OH               H—C—OH                    |——————O
    |                     |                    / H      \
 H—C—OH               H—C—OH                  <  OH   H  >=O
    |                     |                    \        /
   CH2OH                 CH2OH              HO  |——————|
                                                H      OH
```

D-Glucitol (Sorbitol) D-Gluconic acid D-Gluconolactone

glucitol), whilst oxidation with dilute nitric acid or bromine water gives initially the corresponding carboxylic acid, D-gluconic acid, and on further oxidation D-*gluco*-saccharic acid. These polyhydroxy carboxylic acids readily

```
                       CN
                       |
   CO2H              *CHOH              CH=NOH
    |                  |                  |
 H—C—OH             H—C—OH             H—C—OH
    |                  |                  |
HO—C—H             HO—C—H             HO—C—H
    |                  |                  |
 H—C—OH             H—C—OH             H—C—OH
    |                  |                  |
 H—C—OH             H—C—OH             H—C—OH
    |                  |                  |
   CO2H               CH2OH              CH2OH
```

D-*gluco*-Saccharic acid D-Glucose cyanohydrin (Note the new chiral centre) D-Glucose oxime

form lactones (p. 214) (e.g. D-gluconolactone from D-gluconic acid). Oxidation of glucose is easily achieved by Fehling's solution, Benedict's reagent, or ammoniacal silver nitrate.

Few of the addition and condensation reactions characteristic of aldehydes are shown by glucose. Addition of hydrogen cyanide gives two epimeric cyanohydrins, and an oxime can also be formed. The reaction of glucose with thiols gives a normal thioacetal.

Glucose reacts with phenylhydrazine in a way which is typical of aldoses, ketoses and α-hydroxyaldehydes and α-hydroxyketones generally. Initially the phenylhydrazone (p. 102) is formed, but is too soluble to be isolated. A

$$\text{(pyranose form of glucose)} \xrightarrow[H^+]{RSH} \begin{array}{c} CH(SR)_2 \\ H-C-OH \\ HO-C-H \\ H-C-OH \\ H-C-OH \\ CH_2OH \end{array}$$

further reaction with two molecules of phenylhydrazine then occurs, in which one molecule of phenylhydrazine is reduced to phenylamine (aniline) and ammonia and the phenylhydrazone is oxidised to the bis-phenylhydrazone derived from the α-dicarbonyl compound. These compounds, known as

$$\begin{array}{c} CHO \\ H-C-OH \\ HO-C-H \\ H-C-OH \\ H-C-OH \\ CH_2OH \end{array} \xrightarrow{C_6H_5NHNH_2} \begin{array}{c} CH{=}NNHC_6H_5 \\ H-C-OH \\ HO-C-H \\ H-C-OH \\ H-C-OH \\ CH_2OH \end{array} \xrightarrow{C_6H_5NHNH_2} \begin{array}{c} CH{=}NNHC_6H_5 \\ C{=}NNC_6H_5 \\ HO-C-H \\ H-C-OH \\ H-C-OH \\ CH_2OH \end{array} \begin{array}{l} +C_6H_5NH_2 \\ +NH_3 \end{array}$$

D-Glucosazone

osazones are yellow, sparingly soluble solids, which have been extensively used in the past to characterise and identify sugars.* It should be noted that in the formation of the osazone, one of the chiral centres is destroyed so that aldoses which are epimeric at C(2) (e.g. D-glucose and D-mannose; D-allose and D-altrose, etc.) form the same osazone. Since osazones are also formed from ketoses, the same osazone can be obtained from three different sugars (e.g. D-glucose, D-mannose, and D-fructose all give D-glucosazone on treatment with phenylhydrazine).

Although the reaction of glucose with thiols results in the opening of the pyranose ring and formation of derivatives of the open-chain structure, the reaction of glucose with alcohols gives quite different products. The hemi-acetal function present in the pyranose ring structure is converted into the acetal by the normal mechanism (p. 116) and two anomeric glucosides† can be obtained.

* Although osazones are sometimes difficult to recrystallise, and often decompose at the melting point, they frequently have very characteristic crystal shapes, or form distinct crystal clusters, which once provided a valuable means of identification. This is now entirely superseded by chromatographic techniques for the identification of sugars.

† Derivatives of sugars, in which the hydrogen atom of the hydroxyl group of the hemiacetal function is replaced by an alkyl or aryl group, are known generically as 'glycosides'. Such derivatives of glucose, fructose, mannose, etc., are termed 'glucosides', 'fructosides', 'mannosides', etc.

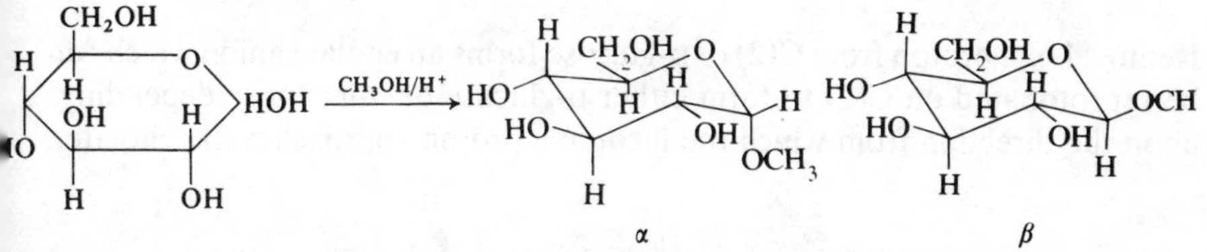

$\xrightarrow{CH_3OH/H^+}$

α β

Methyl-D-glucopyranosides (Methyl-D-glucosides)

No derivative of the open-chain structure is obtained by this method. The glucosides, like normal acetals—are readily hydrolysed by aqueous acid, but are resistant to attack by alkali and although the parent sugar reduces Fehling's solution and ammoniacal silver nitrate, glucosides (and glycosides generally) are not oxidised by these alkaline reagents.

The action of strong alkalis on glucose leads to brown resinous products, but in weakly alkaline solutions, D-glucose rearranges to form a mixture of hexoses, in which D-glucose, D-mannose, and D-fructose predominate. It is very likely that enolisation is responsible for the interconversion, which must occur by way of the open-chain aldehyde.

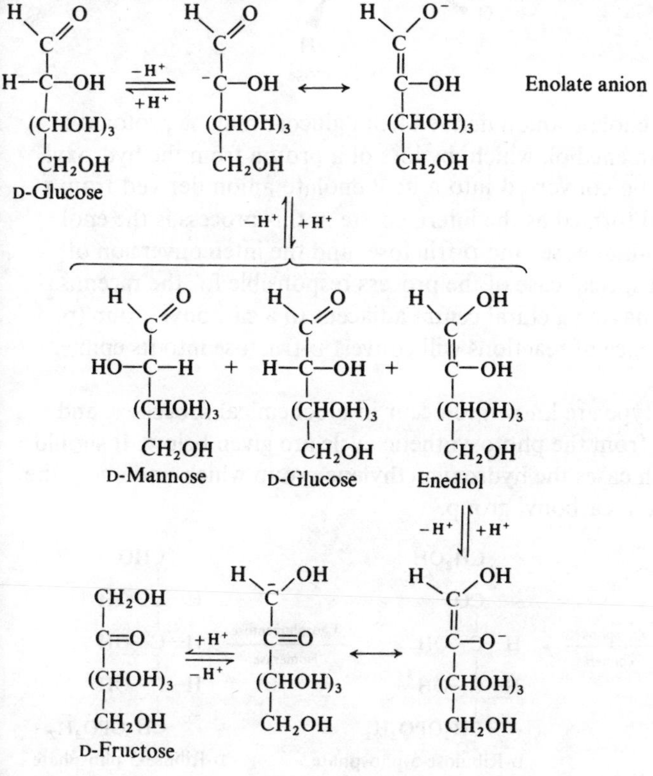

D-Glucose $\underset{+H^+}{\overset{-H^+}{\rightleftharpoons}}$ ⟷ Enolate anion

$-H^+$ ⇅ $+H^+$

D-Mannose + D-Glucose + Enediol

$-H^+$ ⇅ $+H^+$

D-Fructose $\underset{-H^+}{\overset{+H^+}{\rightleftharpoons}}$ ⟷

Removal of a proton from C(2) of D-glucose forms an enolate anion which can be reprotonated on C(2) to form either D-glucose or D-mannose depending upon the direction from which the incoming proton approaches the enolate.

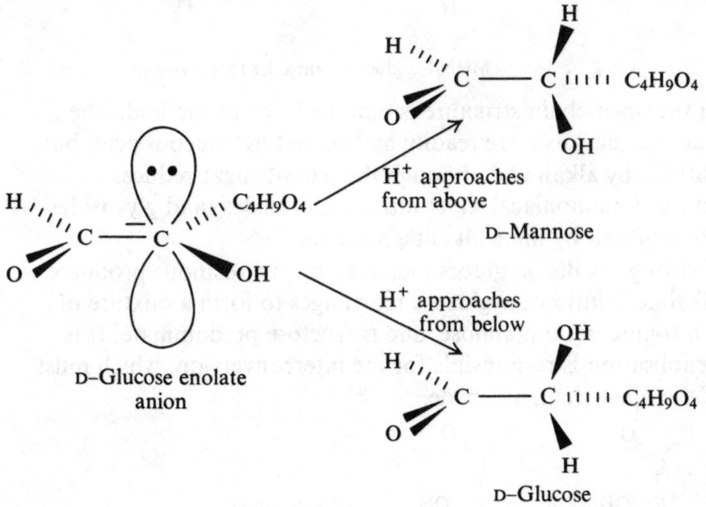

D–Glucose enolate anion

H^+ approaches from above → D–Mannose

H^+ approaches from below → D–Glucose

Alternatively, the enolate anion derived from glucose may be protonated on oxygen to form an enediol, which, by loss of a proton from the hydroxyl group on C(2), can be converted into a new enolate anion derived from D-fructose. The enediol formed as the intermediate in this process is the enol form of D-glucose, D-mannose, and D-fructose, and the interconversion of these sugars is just a special case of the process responsible for the racemisation of enantiomers having a chiral centre adjacent to a carbonyl group (p. 179). A similar sequence of reactions will convert D-fructose into its epimer D-allulose.

Reactions of this type are known to occur in biochemical processes, and two examples taken from the photosynthetic cycle are given below. It should be noted that in both cases the hydroxymethylene group which undergoes the change is adjacent to a carbonyl group.

CH_2OH		CH_2OH		CHO
CO		CO		$H-C-OH$
$HO-C-H$	Phosphoketo-epimerase →	$H-C-OH$	← Phosphopentose isomerase	$H-C-OH$
$H-C-OH$		$H-C-OH$		$H-C-OH$
$CH_2OPO_3H_2$		$CH_2OPO_3H_2$		$CH_2OPO_3H_2$
D-Xylulose-5-phosphate		D-Ribulose-5-phosphate		D-Ribose-5-phosphate

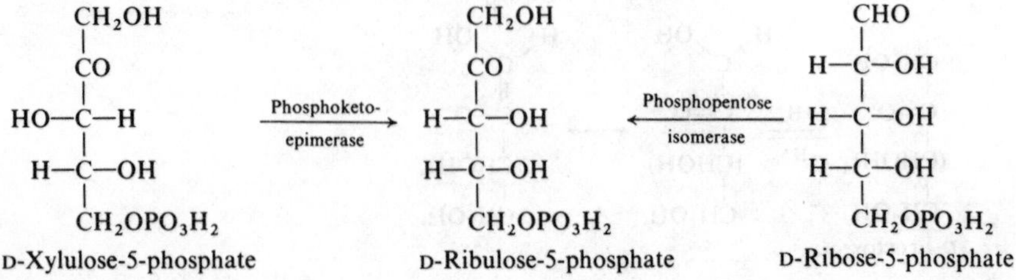

Another reaction observed under alkaline conditions is the degradation of monosaccharides into smaller units by the reverse of the aldol reaction. Typical of this reaction is the cleavage of D-fructose into dihydroxyacetone and D-glyceraldehyde:

$$\begin{array}{l} CH_2OH \\ | \\ C{=}O \\ | \\ HO{-}C{-}H \\ | \\ H{-}C{-}O{-}H \quad \text{Base} \\ | \\ H{-}C{-}OH \\ | \\ CH_2OH \end{array} \longrightarrow \begin{array}{l} CH_2OH \\ | \\ C{-}O^- \\ \| \\ H{-}C{-}OH \end{array} + \begin{array}{l} H{-}C{=}O \\ | \\ H{-}C{-}OH \\ | \\ CH_2OH \end{array}$$

$$\downarrow +H^+$$

$$\begin{array}{l} CH_2OH \\ | \\ C{=}O \\ | \\ CH_2OH \end{array}$$

Compare this mechanism with that of the aldol reaction (p. 109). The enzymic cleavage of fructose-1,6-diphosphate into D-glyceraldehyde-3-phosphate and dihydroxyacetone phosphate is an important step in anaerobic glycolysis (p. 251).

D-Fructose

D-Fructose is one of the most familiar ketoses on account of its occurrence in sucrose (p. 256) from which it may be obtained by acidic or enzymic hydrolysis. Fructose, like glucose, exists predominantly in the form of a cyclic

$$\begin{array}{l} CH_2OH \\ | \\ C{=}O \\ | \\ HO{-}C{-}H \\ | \\ H{-}C{-}OH \\ | \\ H{-}C{-}OH \\ | \\ CH_2OH \end{array} \longrightarrow \begin{array}{l} CH_2OH \\ |\;\; OH \\ C{-} \\ | \\ HO{-}C{-}H \\ | \\ H{-}C{-}OH \quad O \\ | \\ H{-}C{-}OH \\ | \\ CH_2{-} \end{array} \quad \text{or} \quad \begin{array}{l} CH_2OH \\ |\;\; OH \\ C{-} \\ | \\ HO{-}C{-}H \quad O \\ | \\ H{-}C{-}OH \\ | \\ H{-}C{-} \\ | \\ CH_2OH \end{array}$$

$$\equiv \qquad\qquad \equiv$$

D-Fructopyranose [ring: H, C, O, C(OH) CH_2OH; H, C, OH; H; HO, C; OH; H]

D-Fructofuranose [ring: CH_2OH, O, C, H; HO, C(OH)CH_2OH; H, C, C; OH, H] hemiacetal. Crystalline fructose is entirely in the pyranose form, but derivatives of both pyranose and furanose isomers are known. Like glucose, fructose solutions exhibit mutarotation.

Reactions of D-fructose. In many of its reactions fructose behaves similarly to glucose. Treatment with acetic anhydride gives two anomeric penta-acetates, and charring occurs with concentrated sulphuric acid. Reduction gives a mixture of two epimeric polyols since the hydroxymethylene group formed is a new chiral centre. One of these polyols is identical with the compound obtained by reduction of D-glucose and the other is obtained by reduction of D-mannose.

$$
\begin{array}{c}
CH_2OH \\
| \\
C{=}O \\
| \\
HO{-}C{-}H \\
| \\
H{-}C{-}OH \\
| \\
H{-}C{-}OH \\
| \\
CH_2OH
\end{array}
\xrightarrow{\text{Reduction}}
\begin{array}{c}
CH_2OH \\
| \\
H{-}C{-}OH \\
| \\
HO{-}C{-}H \\
| \\
H{-}C{-}OH \\
| \\
H{-}C{-}OH \\
| \\
CH_2OH \\
\text{D-Glucitol} \\
\text{(Sorbitol)}
\end{array}
+
\begin{array}{c}
CH_2OH \\
| \\
HO{-}C{-}H \\
| \\
HO{-}C{-}H \\
| \\
H{-}C{-}OH \\
| \\
H{-}C{-}OH \\
| \\
CH_2OH \\
\text{D-Mannitol} \\

\end{array}
$$

The stereochemical relationship between D-fructose, D-glucose, and D-mannose is also apparent in the formation of D-glucosazone on treatment of D-fructose with phenylhydrazine.

$$
\begin{array}{c}
CH_2OH \\
| \\
C{=}O \\
| \\
HO{-}C{-}H \\
| \\
H{-}C{-}OH \\
| \\
H{-}C{-}OH \\
| \\
CH_2OH
\end{array}
\xrightarrow{+3C_6H_5NHNH_2}
\begin{array}{c}
CH{=}NNHC_6H_5 \\
| \\
C{=}NNHC_6H_5 \\
| \\
HO{-}C{-}H \\
| \\
H{-}C{-}OH \\
| \\
H{-}C{-}OH \\
| \\
CH_2OH \\
\text{D-Glucosazone}
\end{array}
+ C_6H_5NH_2 + NH_3
$$

Oxidation of fructose occurs with breaking of the carbon chain and formation of acids of five or less carbon atoms. Unlike simple ketones, α-hydroxyketones, such as fructose, are readily oxidised even by such mild reagents as Fehling's solution and ammoniacal silver nitrate.

As in the case of glucose, few of the typical derivatives of the carbonyl group are formed; addition of hydrogen cyanide gives two epimeric cyanohydrins, and a *p*-nitrophenylhydrazone is known.

Periodic acid oxidation of sugars and their derivatives. The reaction of periodic acid with *vic*-diols is of great importance in the chemistry of carbohydrates, which contain many such pairs of hydroxyl groups. Whilst the free aldoses and ketoses are completely oxidised to methanoic acid, methanal, and carbon dioxide by an excess of periodic acid, derivatives such as the methyl glycosides undergo oxidative cleavage to products characteristic of the ring size of the starting material. Taking methyl glucoside as an example, furanose and pyranose ring structures are possible, but can be distinguished by the products of periodic acid oxidation. The pyranoside would react with two molecules of periodic acid to give one molecule of methanoic acid by cleavage at the places indicated. The furanose derivative would also react with two

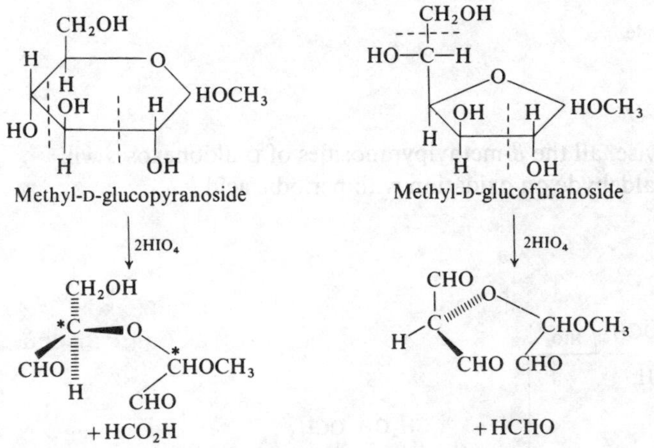

Methyl-D-glucopyranoside $\xrightarrow{2HIO_4}$ dialdehyde $+ HCO_2H$

Methyl-D-glucofuranoside $\xrightarrow{2HIO_4}$ dialdehyde $+ HCHO$

molecules of periodic acid to give one molecule of methanal and a different residual fragment. Methanal and methanoic acid can be readily distinguished, thereby differentiating between the five- and six-membered ring structures.

This example also illustrates another source of structural information. It can be seen that in the dialdehyde obtained from the oxidation of the pyranoside, there are two chiral centres (asterisked) retaining the configurations in the original glycoside. One of these is the penultimate carbon atom of the chain, whose configuration is, by definition, the same for all D-sugars, whilst the other chiral centre in the dialdehyde is the carbon atom of the acetal function, the alternative configurations of which give rise to anomeric glycosides. It is thus possible to correlate the configuration at C(1) of all the anomeric methyl pyranose glycosides derived from D-aldohexoses, since all α-methylpyranosides of D-aldohexoses will give the same dialdehyde on oxidation with

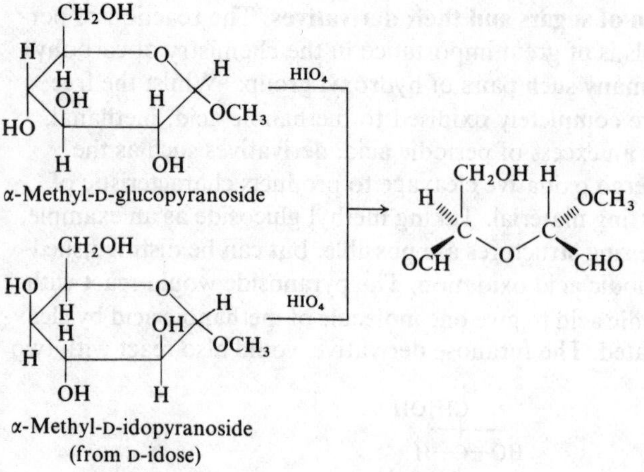

α-Methyl-D-glucopyranoside

α-Methyl-D-idopyranoside
(from D-idose)

periodic acid. Likewise, all the β-methylpyranosides of D-aldohexoses will give the epimeric dialdehyde on oxidation with periodic acid.

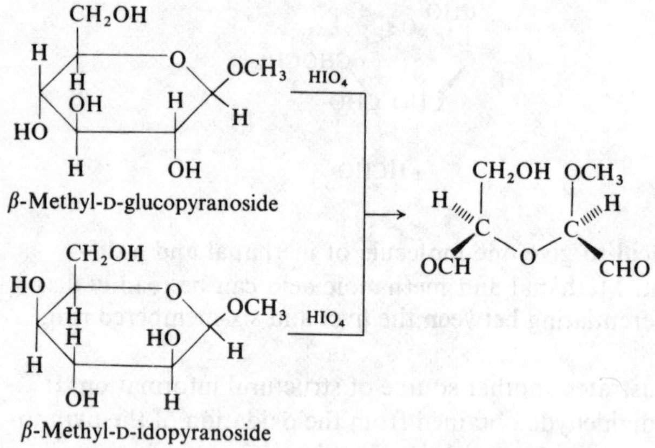

β-Methyl-D-glucopyranoside

β-Methyl-D-idopyranoside

Glycolysis. One important way in which glucose can be used to provide energy is via the process known as glycolysis, which is shown below. The common pathway, into which several different carbohydrate sources feed, starts with glucose-6-phosphate which is formed from free glucose by phosphorylation with adenosine triphosphate (ATP, p. 295). Starch and glycogen (p. 258) are converted enzymically into glucose-1-phosphate by

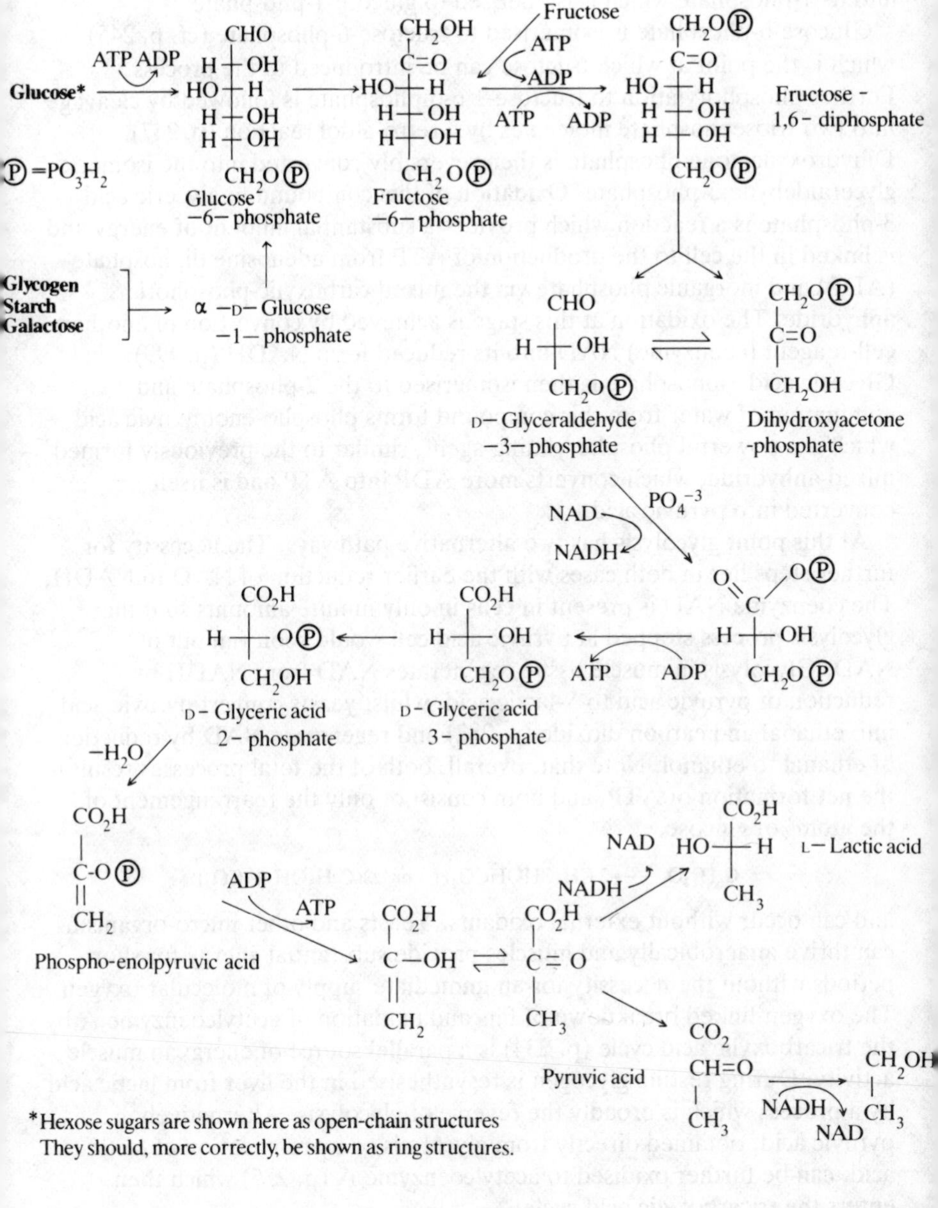

Glucose* → (ATP → ADP)

CHO
H—OH
HO—H
H—OH
H—OH
CH_2OⓅ
Glucose-6-phosphate

→

CH_2OH
C=O
HO—H
H—OH
H—OH
CH_2OⓅ
Fructose-6-phosphate

Fructose (ATP → ADP)

→ (ATP → ADP)

CH_2OⓅ
C=O
HO—H
H—OH
H—OH
CH_2OⓅ
Fructose-1,6-diphosphate

Ⓟ $=PO_3H_2$

Glycogen
Starch
Galactose
→ α-D-Glucose-1-phosphate ↑

CHO
H—OH
CH_2OⓅ
D-Glyceraldehyde-3-phosphate

⇌

CH_2OⓅ
C=O
CH_2OH
Dihydroxyacetone-phosphate

PO_4^{-3}
NAD → NADH

O=C—OⓅ
H—OH
CH_2OⓅ

← (ADP → ATP)

CO_2H
H—OH
CH_2OⓅ
D-Glyceric acid 3-phosphate

←

CO_2H
H—OⓅ
CH_2OH
D-Glyceric acid 2-phosphate

$-H_2O$

CO_2H
C-OⓅ
‖
CH_2
Phospho-enolpyruvic acid

(ADP → ATP) →

CO_2H
C—OH
‖
CH_2

⇌

CO_2H
C=O
CH_3
Pyruvic acid

NADH → NAD

CO_2H
HO—H
CH_3
L-Lactic acid

CO_2
CH=O
CH_3

→ (NADH → NAD)

CH_2OH
CH_3

*Hexose sugars are shown here as open-chain structures
They should, more correctly, be shown as ring structures.

interaction with inorganic phosphate ion and this is isomerised into glucose-6-phosphate. Galactose, from lactose in milk (p. 256), is converted into its 1-phosphate which is isomerised to glucose-1-phosphate.

Glucose-6-phosphate is isomerised to fructose-6-phosphate (cf. p. 245) which is the point at which fructose can be introduced to the process. Further phosphorylation to fructose-1,6-diphosphate is followed by cleavage into two triosephosphate molecules by a retro-aldol reaction (p. 247). Dihydroxyacetone phosphate is then reversibly converted into the isomeric glyceraldehyde-3-phosphate. Oxidation of this compound to glyceric acid 3-phosphate is a reaction which provides a substantial amount of energy and is linked in the cell to the production of ATP from adenosine diphosphate (ADP) and inorganic phosphate via the mixed carboxylic-phosphoric anhydride. The oxidation at this stage is achieved by convertion of another cell-reagent (coenzyme) NAD into its reduced form NADH (p. 299). Glyceric acid 3-phosphate is then isomerised to the 2-phosphate and elimination of water from this compound forms phospho-enolpyruvic acid which is a powerful phosphorylating agent, similar to the previously formed mixed anhydride, which converts more ADP into ATP and is itself converted into pyruvic acid.

At this point glycolysis has two alternative pathways. The necessity for further steps lies in both cases with the earlier reduction of NAD to NADH. The coenzyme NAD is present in cells in only minute amounts so if the glycolysis process stopped at pyruvic acid cell would soon run out of NAD. Glycolysis in muscle tissue regenerates NAD from NADH by reduction of pyruvic acid to *S*-lactic acid, whilst yeasts convert pyruvic acid into ethanal and carbon dioxide (p. 282) and regenerate NAD by reduction of ethanal to ethanol. Note that, overall, both of the total processes result in the net formation of ATP, and both consist of only the rearrangement of the atoms of glucose:

$$C_6H_{12}O_6 \longrightarrow 2\,CH_3CHOHCO_2H \quad \text{or} \quad 2(C_2H_5OH + CO_2)$$

and can occur without external oxidants. Yeasts and other micro-organisms can thrive anaerobically and muscles provide substantial energy for short periods without the necessity for an immediate supply of molecular oxygen. The oxygen-linked breakdown of fats and oxidation of acetylcoenzyme A by the tricarboxylic acid cycle (p. 234) is a parallel source of energy in muscle activity. During resting glycogen is resynthesised in the liver from lactic acid by a process which is broadly the reverse of glycolysis. Alternatively, pyruvic acid, obtained directly from glycolysis or by reoxidation of lactic acid, can be further oxidised to acetylcoenzyme A (p. 235) which then enters the tricarboxylic acid cycle.

Oligosaccharides and polysaccharides

The carbohydrates include many substances more complex than the simple sugars, and numerous naturally occurring carbohydrates are built up by combination of two or more sugar molecules. The names of the various classes of carbohydrate indicate the number of simple sugar (monosaccharide) molecules from which the carbohydrate is constructed (e.g. disaccharides and trisaccharides) whilst the terms oligosaccharide* and polysaccharide* are used to denote compounds derived from small and large numbers of monosaccharide units respectively. Although many common polysaccharides are built up from hexoses, polysaccharides containing tetroses and pentoses are also well known.

The formation of polysaccharides from monosaccharides can be formally regarded as the conversion of the hemiacetal function of a simple sugar into an acetal (a glycoside) by combination with one of the hydroxyl groups of another sugar molecule. The result is a sequence of sugar residues linked by oxygen atoms. Just as the conversion of glucose into methyl glucoside results in two anomeric products (p. 245), so the stereochemistry of the oxygen bridge between sugar residues can differ, and profound differences in biological properties exist between polysaccharides which differ only in the stereochemistry of the ether bridge.

Some of the better known disaccharides are described below:

Maltose

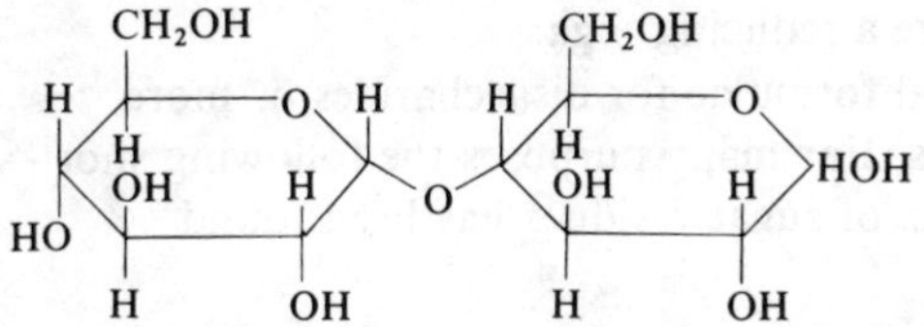

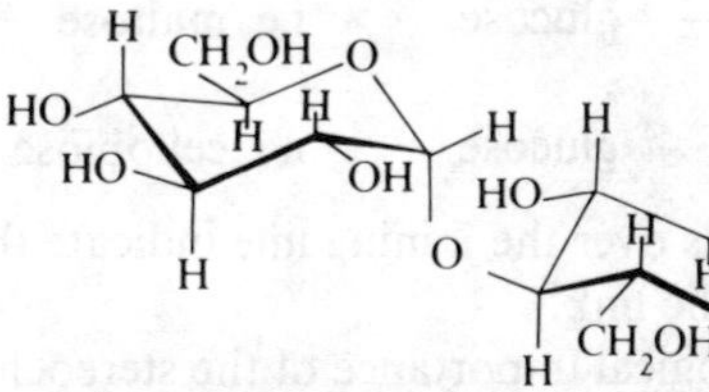

Maltose

Maltose, a disaccharide produced by the enzymic degradation of starch, is formed by the combination of two molecules of D-glucose. Both glucose residues have pyranose rings, and the two residues are linked by combination of

* oligo = a few, poly = many.

the hemiacetal group of one with the hydroxyl group on C(4) of the second. Note that the stereochemistry of the link is that of an α-glucoside.

Since the hemiacetal function of the second (right hand) glucose moiety is not involved in bonding, maltose is a reducing sugar (i.e. it reduces Fehling's solution and ammoniacal silver nitrate), and forms α and β anomers.

Cellobiose

Cellobiose

Cellobiose, a disaccharide obtained by the chemical degradation of cellulose, is constructed from two molecules of D-glucose, linked in a manner very similar to that in maltose, but differing in the stereochemistry of the oxygen bridge, cellobiose being a β-glucoside.* Like maltose, cellobiose has a free hemiacetal function, and is therefore a reducing sugar.

The drawing of full constitutional formulae for disaccharides or more complex oligosaccharides is tedious. For many purposes the following shorthand method of representing chains of sugar residues has been found convenient:

glucose $\xrightarrow{1\alpha \quad\quad 4}$ glucose   i.e. maltose

glucose $\xrightarrow{1\beta \quad\quad 4}$ glucose   i.e. cellobiose

The symbols over the joining line indicate the position and stereochemistry of the glycosidic link.

The biological importance of the stereochemistry of the links between sugar residues in polysaccharides can be judged by a comparison between maltose and cellobiose. Maltose is hydrolysed by maltase, an enzyme present in yeast, which can also hydrolyse many derivatives of α-D-glucose, e.g. α-methyl-D-

* Compare the stereochemistry of the acetal group with that of α-methylglucoside and β-methylgluoside (p. 245).

glucoside. However, maltase is quite unable to hydrolyse cellobiose, even though the molecules are so similar. Cellobiose is hydrolysed to glucose by emulsin, an enzyme occurring in almonds,* which is specific for β-glucosides but has no effect on maltose or other α-glucosides. The reason for this specificity is very probably that, prior to reaction, the enzyme adsorbs the substrate at a special region on the surface of the protein molecule (the 'active site') and this combination of enzyme and substrate is critically dependent on the three-dimensional structure. Alteration of the stereochemistry of the substrate produces a molecule which does not fit the active site of the enzyme and is therefore not susceptible to attack. The biological resolution of racemic mixtures (p. 177) depends upon this stereospecificity.

Gentiobiose (glucose $\xrightarrow{1\beta \quad 6}$ glucose)

Gentiobiose

* The poisonous properties of bitter almonds are due to the presence of 'amygdalin',

$$\text{Glucose} \xrightarrow{1\beta \quad 6} \text{glucose} \xrightarrow{1\beta} \text{OCH(CN)C}_6\text{H}_5 \quad \text{Amygdalin}$$

$$\downarrow \text{H}_2\text{O} \quad \text{emulsin}$$

$$\text{D-Glucose} + \text{C}_6\text{H}_5\text{CH(OH)CN} \longrightarrow \text{C}_6\text{H}_5\text{CHO} + \text{HCN}$$

a glycoside of gentiobiose and benzaldehyde cyanohydrin. Hydrolysis of the β-glucosidic links by emulsin liberates the cyanohydrin, which decomposes to benzaldehyde and hydrogen cyanide. Sweet almonds contain emulsin but no amygdalin and hence are not toxic.

Gentiobiose, also formed from two molecules of D-glucose, is an example of a disaccharide in which the hydroxyl group on C_6 of a glucose unit is involved in the glycosidic link. Gentiobiose is a reducing sugar, and, since it is a β-glucoside, is hydrolysed by emulsin.

Lactose (galactose $\xrightarrow{1\beta \quad 4}$ glucose)

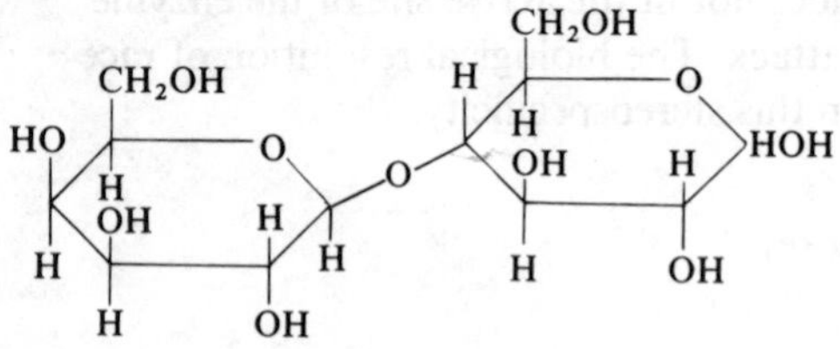

Lactose is a disaccharide, present in milk, formed from one molecule of D-galactose and one of D-glucose joined by a β linkage. It is not hydrolysed by either emulsin or maltase, since it is not a D-glucoside, but can be cleaved by β-galactosidases, which are specific for derivatives of β-D-galactose.

Sucrose (glucose $\xrightarrow{1\alpha \quad 2\beta}$ fructose)

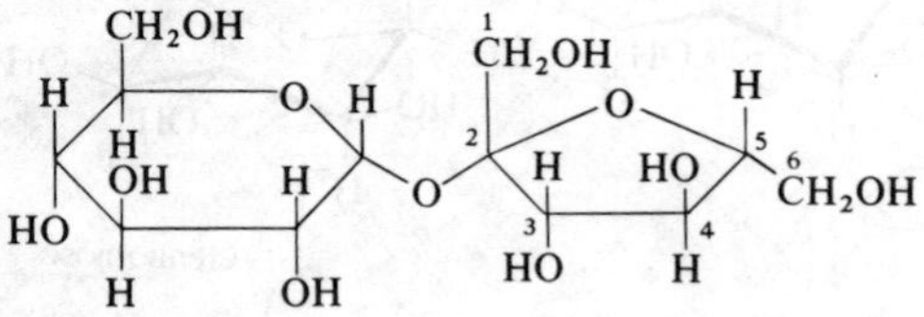

Sucrose (cane sugar) is commercially the most important of the simple carbohydrates, and is extracted from the juice of sugar cane or sugar beet. It is formed by combination of one molecule of D-glucose with one molecule of D-fructose, but unlike previous examples, both hemiacetal groups are involved in this linkage. Since acetals and glycosides are stable to alkali and no aldehyde or ketone function (or its equivalent) is present in the molecule, sucrose is a non-reducing sugar, being unaffected by Fehling's solution or ammoniacal silver nitrate. However, boiling for a few minutes with dilute mineral acid cleaves the glycosidic linkage liberating free glucose and fructose, which will then readily reduce these reagents.

Since the glycosidic linkage involves both hemiacetal functions, four stereoisomers are possible:

glucose $\overset{1\alpha\qquad 2\alpha}{\rule{3em}{0.4pt}}$ fructose

glucose $\overset{1\beta\qquad 2\alpha}{\rule{3em}{0.4pt}}$ fructose

glucose $\overset{1\alpha\qquad 2\beta}{\rule{3em}{0.4pt}}$ fructose

glucose $\overset{1\beta\qquad 2\beta}{\rule{3em}{0.4pt}}$ fructose

Hydrolysis of sucrose by enzymes of known stereospecificity shows that sucrose is an α-glucoside and a β-fructoside.

Trehalose (glucose $\overset{1\alpha\qquad 1\alpha}{\rule{3em}{0.4pt}}$ glucose)

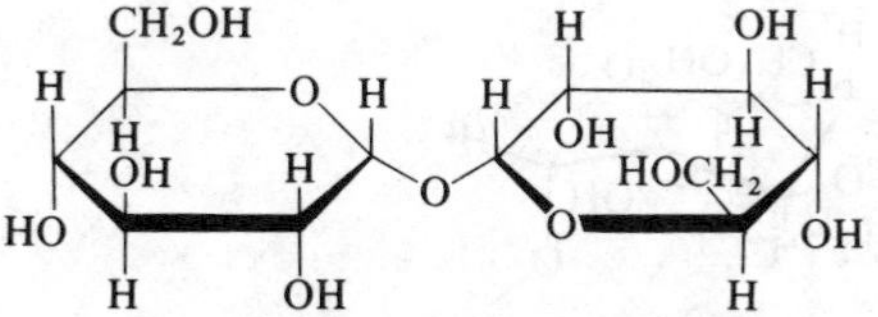

Trehalose is a non-reducing sugar formed from two molecules of D-glucose employed as a reserve carbohydrate by insects.

Trisaccharides, tetrasaccharides, and other oligosaccharides are known, but follow the same sort of patterns as those outlined above. Thus cellotriose (glucose $\overset{1\beta\qquad 4}{\rule{3em}{0.4pt}}$ glucose $\overset{1\beta\qquad 4}{\rule{3em}{0.4pt}}$ glucose) is a reducing sugar hydrolysed by emulsin, whereas raffinose (galactose $\overset{1\alpha\qquad 6}{\rule{3em}{0.4pt}}$ glucose $\overset{1\alpha\qquad 2\beta}{\rule{3em}{0.4pt}}$ fructose) is a non-reducing sugar, unaffected by emulsin.

Cellulose, $(C_6H_{10}O_5)_n$ is the principal polysaccharide of the cell walls of higher plants, and the rigidity of plant tissue is due to its presence. Cellulose consists of very long chains of D-glucose residues linked (1β–4) as in cellobiose (p. 254) with the resultant molecule having a relative molecular mass of

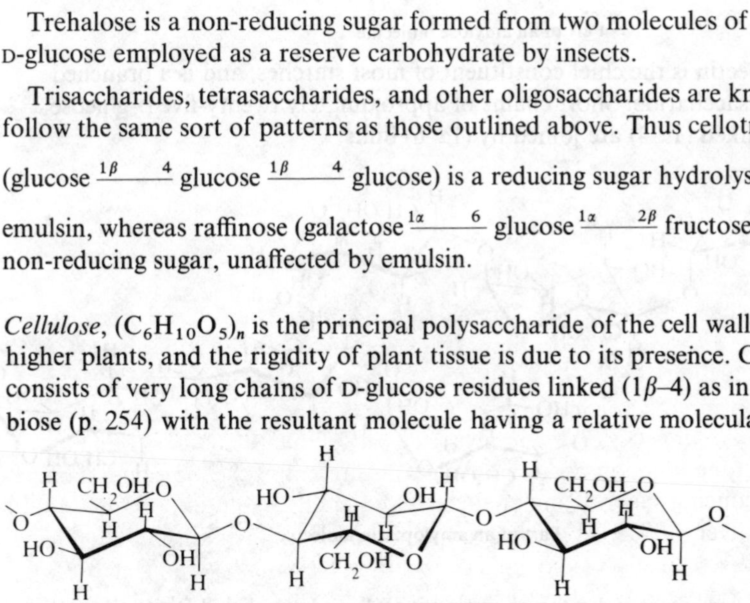

Part of a cellulose molecule

10^5–10^6 depending upon the plant source. Experimental problems of handling molecules of this size make it difficult to decide whether cellulose is an open-chain polyglucose or a giant loop, possibly with cross links.

Starch is the reserve polysaccharide of plants, being deposited in characteristic granules. These starch granules contain two components, amylose and amylopectin, both of which are built entirely of D-glucose residues.

Amylose, the minor constituent of starch, is the polysaccharide which gives the characteristic blue colour with iodine. It consists of a long chain of about three hundred glucose residues linked (1α–4) as in maltose (p. 253).* This long chain is thought to be coiled up into a helix in which there are about six glucose residues per turn (i.e. a helix of approximately fifty turns). The blue iodine complex is considered to be formed by insertion of iodine atoms or molecules into the axial cavity of the helix.

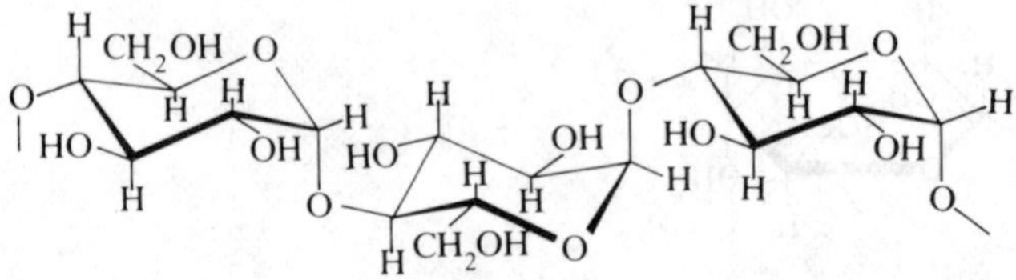

Part of an amylose molecule

Amylopectin is the chief constituent of most starches, and is a branched chain polysaccharide. Short chains of approximately twenty-five D-glucose residues linked (1α-4) are joined by (1α–6) links.

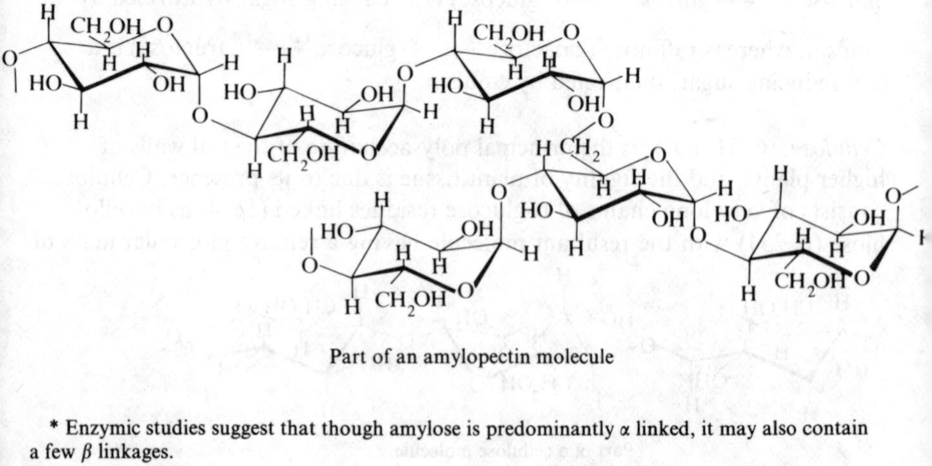

Part of an amylopectin molecule

* Enzymic studies suggest that though amylose is predominantly α linked, it may also contain a few β linkages.

This type of structure can best be represented by the diagram below, where the horizontal lines represent chains of (1α–4)-linked glucose units and the vertical sections represent (1α–6) links between adjacent chains

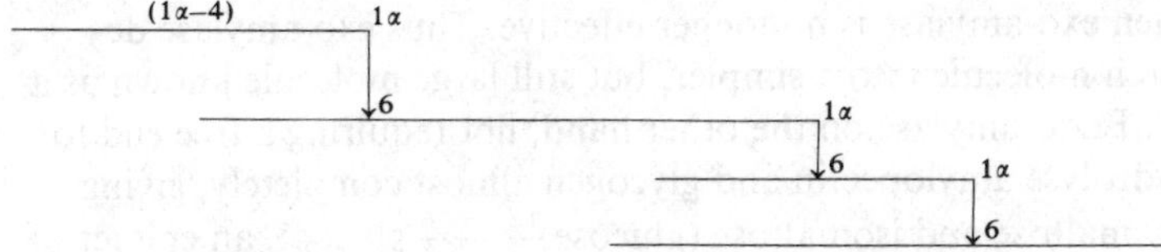

Glycogen (liver starch) is the principal reserve polysaccharide of animals, being deposited in the liver and muscles. Its structure is similar to that of amylopectin, but having shorter chains of approximately twelve (1α–4)-linked glucose residues with (1α–6) cross-links.

The storage of glucose by conversion into these insoluble, polymeric forms avoids the problems of high osmotic pressure which would arise from the storage of comparable quantities of highly soluble simple sugars.

Enzymic degradation of starch and cellulose

Cellulose, amylose, amylopectin, and glycogen are all polysaccharides constructed solely from D-glucose units, differing only in the position and stereochemistry of the glycosidic links. Comparison of the enzymic degradation of these compounds provides an indication of the remarkable specificity of enzymes.

Cellulose can be degraded by a group of enzymes known as 'cellulases', specific for (1β–4)-linked D-glucose polymers. Very few of the higher animals secrete cellulases in the digestive tract, and herbivores, for which cellulose is a major constituent of the diet, rely on symbiotic, cellulase-containing, microorganisms, to degrade cellulose in the food.

A group of enzymes known as 'amylases' are responsible for the hydrolysis of starch, and these are mostly specific for α-linked D-glucose polymers but have no action on cellulose. Several types of amylase are known, with varying substrate specificity. Exo- and endo-amylase specifically catalyse the hydrolysis of (1α–4)-linked D-glucose chains to maltose, and differ in the position of attack. Exo-amylase degrades the chain starting from a free end, whereas endo-amylase is able to attack in the middle of a chain. Either of these enzymes will degrade amylose extensively, but complete degradation requires the presence of yet another enzyme, 'Z-enzyme', which is known to be specific for certain types of β-glucoside, indicating the presence of a small number of β-links in the amylose molecule.

If exo-amylase acts on glycogen or amylopectin, the free ends of the (1α–4)-linked glucose chains are progressively degraded until a (1–6) link is encountered, when exo-amylase is no longer effective. Thus exo-amylase degrades the starch molecule into a simpler, but still large molecule known as a 'limit dextrin'. Endo-amylase on the other hand, not requiring a free end for attack, can hydrolyse amylopectin and glycogen almost completely, giving predominantly maltose and isomaltose (glucose $\stackrel{1\alpha \qquad 6}{\text{———}}$ glucose, an epimer of gentiobiose, p. 255). Thus complete enzymic degradation of starch into glucose requires exo-amylase and endo-amylase, which degrade the polymer into disaccharides, and Z-enzyme, maltase, and amylo-1,6-glucosidase, which acts on isomaltose, to convert the disaccharides into glucose.

Numerous other polysaccharides are known in nature, consisting of chains of condensed sugar units of varying types. Araban (poly-L-arabinose), found in association with pectin, and xylan (poly-D-xylose), found in woody plant tissue, are examples of polypentoses. Starch and cellulose are examples of polyhexoses, but many other types are known. Some micro-organisms produce dextrans ((1α–6)-linked poly-D-glucopyranose), and mannans (D-mannose chains) are found in the wood of some conifers. Galactans (poly-D-galactose) are also known. Inulin is a (2β–1)-linked poly-D-fructofuranoside found in the tubers of dahlias and other plants.

Besides polymers of simple carbohydrates, chains of modified sugars are known to occur widely. Pectins, which are constituents of the cell wall of plants, consist of chains of D-glucuronic acid partly in the form of its methyl ester; and chitin, a polysaccharide found in the shells of lobsters, crabs, and

D-Glucuronic acid

N-Acetylglucosamine

cockroaches, is derived from *N*-acetylglucosamine. The cell walls of Gram-positive micro-organisms contain large amounts of a muropeptide, the structure of which consists of chains of alternate (1β–4)-linked residues of *N*-acetylglucosamine and *N*-acetylmuramic acid cross-linked by peptide chains (p. 268).

CHO
CH_3 H—C—$NHCOCH_3$
H—C——O—C—H
CO_2H H—C—OH
H—C—OH
CH_2OH

CH_2OH
H O
H
O H HOH
HO
H $NHCOCH_3$
CH_3CH
CO_2H

N-Acetylmuramic acid (derived from *N*-acetylglucosamine and L-lactic acid).

Problems

1. Give the Fischer projections of L-sorbose and L-glucose.
2. Into which hexoses will D-galactose be transformed in weakly alkaline solution?
3. Draw the ring structures for all furanose and pyranose forms of D-glucose.
4. A non-reducing sugar, *A*, $C_{18}H_{32}O_{16}$, on acidic hydrolysis gives D-glucose and D-fructose in molecular proportions 2:1. Cautious hydrolysis of *A* gives D-glucose and $C_{12}H_{22}O_{11}$, a reducing sugar. Give three structures for *A* which would satisfy this data. What other information is required before the structure of *A* can be described unambiguously? How many different compounds $C_{18}H_{32}O_{16}$ are there which would fit the degradative data above?
5. Give mechanistic explanations for catalysis of the mutarotation of glucose (p. 242) by both acids and bases.
6. The following enzyme-catalysed reactions occur during the alcoholic fermentation of starch:

$CH_2OPO_3H_2$
C=O
HO—C—H
H—C—OH
H—C—OH
$CH_2OPO_3H_2$

D-Fructose-1,6-diphosphate

Aldolase

$CH_2OPO_3H_2$
C=O
CH_2OH

Dihydroxyacetone phosphate

Phosphotriose isomerase

CHO
H—C—OH
$CH_2OPO_3H_2$

D-Glyceraldehyde-3-phosphate

Rationalise both of these transformations mechanistically.

18

Amino acids and proteins

Amino acids

Numerous types of amino acid are known, but of these the biologically significant ones are predominantly primary α-amino acids along with a few secondary α-amino acids; β-, γ-, and δ-amino acids are of much less importance. The majority of naturally occurring amino acids are of the type $RCH(NH_2)CO_2H$, and attention will be confined principally to this class of compound.

Preparation of α-amino acids can be achieved either by the action of ammonia on the appropriate halogen-substituted acid, or by hydrolysis of the amino-nitrile, which can be derived from the lower aldehyde as shown in the scheme below.

$$RCHClCO_2H \xrightarrow{NH_3} RCH(NH_2)CO_2^- \overset{+}{N}H_4 \longrightarrow RCH(NH_2)CO_2H$$

$$RCHO \xrightarrow{HCN/CN^-} RCH(CN)OH \xrightarrow{NH_3} RCH(CN)NH_2 \xrightarrow{H^+/H_2O} RCH(NH_2)CO_2H$$

$$RCHO \xrightarrow{\overset{+}{N}H_4\ CN^-} RCH(CN)NH_2$$

(Displacement of the hydroxyl group of the cyanohydrin by ammonia is possible only on account of the adjacent powerfully electronegative cyano group.) Numerous other preparative routes are available, but need not be considered here. Where enantiomerism is possible, most syntheses give racemic products.

A list of the biologically important amino acids is given below.

Properties of amino acids. Amino acids are normally solids of high melting point, very soluble in water, and insoluble in organic solvents. The molecule $RCH(NH_2)CO_2H$ contains both an acidic and a basic group and in the solid

Some common amino acids

		Abbreviation
Glycine	$CH_2(NH_2)CO_2H$	(Gly)
*Alanine	$CH_3CH(NH_2)CO_2H$	(Ala)
*Valine	$(CH_3)_2CHCH(NH_2)CO_2H$	(Val)
Leucine	$(CH_3)_2CHCH_2CH(NH_2)CO_2H$	(Leu)
Isoleucine	$CH_3CH_2CH(CH_3)CH(NH_2)CO_2H$	(Ileu)
Aspartic Acid	$HO_2CCH_2CH(NH_2)CO_2H$	(Asp)
Glutamic Acid	$HO_2CCH_2CH_2CH(NH_2)CO_2H$	(Glu)
Asparagine	$H_2NCOCH_2CH(NH_2)CO_2H$	(Asp—NH_2)
*Glutamine	$H_2NCOCH_2CH_2CH(NH_2)CO_2H$	(Glu—NH_2)
Ornithine	$H_2N(CH_2)_3CH(NH_2)CO_2H$	(Orn)
Lysine	$H_2N(CH_2)_4CH(NH_2)CO_2H$	(Lys)
Arginine	$(HN{=})(H_2N)C{-}NH(CH_2)_3CH(NH_2)CO_2H$	(Arg)
Cysteine	$HSCH_2CH(NH_2)CO_2H$	(Cy—SH)
Cystine	$S{-}CH_2CH(NH_2)CO_2H$ | $S{-}CH_2CH(NH_2)CO_2H$	(Cy—S—S—Cy)
Methionine	$CH_3SCH_2CH_2CH(NH_2)CO_2H$	(Met)
*Serine	$HOCH_2CH(NH_2)CO_2H$	(Ser)
Threonine	$CH_3CH(OH)CH(NH_2)CO_2H$	(Thr)
Hydroxylysine	$H_2NCH_2CH(OH)CH_2CH_2CH(NH_2)CO_2H$	(Lys—OH)
*Phenylalanine	$C_6H_5CH_2CH(NH_2)CO_2H$	(Phe)
†Diaminopimelic acid	$HO_2CCH(NH_2)CH_2CH_2CH_2CH(NH_2)CO_2H$	

* Both D and L-forms occur naturally.
† DD, LL, and *meso*-forms all occur naturally in the muropeptide of bacterial cell walls.

state and neutral solution the species present is not $RCH(NH_2)CO_2H$ but the 'internal salt' $RCH(\overset{+}{N}H_3)CO_2^-$. This dipolar type of structure (known as a 'zwitterion', cf. urea, p. 151) is responsible for the low solubility in organic solvents and high melting point (a characteristic of ionic solids), though hydrogen bonding may also contribute to these properties to some extent. The presence of both basic and acidic groups means that in an aqueous solution of an amino acid, the species actually present is *p*H-dependent. At low *p*H the

*p*H	PREDOMINANT SPECIES	NET CHARGE
0	$RCH(\overset{+}{N}H_3)CO_2H$	+1
7	$RCH(\overset{+}{N}H_3)CO_2^-$	0
14	$RCH(NH_2)CO_2^-$	−1

carboxylic acid group will be undissociated and the amino group protonated. At high *p*H the amino acid will be in the form of an aminocarboxylate anion. At intermediate *p*H the zwitterion will be the principal species in solution,

Tyrosine	$HO-C_6H_4-CH_2CH(NH_2)CO_2H$	(Tyr)
Proline	H_2C-CH_2 / $H_2C\quad CHCO_2H$ / N / H	(Pro)
Hydroxyproline	$HOHC-CH_2$ / $H_2C\quad CHCO_2H$ / N / H	(Hypro)
Histidine	$N-C-CH_2CH(NH_2)CO_2H$ / $HC\quad CH$ / N / H	(His)
Tryptophan	$C-CH_2CH(NH_2)CO_2H$ / CH / N / H	(Try)
Creatine	$HN\quad CH_3$ / $C-N$ / $H_2N\quad CH_2CO_2H$	

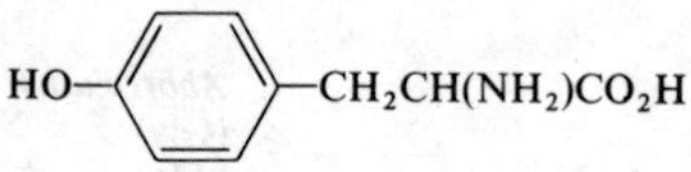

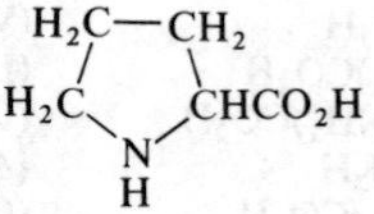

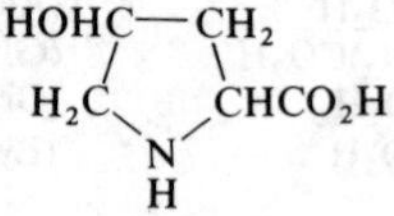

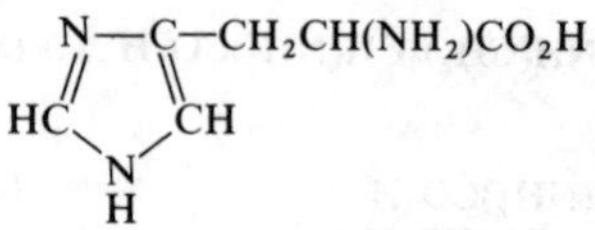

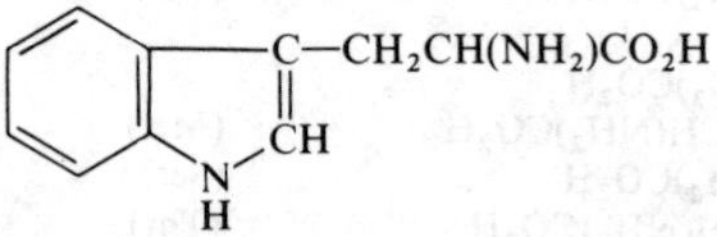

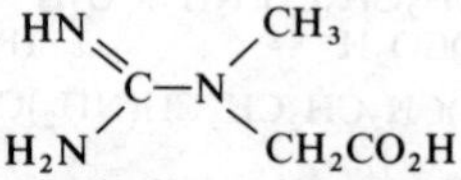

and the species $RCH(NH_2)CO_2H$ is never present to any significant extent. The *p*H of a solution of an amino acid at which the average charge per molecule is zero is known as the 'isoelectric point'.* It does not necessarily follow that at the isoelectric point all the molecules are in the zwitterion form as an equilibrium with equal concentrations of protonated and deprotonated species will exist but the zwitterion is the predominant solute.

Amino acids of the structure $RCH(NH_2)CO_2H$ can exist as enantiomers:

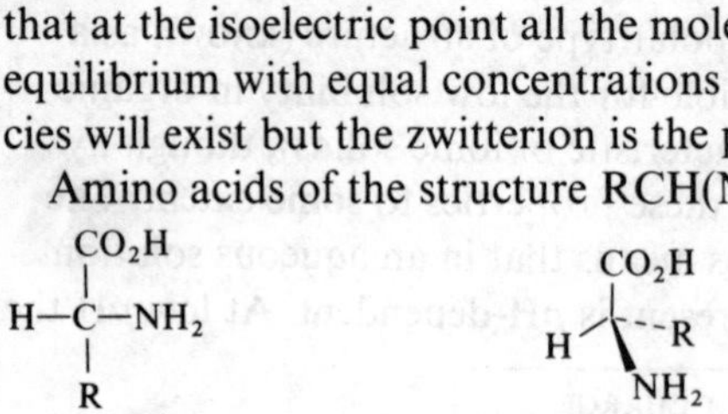

CO_2H / $H-C-NH_2$ / R $\qquad$ CO_2H / H R / NH_2 $\qquad$ D

CO_2H / H_2N-C-H / R $\qquad$ CO_2H / H_2N R / H $\qquad$ L

* A monoamino monocarboxylic acid has two pK_a values corresponding to dissociation of the $-\overset{+}{N}H_3$ and CO_2H groups in the cation $RCH(\overset{+}{N}H_3)CO_2H$. For these simple acids, the isoelectric point is given by $pH_{\text{isoelectric}} = \dfrac{pK_1 + pK_2}{2}$

The Fischer projections of the two possible structures are shown above, and the amino acid which, when drawn with the carboxylic acid group at the top and the R group at the bottom of the carbon chain, has the amino group on the right-hand side is known as D. This convention is the same as that applying to α-hydroxy acids (p. 212) but different from that applying to carbohydrates, which can lead to confusion (p. 184). For the purposes of designating chirality by *R* and *S* (p. 183), although the priority sequence of the groups may, in theory, be altered by sufficiently wide changes in the structure of the side chain, in practice the side chains of most biologically important amino acids have priorities with respect to NH_2 and CO_2H which makes D ≡ *R* and L ≡ *S* (cysteine and cystine are exceptions).

The great majority of naturally occurring chiral α-amino acids are L. Some D-amino acids occur naturally in fungal peptides with antibiotic activity, and also in the muropeptide of the cell wall of Gram-positive bacteria. An enzyme which specifically catalyses the oxidation of D-amino acids occurs in the liver of higher animals.

Reactions of amino acids. The primary amine function of α-amino acids of the type $RCH(\overset{+}{N}H_3)CO_2^-$ has all the reactions characteristic of such a group (p. 84). Reaction of amino acids with mineral acids forms salts, although in fact it is the carboxylate anion of the zwitterion which has been protonated. Treatment with nitrous acid gives the corresponding hydroxy-compound. Amino acids can be acylated and alkylated, but for these reactions to occur, the amino acid must be in a form with a free $—NH_2$ group, i.e. reaction must be conducted under alkaline conditions as the $—\overset{+}{N}H_3$ group is not nucleophilic.

$$RCH(\overset{+}{N}H_3)CO_2^- \xrightarrow{HCl} RCH(\overset{+}{N}H_3)CO_2H\ Cl^-$$

$$RCH(\overset{+}{N}H_3)CO_2^- \xrightarrow[Na_2CO_3]{(CH_3CO)_2O} RCH(NHCOCH_3)CO_2^-\ Na^+$$

$$RCH(\overset{+}{N}H_3)CO_2^- \xrightarrow{HNO_2} RCH(OH)CO_2H$$

$$RCH(\overset{+}{N}H_3)CO_2^- \xrightarrow[K_2CO_3]{CH_3I} RCH(^+NH_2CH_3)CO_2^- \xrightarrow[K_2CO_3]{CH_3I} RCH(^+NH(CH_3)_2)CO_2^- \xrightarrow[K_2CO_3]{CH_3I} RCH(^+N(CH_3)_3)CO_2^- \quad \text{A betaine}$$

Alkylation ultimately gives zwitterionic compounds containing quaternary ammonium cations and carboxylate anions in the same molecule, which are known as 'betaines' the name betaine also being applied specifically to the simplest member of the group, $(CH_3)_3\overset{+}{N}CH_2CO_2^-$.

Primary α-amino acids react readily with methanal (formaldehyde) to give the corresponding methylene-imine quantitatively. This reaction is of significance in the estimation of amino acids, since although amino acids cannot be titrated against base or acid, being salt-like, the methylene-imino group is only very feebly basic, so that the products of reaction with methanal have the acidity of normal carboxylic acids, and can be titrated with alkali using normal indicators.

$$\underset{\substack{|\\ {}^{+}NH_3}}{RCHCO_2^-} \xrightarrow{HCHO} \underset{\substack{|\\ NH\\ |\\ CH_2OH}}{RCHCO_2H} \xrightarrow{-H_2O} \underset{\substack{|\\ N\\ \|\\ CH_2}}{RCHCO_2H}$$

Enzymic oxidation of α-amino acids leads to α-keto acids (p. 88).

The carboxyl group of α-amino acids shows some of the normal reactions, forming a series of salts, e.g. $RCH(NH_2)CO_2^-K^+$. Chelate complexes are formed with a number of metal ions, the cupric complexes being of use in the isolation of amino acids.

$$\underset{\substack{|\\ {}^{+}NH_3}}{RCHCO_2^-} + Cu^{+2} \longrightarrow \left[\begin{array}{ccc} & O & \\ & \| & \\ & C & \\ RCH & & O^- \\ | & & | \\ H_2\overset{+}{N} & \text{——} & Cu \end{array}\right]^+$$

In some reactions, the conditions must be adjusted to ensure that the carboxylate group in the amino acid is protonated, when reactions specifically require the presence of a $—CO_2H$ group. Esterification is possible in the presence of enough mineral acid to ensure that the carboxylate anion is completely protonated; and the salts obtained after esterification may be converted into the free amino esters by treatment with weak bases. The amino

$$\underset{\substack{|\\ {}^{+}NH_3}}{RCHCO_2^-} \xrightarrow[HCl]{CH_3OH} \underset{\substack{|\\ {}^{+}NH_3\ \ Cl^-}}{RCHCO_2CH_3} \xrightarrow{NaHCO_3} \underset{\substack{|\\ NH_2}}{RCHCO_2CH_3}$$

esters are unstable and rapidly condense to cyclic diamides known as diketopiperazines (cf. lactides, p. 213). Because of the ease with which the activated carbonyl group would react with the amino function, no simple anhydrides or acyl chlorides can be formed.

$$RCH(CO_2CH_3)(NH_2) + H_2N\text{–}CHR\text{–}CO_2CH_3 \longrightarrow \text{RCH–C(=O)–NH–CHR–C(=O)–NH (ring)} + 2CH_3OH$$

A diketopiperazine

Decarboxylation of amino acids can be achieved by heating the alkali metal salts with soda lime (p. 222) giving the corresponding primary amines in very poor yield. Enzymic decarboxylation occurs during putrefaction of meat when degradation of protein and decarboxylation of the amino acids ornithine and lysine give rise to the evil smelling 1,4-diaminobutane and 1,5-diaminopentane known as putrescine and cadaverine respectively.

Primary α-amino acids react with ninhydrin* to give an intense purple colour. The reaction occurs in two stages, the amino acid being initially oxidised to the lower aldehyde or ketone with liberation of ammonia and carbon dioxide. The ammonia then reacts with the compound produced by reduction of ninhydrin and unchanged ninhydrin to give the purple compound. The quantitative formation of this substance coupled with the intensity of its colour makes this a valuable reaction, which has been extensively used for the qualitative detection of amino acids (e.g. as a chromatographic spray) and for quantitative estimation by spectrophotometric means. The intensity

$$\text{Ninhydrin} + RCH(\overset{+}{N}H_3)CO_2^- \longrightarrow \text{(2-hydroxy-2-H-indane-1,3-dione)} + RCHO + CO_2 + NH_3$$

Ninhydrin

$$\text{Ninhydrin} + NH_3 + \text{(2-hydroxy-2-H-indane-1,3-dione)} \longrightarrow \text{Ruhemann's purple}$$

Ruhemann's purple

* Ninhydrin is also known as triketohydrindene hydrate and indanetrione hydrate. It is the *gem*-diol formed by hydration of a triketone.

of the colour produced makes the method correspondingly sensitive. It should be noted that only primary α-amino acids can give this test with ninhydrin. β-, γ-, and δ-amino acids are not oxidised in the first step, and secondary and tertiary α-amino acids are oxidised but do not liberate ammonia. Secondary α-amino acids such as proline and hydroxyproline give much fainter yellow colours with ninhydrin and cannot be estimated in this way.

Peptides and proteins

Peptides and proteins are groups of compounds of similar structure, differing only in the molecular size. Both are polyamides derived from α-amino acids having the general structure:

$$H_2N{-}\underset{\displaystyle R'}{\overset{}{CH}}{-}CO\left(NH{-}\underset{\displaystyle R}{CH}{-}CO\right)_n NH{-}\underset{\displaystyle R''}{CH}{-}CO_2H$$

Where the polyamide is constructed from relatively few amino acids, the term 'peptide' is employed, with a prefix indicative of the number of amino acid residues (e.g. dipeptide, tripeptide, octapeptide). The term 'protein' is usually employed where the number of amino acids is very large, and proteins are known which have relative molecular masses up to 10^7 or 10^8, though there is no clearly defined lower limit for the term.

Proteins are of great biological importance, being a major constituent of the soft structural tissue of animals. Proteins, known as enzymes, act as the catalysts for cell reactions, and numerous polypeptide hormones are known. Metabolic activity in the cell is controlled by nucleoproteins, and soluble proteins in the blood are concerned with oxygen transport (haemoglobin) and immune response amongst many other functions.

The only chemical reaction which will be considered here is hydrolysis, which can be achieved by enzymic or 'chemical' means. Hot dilute mineral acid will slowly cleave the amide links to give random degradation, ultimately resulting in the formation of simple amino acids. Controlled acidic hydrolysis will degrade a protein into a mixture of peptides. Enzymic hydrolysis is also possible, and the proteolytic enzymes vary greatly in their specificity. Some, like papain or ficin, are virtually non-specific and degrade proteins to free amino acids, whilst others like trypsin, chymotrypsin and pepsin will hydrolyse only particular links in protein molecules (cf. maltase, emulsin, etc. pp. 254, 259). Thus pepsin will cleave the amide link between the carbonyl group of a dicarboxylic L-amino acid and the amino group of an aromatic L-amino acid, provided that the second carboxylic acid group of the former is uncombined. Chymotrypsin is less specific and cleaves the amide link on the carbonyl side of aromatic L-amino acids. Trypsin is specific for amide links involving the carboxyl group of lysine or arginine. In all these

Pepsin cleaves here | Chymotrypsin cleaves here

$$-NH-CH(CH_2)_nCO_2H-CO-\,|\,-NH-CH(CH_2C_6H_4X)-CO-\,|\,-NH-CH(R)-CO-$$

$n = 1$ (Aspartic acid) or 2 (Glutamic acid) | X = H (Phenylalanine) or OH (Tyrosine)

cases the enzyme not only requires the presence of a particular amino acid, but will hydrolyse the link on only one side of the acid, not on both. The specificity of these enzyme-catalysed reactions undoubtedly involves interaction of the side-chain functionality of the polypeptide with receptor sites on the proteolytic enzyme. Other enzymes with differing specificity are known, carboxypeptidases degrade peptides or proteins with a free terminal carboxylic acid group by stepwise removal of single amino acids, whilst aminopeptidases effect a similar stepwise degradation from the end with a free amino group. Neither of these groups of enzymes is effective in the hydrolysis of peptides or proteins with cyclic structures.

The investigation of the structure of proteins utilises these and other methods of degradation, and a number of techniques have been devised to identify the terminal amino acids. One, widely employed to identify the

$$H_2N-CH(R)-CO-NH-CH(R')-CO-NH-CH(R'')-CO_2H$$

$$\downarrow \; F-C_6H_3(NO_2)_2$$

$$O_2N(NO_2)C_6H_3-NH-CH(R)-CO-NH-CH(R')-CO-NH-CH(R'')-CO_2H$$

$$\downarrow \; H^+/H_2O$$

$$O_2N(NO_2)C_6H_3-NHCH(R)CO_2H + R'CH(NH_2)CO_2H + R''CH(NH_2)CO_2H$$

amino-terminal amino acid of a chain, is to react the polypeptide with 2,4--dinitrofluorobenzene, which converts the free amino group into its 2,4-dinitrophenyl derivative (p. 57). Subsequent hydrolysis of the polypeptide gives normal amino acids, with the exception of the terminal *N*-aryl amino acid, which can be separated and identified chromatographically.

The primary structures of two biologically important peptides are given below. Glutathione is a tripeptide, required as a cofactor in some enzymic oxidations. It is unusual in having the glutamic acid residue linked through

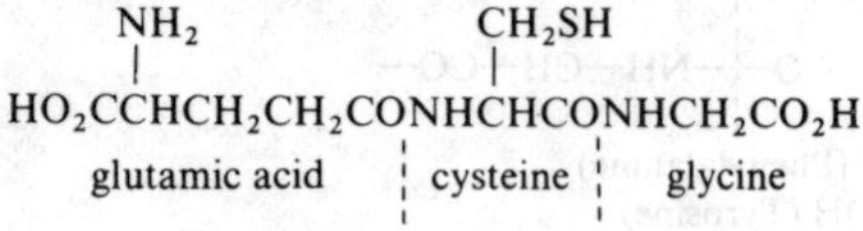

```
      NH2                 CH2SH
      |                   |
HO2CCHCH2CH2CONHCHCONHCH2CO2H
  glutamic acid   | cysteine |  glycine
```

Glutathione

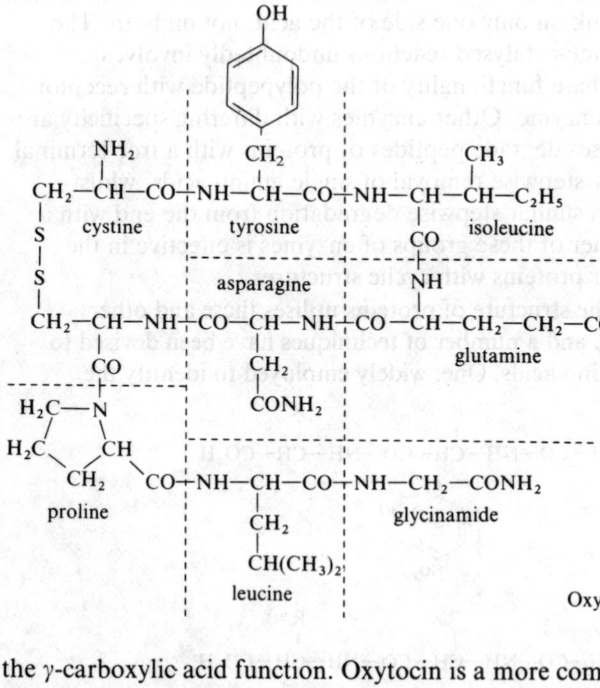

```
                              OH
                              |
                           (C6H4)
           NH2                |                CH3
           |                  CH2              |
   CH2—CH—CO—|—NH—CH—CO—|—NH—CH—CH—C2H5
   |   cystine       tyrosine     |    isoleucine
   S                              CO
   |  - - - - - - - - - - - - - - + - - - - - - - - -
   S            asparagine        NH
   |                              |
   CH2—CH—NH—|—CO—CH—NH—|—CO—CH—CH2—CH2—CONH2
        |             |                glutamine
        CO            CH2
 - - - -+- - - -      |
   H2C—N              CONH2
   /     \
 H2C      CH    - - - - - - - - - - - - - - - - - - -
   \    /   \
    CH2      CO—|—NH—CH—CO—|—NH—CH2—CONH2
   proline           |          glycinamide
                     CH2
                     |
                     CH(CH3)2
                     leucine
```

Oxytocin

the γ-carboxylic acid function. Oxytocin is a more complex octapeptide hormone secreted by the posterior lobe of the pituitary gland, which stimulates contraction of the uterus and initiates lactation at the end of pregnancy. A very similar octapeptide hormone vasopressin, also secreted by the posterior pituitary, is the antidiuretic hormone, and differs from oxytocin in the replacement of L-isoleucine by L-phenylalanine, and L-leucine either by L-lysine (in hog vasopressin) or L-arginine (in beef vasopressin). The primary structure of beef insulin, which contains fifty-one amino acid residues, is shown below. The amino-terminal end of the peptide chain is shown by the symbol H· (e.g. H·Gly· below means H_2NCH_2CO—) and the carboxylic acid end by ·OH (e.g. ·Ala·OH means —$NHCH(CH_3)CO_2H$).

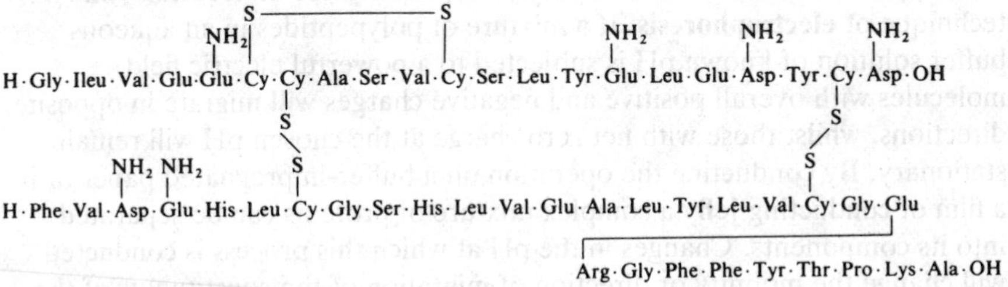

H·Gly·Ileu·Val·Glu·Glu·Cy·Cy·Ala·Ser·Val·Cy·Ser·Leu·Tyr·Glu·Leu·Glu·Asp·Tyr·Cy·Asp·OH

H·Phe·Val·Asp·Glu·His·Leu·Cy·Gly·Ser·His·Leu·Val·Glu·Ala·Leu·Tyr·Leu·Val·Cy·Gly·Glu

Arg·Gly·Phe·Phe·Tyr·Thr·Pro·Lys·Ala·OH

Beef insulin (for key to abbreviations see p. 263)

The structure and physical properties of proteins

Much chemical and biochemical interest is focused upon the relationship between the structure and function of proteins. Peptides and proteins may have both basic (—NH_2, CO_2^-) and acidic (—$\overset{+}{N}H_3$, CO_2H) groups in the molecule either as the terminal groups of the polyamide chain or due to inclusion of polyfunctional amino acids such as lysine or glutamic acid. The net charge on the molecule will vary with the *p*H of its environment, just as with the simple amino acids (p. 263). Thus under conditions of low *p*H the protein will be positively charged, whilst at high *p*H it will be negatively charged. At some intermediate *p*H the net or average charge on the molecule will be zero and this *p*H is referred to as the **isoelectric point** (cf. p. 264).

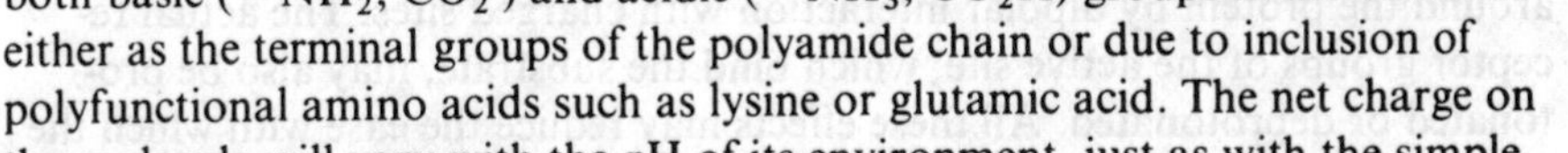

	*p*H 0	Isoelectric point	*p*H 14
Top groups	$\overset{+}{N}H_3$ $\overset{+}{N}H_3$ CO_2H	$\overset{+}{N}H_3$ NH_2 CO_2^-	NH_2 NH_2 CO_2^-
Protein	$H_3\overset{+}{N}$— Protein —CO_2H	$H_3\overset{+}{N}$— Protein —CO_2^-	H_2N— Protein —CO_2^-
Bottom groups	CO_2H $\overset{+}{N}H_3$	CO_2^- $\overset{+}{N}H_3$	CO_2^- NH_2
Net charge	+4	0	−3

It is not very surprising that the properties of protein solutions show marked changes on passing through the isoelectric point, since the solvation of molecules or interaction of adjacent protein molecules is clearly likely to be affected by the charge distribution and total charge on the molecule. The viscosity of gelatin solution passes through a minimum at *p*H 4·7 (the isoelectric point) and the solubilities of insulin and casein are lowest at their isoelectric points (5·3 and 4·7 respectively).

The variation of the total charge on a polypeptide or protein with change in the pH of the medium can be used to separate these molecules by the technique of **electrophoresis**. If a mixture of polypeptides in an aqueous buffer solution of known pH is subjected to a powerful electric field, molecules with overall positive and negative charges will migrate in opposite directions, whilst those with net zero charge at the chosen pH will remain stationary. By conducting the operation on a buffer-impregnated paper or in a film of conducting jelly a complex mixture of proteins can be separated into its components. Changes in the pH at which this process is conducted will change the mobility or direction of migration of the constituents of the mixture.

The activity of many enzymes is known to be *p*H-dependent, activity being at a maximum at a particular *p*H. It is thought that enzymes adsorb their substrates at a special region of the molecule known as the 'active site'. Changes in *p*H will change the distribution of charges on the molecule, which in its turn will affect hydration, either by varying the number of hydrogen-bonding groups or by changing the extent to which water molecules cluster around the protein by dipolar interaction with charged sites. The actual receptor groups of the active site, which bind the substrate, may also be protonated or deprotonated. All these effects may reduce the ease with which the enzyme adsorbs its particular substrate and so decrease the catalytic activity.

Quite apart from the effect of environment upon the biological activity of proteins, it is known that their structure is intimately related to their function. It is customary to divide the structural features of proteins into a number of categories. The **primary structure** of a protein is the sequence of amino acid residues, as determined by chemical analysis, and this chain may be coiled or arranged in a particular fashion as a result of hydrogen bonding between amide groups. Those features of the protein structure which result from such interaction between peptide (amide) links are classed as **secondary structure**. Further folding of the secondary structure may result from the interaction of functional groups on the side chains of the amino acids (e.g. —SH, NH_2, CO_2H, —OH, etc.) and this constitutes the **tertiary structure**. Finally, proteins may have a **quaternary structure** resulting from the interaction of several protein molecules to give groups or clusters which may possess a high degree of symmetry, sometimes directly observable by the electron microscope.

Numerous polypeptides and proteins have been studied by X-ray crystallography, and some features of their structure have been recognised. Two types of regular secondary structure are found to occur frequently, although numerous more disorderly arrangements are also known. In the 'α' form, the polyamide chain is coiled into a helix, in which adjacent turns of the coil (approximately 3·6 amino acid residues per turn) are joined by hydrogen bonds between neighbouring amide groups. The helix is dextral, which ensures that the bulky side-chains of the L-amino acids project away from the spiral polyamide chain. In the 'β' structure the polyamide chains are arranged adjacent in an anti-parallel fashion to form a sheet of polypeptide chains cross linked by hydrogen bonds. In this array the side chains of the amino acids lie alternately above and below the plane of the sheet. Sections of β structure can be formed within a single protein molecule by pleated folding of the polypeptide chain. Proteins that have extensive β structure, such as silk fibroin, are not readily extensible since the polypeptide chains are already fully extended, but those with the α structure predominating (e.g. hair, wool) are elastic, since mechanical stress can be relieved by conversion of the helical polypeptide chain into the extended conformation of the β structure.

The tertiary structure of proteins, involving interaction of the side chains of amino acids, does not lead to regularity such as described above. Besides hydrogen bonding, the formation of disulphide links is an important factor in the stabilisation of tertiary structure. Insulin has three such disulphide bridges, two of which link the two separate polypeptide chains in the molecule. The tertiary structure often holds the protein molecule in a conformation in which the hydrophilic groups (such as OH, NH_2, CO_2H) are exposed on the surface and the hydrophobic groups (alkyl and aryl side chains) are enclosed within the centre of the molecule.

Quaternary structure of proteins varies widely. Some electron micrographs reveal clearly the aggregation of protein molecules, whose fine structure cannot be resolved. One common form of quaternary structure found in fibrous proteins (wool, hair), consists of six protein chains, each individually in the form of an α helix, being coiled around a central helical protein molecule to give a rope-like structure.

The biological activity of proteins is often intimately connected with the higher orders of structure, and living organisms are able to synthesise proteins in the desired conformations, which frequently are metastable (i.e. not the most stable structure possible). Under the influence of heat, extremes of pH, or many chemical reagents, proteins will often lose their biologically desirable conformation, being converted into random and disoriented structures with consequent loss of biological activity. This process, known as denaturation, is most familiar in the change of texture of egg white on heating, and is also responsible for the change in texture of meat during cooking. In

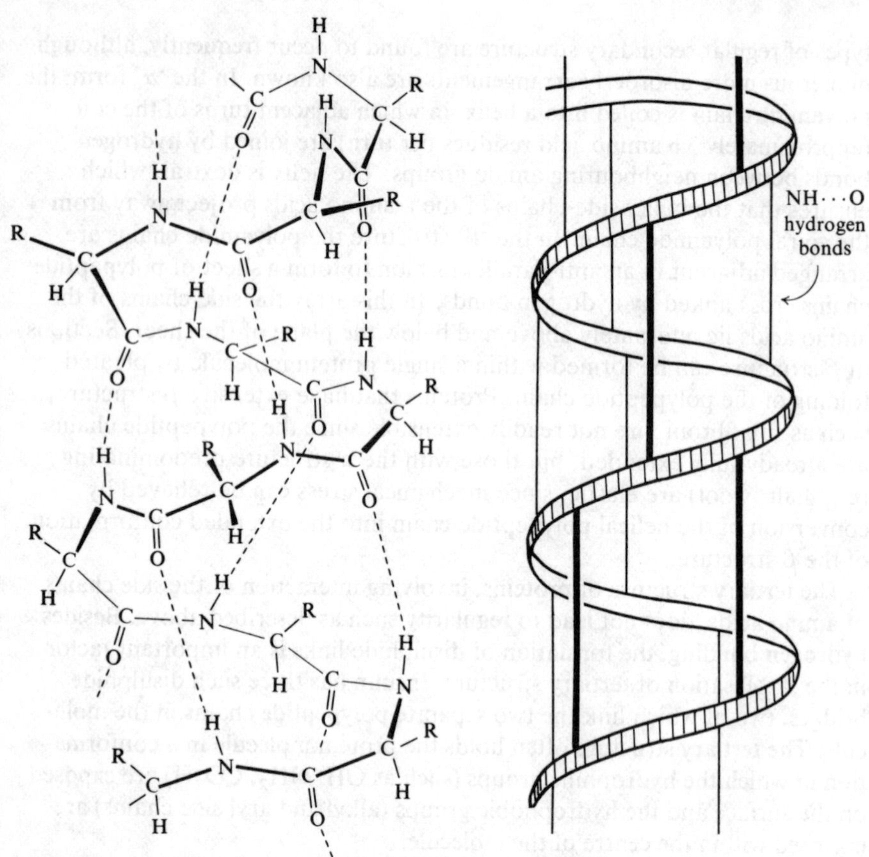

A protein chain in the 'α' conformation.

A protein in the 'β' conformation

the latter case, cooking causes a marked increase in the digestibility of the meat, since denaturation exposes linkages in the protein, which in the raw state are not so readily accessible to the proteolytic enzymes of the digestive tract. In these denaturations, unfolding of the protein chains exposes the hydrophobic groups normally enclosed within the central part of the molecular shape, and interaction between the exposed hydrophobic sections of neighbouring molecules contributes to the coagulation of denatured protein.

Although of no biological significance, the processes of hair-waving exemplify interference with secondary and tertiary structure in different ways. Water-waving utilises the ability of water to penetrate the protein tissue, which is softened by the breaking of hydrogen bonds between amide groups in the protein and formation of new hydrogen bonds to water molecules. On drying, new hydrogen bonds are reformed within the protein, which preserves the shape resulting from external constraint. In permanent waving a similar result is achieved by the reduction of disulphide bridges to thiol groups, followed by reoxidation (by the setting lotion) to form a new set of disulphide links.

Problems

1. How would glutathione and oxytocin be affected by pepsin, chymotrypsin, and aminopeptidase?
2. A compound $C_{12}H_{17}N_3O_3$ on acidic hydrolysis gives L-tyrosine, L-alanine, and ammonia. Give all possible structures for this compound. How might you attempt to distinguish between these structures?

19

Aromatic compounds, nucleic acids, and nucleotide coenzymes

Although all of the aromatic compounds previously described have been derivatives of benzene, benzene is only one of a large number of compounds which exhibit certain characteristic chemical and physical properties. The common feature of these compounds is that they contain a planar, cyclic, conjugated π bond system of $(4n + 2)$ electrons. The detailed chemistry of these compounds will not be covered but representatives of the various groups will be described.

The carbocyclic (i.e. a ring of carbon atoms) cyclopropenium cation is an example of an aromatic 2π electron system. The next group, in which examples are more plentiful, have 6π electrons and includes benzene, cyclopentadienide anion, and tropylium cation. The formation of an aromatic system is energetically very favourable, and as a result the hydrocarbon cyclopentadiene is sufficiently acidic to evolve hydrogen with alkali metals forming the stable anion. In fact, cyclopentadiene ($pK_a \sim 16$) is almost as acidic as water ($pK_a = 15.75$). (In the same circumstances, alkanes and cycloalkanes are inert.) Salts of the tropylium cation, e.g. the perchlorate, are known and are fully

Carbocyclic aromatic systems

No. of π electrons $(4n + 2)$

2 $(n = 0)$

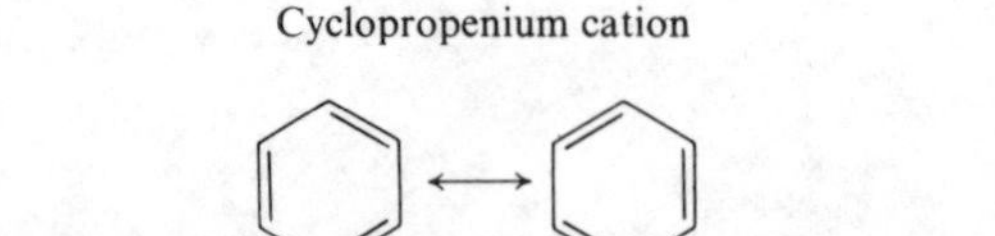

HC=CH, C(+)H ⟷ HC—CH(+), =CH ⟷ HC(+)—CH, =CH

Cyclopropenium cation

6 $(n = 1)$

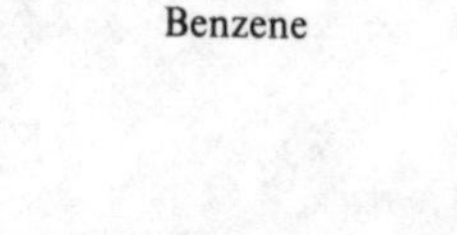

Benzene ⟷ Benzene

Benzene

6 ($n = 1$)

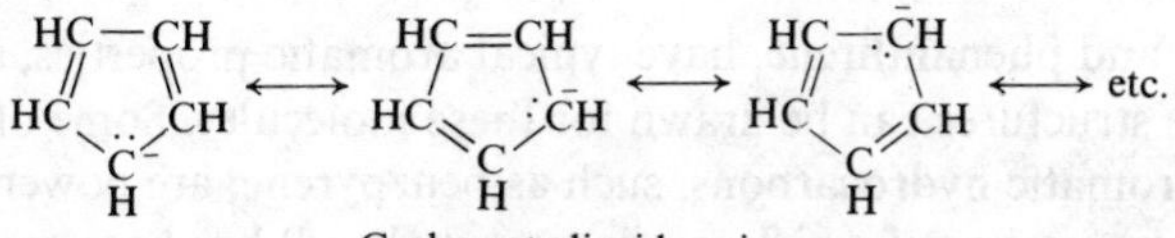

Cyclopentadienide anion

6 ($n = 1$)

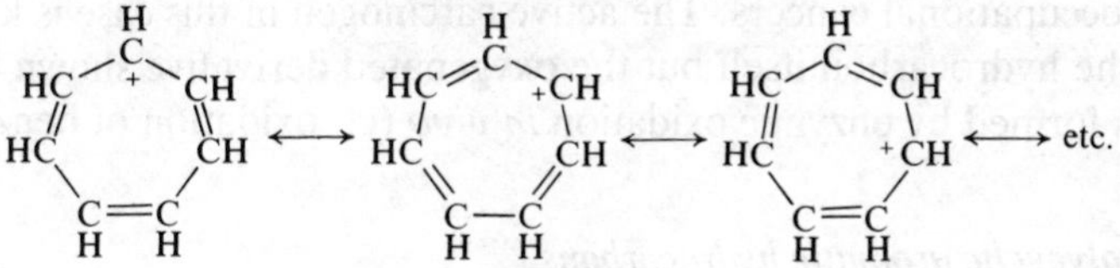

Tropylium cation

18 ($n = 4$)

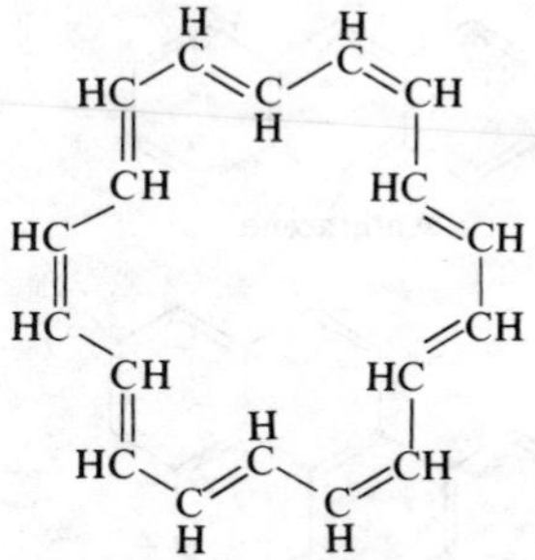

[18] Annulene

ionic. Monocyclic systems with larger numbers of π electrons are also known, e.g. [18] annulene.

Cyclic systems with ($4n$) π electrons are not aromatic. Cyclobutadiene is a very reactive alkene and cyclo-octatetraene is non-planar, having the folded structure illustrated, and shows alkene-like reactions.

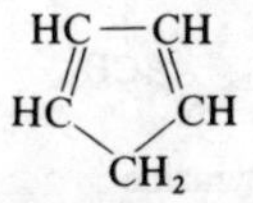

Cyclopentadiene

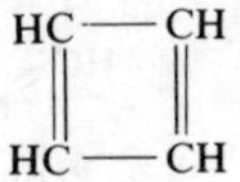

Cyclobutadiene

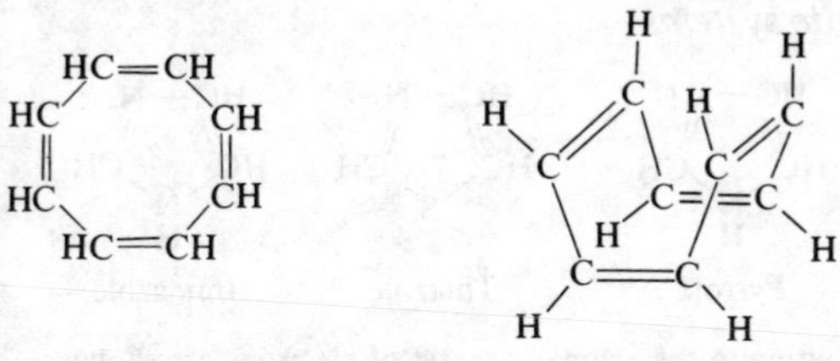

Cyclo-octatetraene

A large number of aromatic compounds are derived from systems of 'fused' benzene rings. The simpler members of this series, naphthalene, anthracene, and phenanthrene, have typical aromatic properties, and several Kekulé-type structures can be drawn for these molecules. Some of the higher polycyclic aromatic hydrocarbons, such as benzpyrene, are powerful carcinogens, the presence of which in oils (e.g. shale oil) has been responsible for certain occupational cancers. The active carcinogen in this case is known to be not the hydrocarbon itself but the oxygenated derivative shown below, which is formed by enzymic oxidation *in vivo* (cf. oxidation of benzene, p. 39).

Fused polycyclic aromatic hydrocarbons

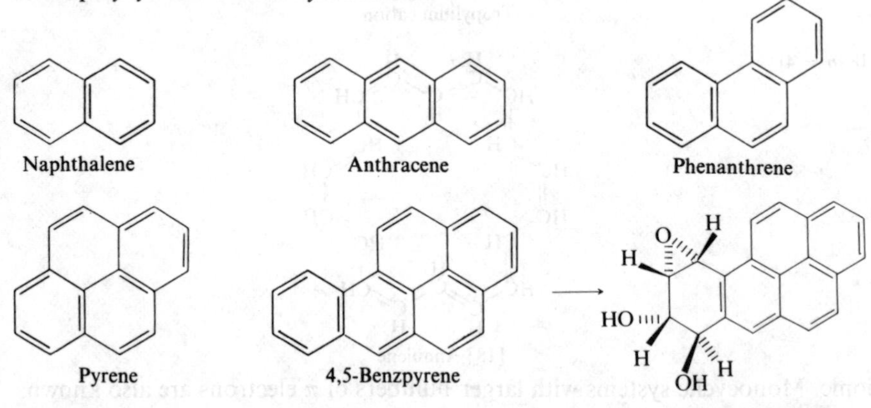

Benzenoid heterocyclic aromatic systems

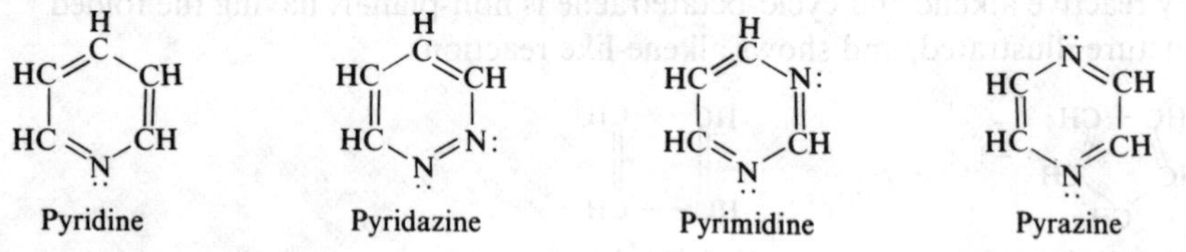

Non-benzenoid heterocyclic aromatic systems

Furan Thiophen Pyrrole Thiazole Iminazole

In the diagrams above, lone pairs contributing to the aromatic sextet of electrons are shown inside the ring, lone pairs not involved in the aromatic sextet are shown outside.

Heterocyclic systems. Aromaticity is not restricted to carbocyclic compounds, and replacement of some of the carbon atoms in the compounds mentioned above by other atoms gives new aromatic systems, provided that the π electron system is unchanged. Replacement of CH groups in benzene by the isoelectronic (i.e. containing the same number of electrons) nitrogen atom results in a series of heterocyclic aromatic compounds pyridine, pyridazine, pyrimidine, and pyrazine, and even further replacement is possible. In all these compounds the cyclic 6π electron system—the 'aromatic sextet'—utilises one of the electrons from each carbon and nitrogen atom leaving a lone pair in an sp^2 orbital on each nitrogen, where in benzene there would be a C—H bond. As a result these heterocyclic compounds are feebly basic, the basicity of a lone pair in an sp^2 orbital being markedly less than that of a lone pair in an sp^3 orbital (cf. the acidity of C—H in alkanes and alkynes, p. 135). The cyclopentadienide anion can also be regarded as the carbocyclic parent of a series of heterocyclic aromatic compounds. Furan and thiophen have an aromatic sextet in which one electron comes from each of the four carbon atoms (i.e. the two double bonds) and two are contributed from a lone pair on oxygen or sulphur. In pyrrole, the lone pair on the nitrogen atom is used in this way, so that pyrrole is non-basic, having no lone pair on nitrogen available for combination with a hydrogen ion (cf. pyridine where the lone pair is not involved in the aromatic sextet). Further replacement of the carbon atoms of these rings leads to more complex heteroaromatic species such as iminazole (occurring in the amino acid histidine, p. 264) and thiazole. Note that in thiazole the lone pair of the nitrogen atom is in an sp^2 orbital corresponding to a C—H bond in thiophen and is therefore not involved in the aromatic sextet. Thiazole is therefore basic. In iminazole the two nitrogen atoms have quite different utilisation of their electrons, one is basic as in pyridine, and the other is non-basic, as in pyrrole, since its lone pair is part of the aromatic sextet.

As with benzenoid (benzene-like) aromatic systems, so with non-benzenoid systems, fused polycyclic aromatic compounds are possible. Quinoline and isoquinoline are derived by fusion of benzene and pyridine rings. Indole, the parent from which the amino-acid tryptophane (p. 264) is derived, consists of benzene and pyrrole rings fused together, and pteridine, purine, and alloxazine are more complex examples of heteroaromatic systems obtained in this way. Alloxazine may be considered as a heterocyclic analogue of anthracene.

Fused heterocyclic aromatic systems

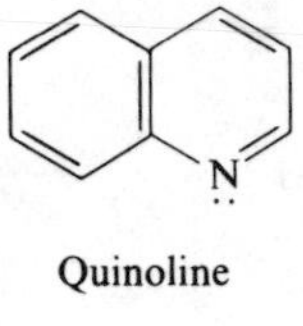

Quinoline

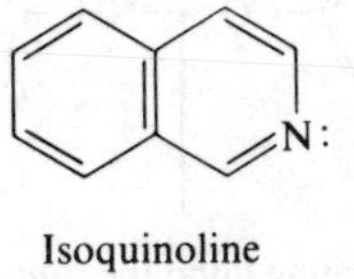

Isoquinoline

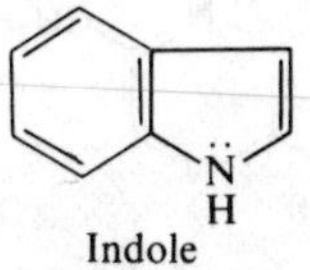

Indole

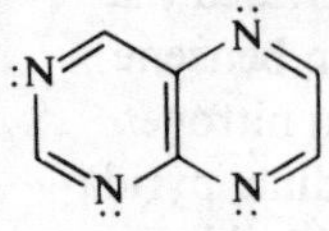

Pteridine 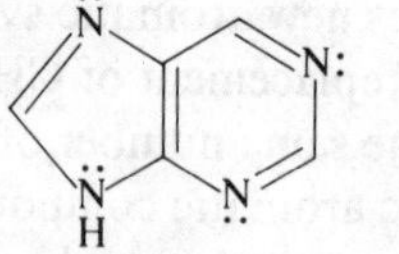 Purine 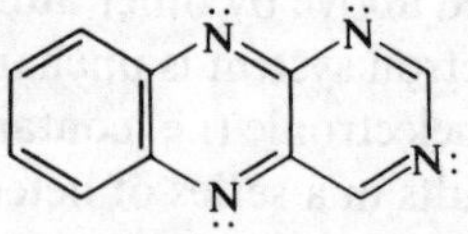 Alloxazine

Although the parent heterocyclic compounds do not occur naturally, their derivatives are widespread in nature and of considerable importance. Nicotinamide, the amide of nicotinic acid, and pyridoxal (Vitamin B_6) are both derived from pyridine, and both are vitamins of the B group. Nicotinamide is

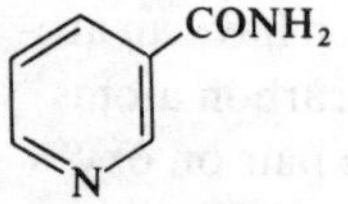

Nicotinamide 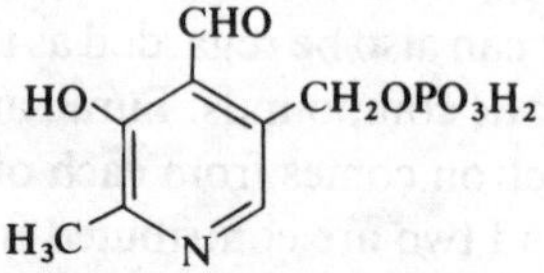 Pyridoxal phosphate

an important part of the coenzymes NAD and NADP (p. 299) whilst pyridoxal phosphate is a cofactor required for decarboxylation and transamination of amino acids. Pyrimidine bases are significant as part of the nucleic acid structure (p. 288) and are also found in a number of coenzymes. Vitamin B_1 (thiamine) is a derivative of pyrimidine and thiazole which is required, as

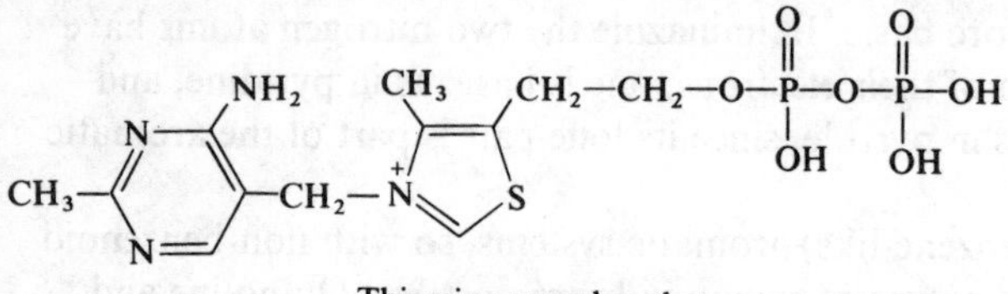

Thiamine pyrophosphate

its pyrophosphate, as a cofactor in the enzymic decarboxylation of α-keto-acids. Simple pyrazine and pyridazine derivatives are of no biological significance, but the pteridine system is found in folic acid (Vitamin B_{10}).

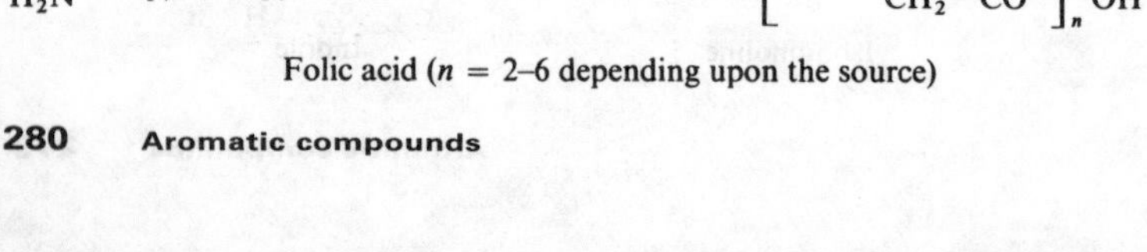

Folic acid (n = 2–6 depending upon the source)

The tetrahydro-derivative of folic acid (see below) is required as a cofactor in biological syntheses in which single carbon units undergo oxidation or reduction and transfer onto substrate molecules as —CH_3, —CH_2OH, or —CHO groups. Riboflavin (Vitamin B_2) is derived from alloxazine and occurs as the prosthetic group* in a number of dehydrogenases (enzymes which oxidise by removal of hydrogen).

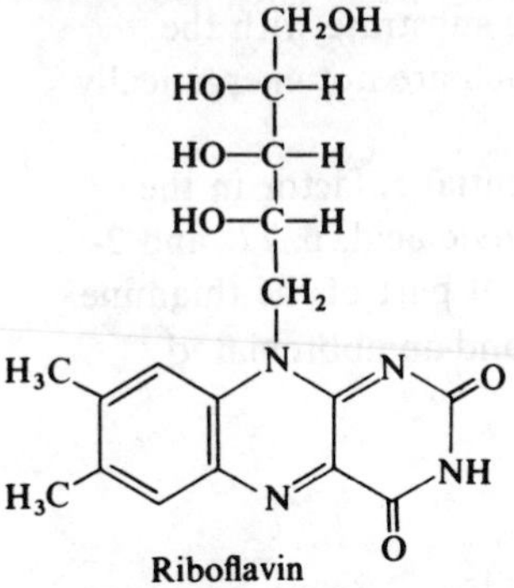

Riboflavin

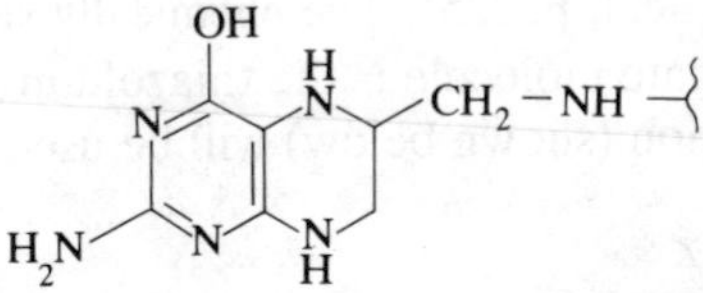

Tetrahydrofolic acid (part structure)

Purine derivatives are important in the nucleic acids, and purine is the skeleton of uric acid, the principal end product of nitrogen metabolism in terrestrial invertebrates and reptiles. Caffeine, the principal stimulant and

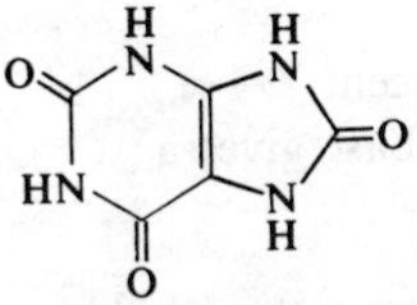

Uric acid

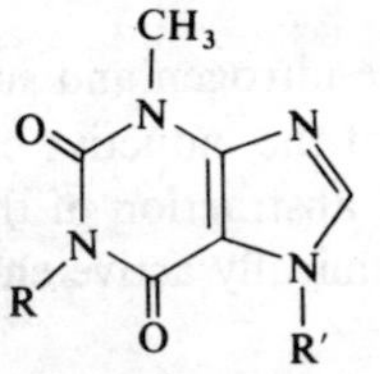

Caffeine R=R′=CH_3
Theophylline R=CH_3 R′=H
Theobromine R=H R′=CH_3

diuretic of coffee, is a simple purine derivative, and theophylline and theobromine which occur in tea and cocoa respectively have very similar structures.

Many of the drugs used in medicine are derivatives of these and other heteroaromatic systems.

The mechanism of action of coenzymes

It is frequently found in cell-reactions that, in addition to the enzyme and substrate, a third substance is required for reaction to occur. These

* A prosthetic group is the non-protein portion of an enzyme, which is frequently chemically involved in the enzymically catalysed reaction.

compounds, known as coenzymes, are really reagents in the enzyme-catalysed reactions. Many coenzymes have elaborate constitutions in which heterocyclic sections are of major importance in the chemical behaviour.

Although the chemical role of all coenzymes is not yet fully understood, the way in which certain coenzymes function is known in some detail. The chemical behaviour of two coenzymes will be described here with particular reference to the way in which combination of the substrate with the coenzyme makes available reaction pathways which are not energetically feasible for the isolated substrate molecule.

Thiamine pyrophosphate (TPP, p. 280) is an essential cofactor in the enzymic decarboxylation of α-oxo acids (e.g. pyruvic acid, p. 77, and 2-oxoglutaric acid, p. 235). The chemically significant part of the thiamine pyrophosphate molecule is the thiazolium ring, and an abbreviated representation (shown below) will be used here.

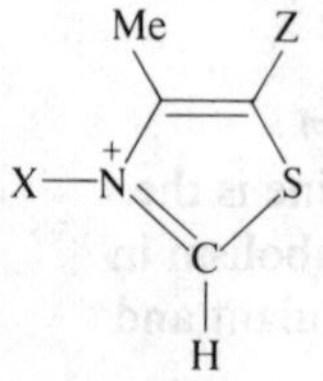

The CH group between the nitrogen and sulphur atoms is weakly acidic on account of the inductive effect of the adjacent electronegative atoms, and abstraction of this proton by a base gives a zwitterion which is the chemically active entity.

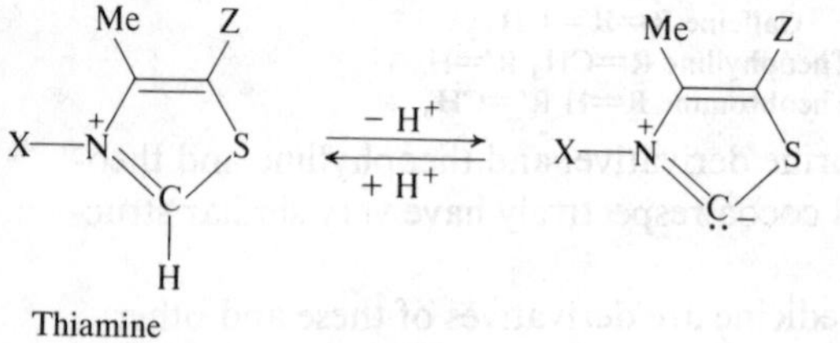

$$\xrightleftharpoons[+\,H^+]{-\,H^+}$$

Thiamine pyrophosphate

(Note that the lone pair of the carbanion is in an sp^2 orbital lying in the plane of the ring and perpendicular to the aromatic π-orbitals, so that the negative charge cannot be delocalised by conjugation with the aromatic system.) This carbanion can react with a carbonyl group in a fashion identical with that described for the aldol reaction (p. 109). Reaction with pyruvic acid will proceed as follows:

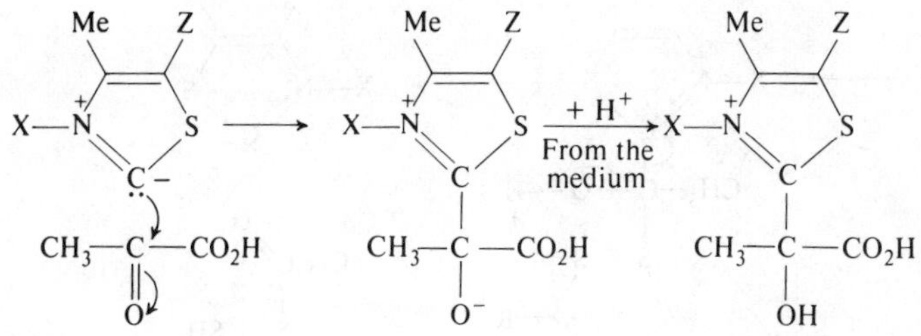

The addition compound (I) of pyruvic acid and TPP so obtained has a carboxylic acid group and an electronegative unsaturated system in precisely the relationship necessary to assist decarboxylation (p. 222) to form the intermediate compound (II). Compound (II) can then undergo protonation and cleavage to generate ethanal.

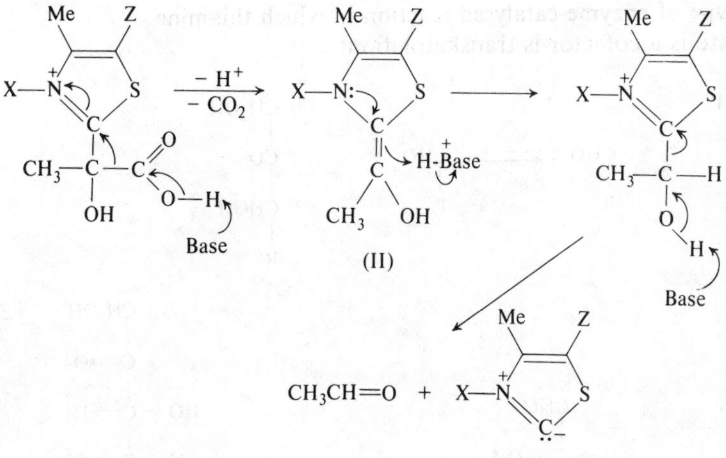

The overall reaction is then:

$$CH_3COCO_2H \longrightarrow CH_3CHO + CO_2$$

which occurs as one of the final steps in the alcoholic fermentation of glucose by yeast (p. 25). Alternatively, compound (II) can interact with thioctic acid (p. 77), which acts as an oxidising agent and removes the side chain as an ethanoyl (acetyl) group.

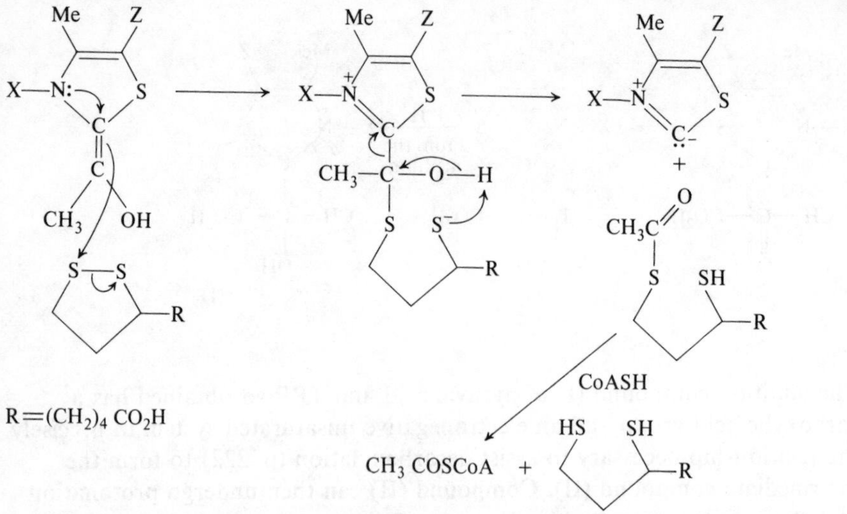

$R{=}(CH_2)_4CO_2H$

CoASH

$CH_3COSCoA$ +

Subsequently the thioester of the reduced thioctic acid transfers its acyl group to coenzyme A by a typical transesterification mechanism (p. 142). The overall reaction is now the oxidative decarboxylation of pyruvic acid (summarised on p. 77).

Another type of enzyme-catalysed reaction in which thiamine pyrophosphate is a cofactor is transketolation:

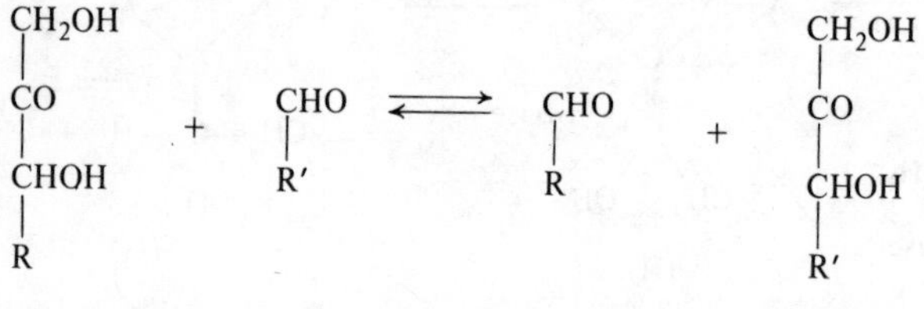

$$CH_2OH{-}CO{-}CHOH{-}R \; + \; R'{-}CHO \;\rightleftharpoons\; R{-}CHO \; + \; CH_2OH{-}CO{-}CHOH{-}R'$$

E.g.

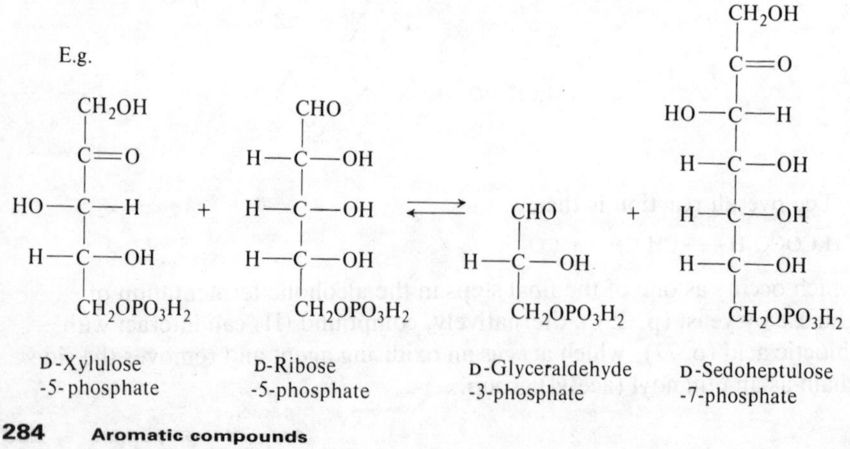

$$\underset{\text{D-Xylulose-5-phosphate}}{CH_2OH{-}C(=O){-}CH(OH)_{\text{(HO–C–H)}}{-}CH(OH){-}CH_2OPO_3H_2} \; + \; \underset{\text{D-Ribose-5-phosphate}}{CHO{-}[CH(OH)]_3{-}CH_2OPO_3H_2} \;\rightleftharpoons\; \underset{\text{D-Glyceraldehyde-3-phosphate}}{CHO{-}CH(OH){-}CH_2OPO_3H_2} \; + \; \underset{\text{D-Sedoheptulose-7-phosphate}}{CH_2OH{-}C(=O){-}CH(OH)_{\text{(HO–C–H)}}{-}[CH(OH)]_3{-}CH_2OPO_3H_2}$$

The transfer of a two-carbon fragment can be rationalised in the following way:

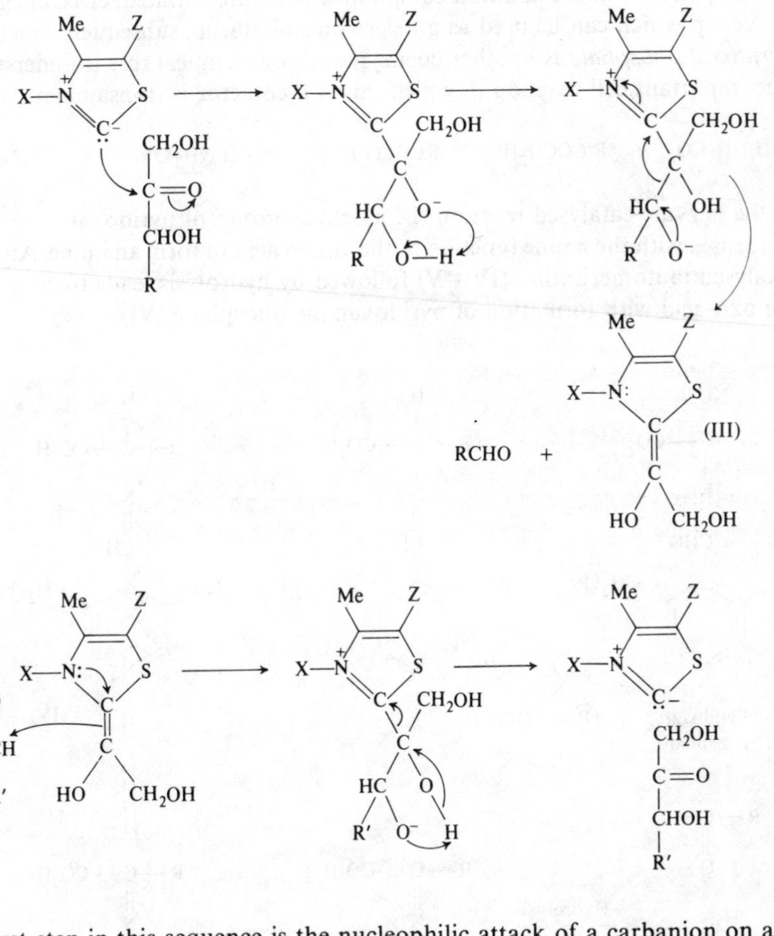

The first step in this sequence is the nucleophilic attack of a carbanion on a carbonyl group. Proton exchange is then followed by a cleavage similar to the reverse of the aldol reaction (p. 109). The breaking of the C——C bond is assisted by the adjacent $C{=}\overset{+}{N}$ group which acts as an 'electron sink' and accommodates the electrons displaced during heterolysis of the C——C bond The intermediate (III) can now react with another aldehyde group by the reverse of the above process to regenerate the conjugate base of the coenzyme after transfer of the two-carbon fragment.

The mode of action of thiamine pyrophosphate in all of these processes is dependent upon two factors:

(i) the ready formation of a carbanion, which can react with carbonyl groups;

(ii) the presence in the addition compounds of an unsaturated, electronegative group which can be used as an electron sink during subsequent reactions.

Pyridoxal phosphate is another coenzyme whose chemical role is understood. One important cell reaction in which this is a cofactor is transamination:

$$\mathrm{RCH\,\overset{+}{N}H_3CO_2^-} + \mathrm{R'COCO_2H} \rightleftharpoons \mathrm{RCOCO_2H} + \mathrm{R'CH\,\overset{+}{N}H_3CO_2^-}$$

In the enzyme-catalysed reaction the aldehyde group of pyridoxal condenses with the amine function of the amino acid to form an imine. An acid-catalysed tautomerisation (IV ⇌ V) followed by hydrolysis leads to release of the oxo-acid with formation of pyridoxamine phosphate (VI).

Pyridoxal phosphate

Ⓟ $= -OPO_3H_2$

$+H^+$

(IV)

(V)

Hydrolysis $-H^+$

(VI)

Pyridoxamine phosphate can then react with a molecule of another oxo-acid by the reverse of the mechanism shown above to regenerate pyridoxal phosphate and form a new amino acid.

$$R'{-}C({=}O){-}CO_2H \; + \; \text{pyridoxamine phosphate } (H_2N{-}CH_2{-}) \longrightarrow R'{-}C(CO_2H){=}N{-}CH_2{-}\text{pyr} \xrightarrow[\text{tautomerisation}]{\text{Acid catalysed}} R'{-}CH(CO_2H){-}N{=}CH{-}\text{pyr}$$

$$\xrightarrow{\text{Hydrolysis}} \text{pyridoxal phosphate (CHO, HO, } CH_2Ⓟ, CH_3, N) + R'CH(\overset{+}{N}H_3)CO_2^-$$

Another cell reaction for which pyridoxal phosphate is required is amino acid epimerisation:

$$\text{L-}C_6H_5CH_2CH(\overset{+}{N}H_3)CO_2^- \rightleftharpoons \text{D-}C_6H_5CH_2CH(\overset{+}{N}H_3)CO_2^-$$

and the tautomeric change (IV) ⇌ (V) in the scheme above provides a reasonable pathway for inverting the configuration of the chiral centre of an amino acid. Pyridoxal phosphate is also involved in the decarboxylation of amino acids, and the protonated imine (IV) can be seen to have the electronic features necessary to assist decarboxylation.

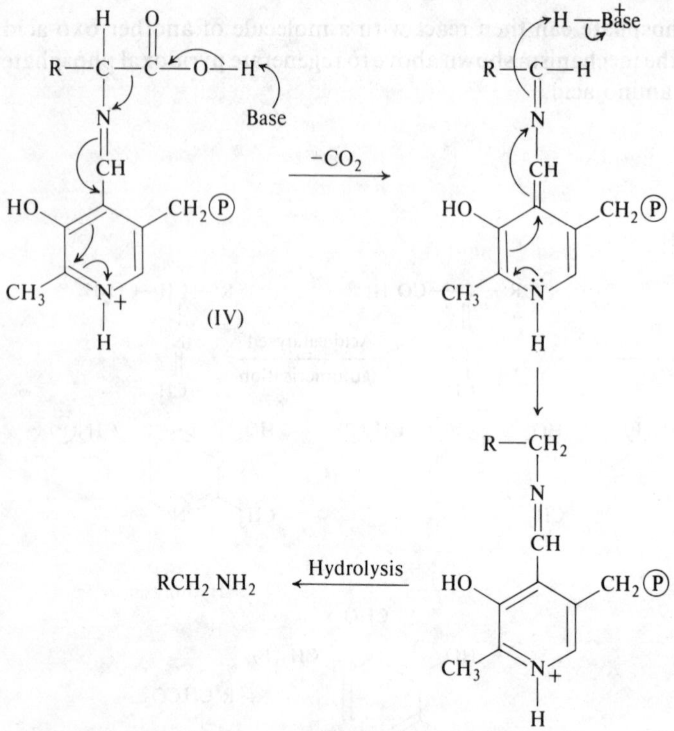

It can be seen that, just as thiamine pyrophosphate uses the carbanion of its conjugate base to become attached to a molecule in order to make energetically accessible reaction pathways not possible in the isolated substrate molecule, so pyridoxal phosphate uses its aldehyde function to link onto amino groups with the same effect. It appears that pyridoxal phosphate is bound to the enzymes by condensation of this aldehyde group with the terminal amino group of a lysine residue (p. 263) in the protein chain of the enzyme. Strictly, the reaction schemes shown above should have this imine function in place of the aldehyde group.

Nucleic acids

Nucleic acids are acidic macromolecules ('macro' = large) found originally in the nucleus of cells but occurring also in the cytoplasm. Nucleic acids occur as 'nucleoproteins' in combination with protein, and it is known that viruses, which in some cases can be obtained as crystalline substances, are largely nucleoprotein. The biochemical role of nucleic acids and nucleoproteins is outside the scope of this text, but their importance is shown by the fact that they are responsible for the transmission of hereditary characteristics and control of protein synthesis in the cell.

Separation of the nucleic acids from other cell constituents gives the purified acids as fibrous precipitates. Hydrolysis of purified nucleic acid gives three types of product, these being a group of four basic compounds, a sugar, and phosphoric acid. Two types of nucleic acid are known, distinguishable primarily by the sugar obtained on hydrolysis. Ribonucleic acid (RNA) gives D-ribose, whilst deoxyribonucleic acid (DNA) gives 2-deoxy-D-ribose. The group of bases obtained also varies, both RNA and DNA giving adenine,

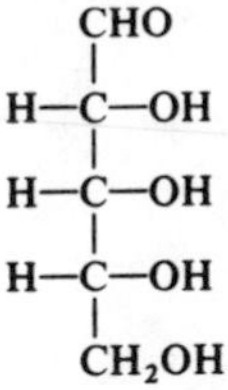
D-Ribose

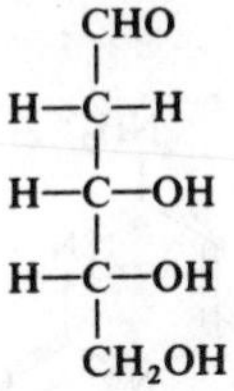
2-Deoxy-D-ribose

guanine (derived from purine, p. 280) and cytosine (a pyrimidine), but RNA contains uracil as its fourth base whilst DNA contains thymine. Enzymic hydrolysis breaks the nucleic acids down into fragments known as nucleosides (composed of one molecule of base combined with one molecule of sugar) and nucleotides (containing one molecule each of base, sugar, and phosphoric

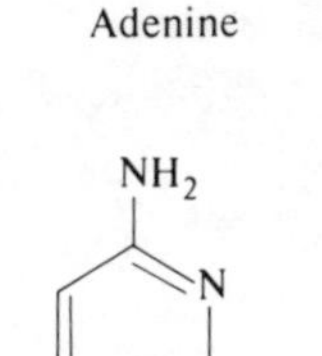
Adenine

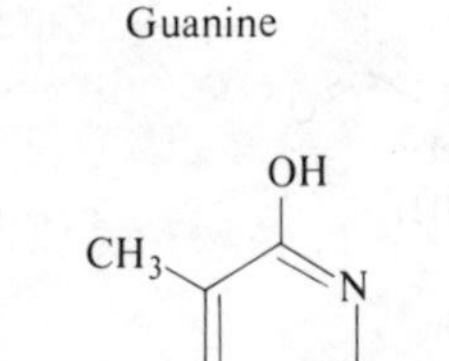
Guanine

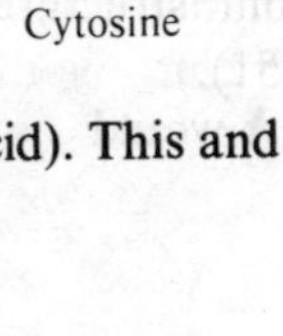
Cytosine

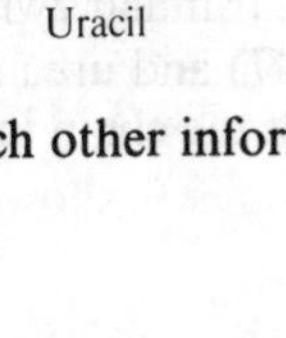
Uracil

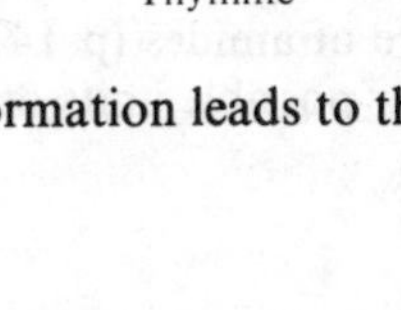
Thymine

acid). This and much other information leads to the conclusion that the nucleic acid consists of a chain of alternate sugar and phosphate residues, with the bases attached to the sugar units:

—Phosphate—Sugar—Phosphate—Sugar—Phosphate—
(with Base attached below each Sugar)

The structure of the repeating unit in this chain can be studied in the nucleotides, which have been shown to contain the sugar in a furanose ring (p. 241). Thus the structure of the ribonucleotides derived from adenine and cytosine are:

NH₂ ... $H_2O_3POCH_2$ ... H H H H OH OH

NH_2 ... $H_2O_3POCH_2$ ... O ... H H H H OH OH

The pyrimidine residue in the cytosine nucleotide may appear to have lost its aromatic character, but the pyrimidine bases are probably more accurately

H_2N ... N ... N ... OH — Tautomerism ⇌ — H_2N ... N ... N H ... O ⟷ H_2N ... N ... N^+ H ... O^-

(VII) (VIII) (IX)

represented by the zwitterionic structure (IX) rather than either of the uncharged tautomeric structures (VII) and (VIII). (Similar zwitterionic structures are important in the structure of amides (p. 147) and urea (p. 151).)

Thus a sequence of the four possible units of the chain in RNA would have the structure:

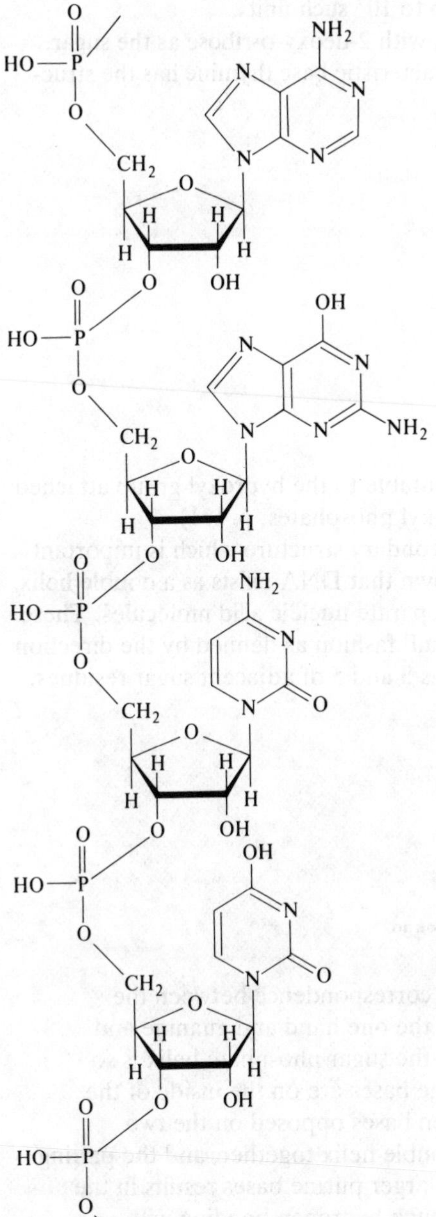

and the whole molecule may contain up to 10^7 such units.

A similar sequence is found in DNA, with 2-deoxy-D-ribose as the sugar. The nucleotide unit containing the characteristic base thymine has the structure:

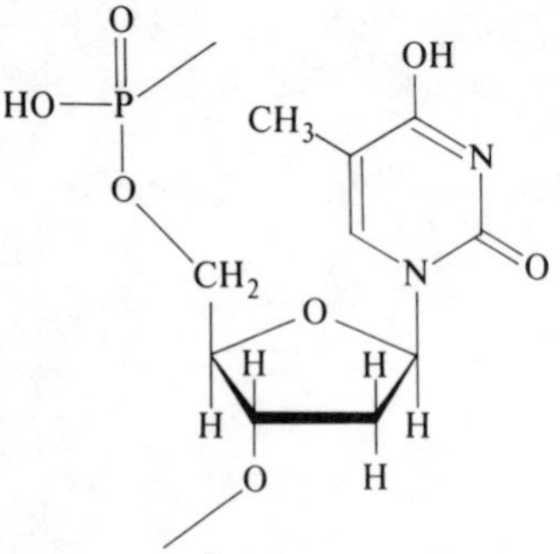

The acidity of these molecules is attributable to the hydroxyl group attached to the phosphorus atom (cf. *pK* of dialkyl phosphates, p. 161).

Nucleic acids, like proteins, have secondary structure which is important in the biological function, and it is known that DNA exists as a double helix resulting from the association of two separate nucleic acid molecules. These two molecules associate in a 'head to tail' fashion as defined by the direction of the phosphate link between positions 3 and 5 of adjacent sugar residues.

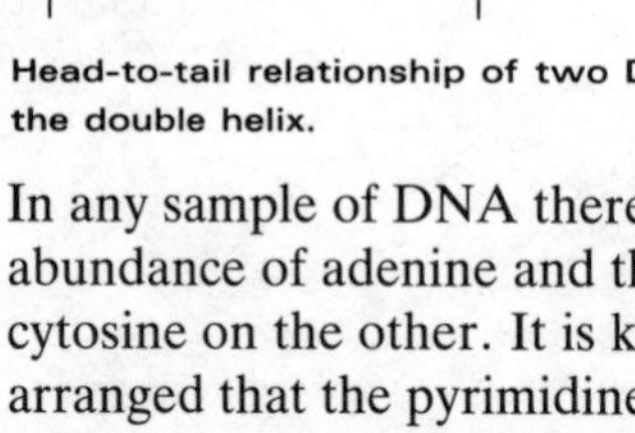

| 5	| 3
Sugar—Base	Base—Sugar
| 3	| 5
Phosphate	Phosphate
| 5	| 3
Sugar—Base	Base—Sugar
| 3	| 5
Phosphate	Phosphate
|	|

Head-to-tail relationship of two DNA molecules in the double helix.

In any sample of DNA there is a close correspondence between the abundance of adenine and thymine on the one hand and guanine and cytosine on the other. It is known that the sugar phosphate helix is so arranged that the pyrimidine and purine bases are on the inside of the spiral, when hydrogen bonding between bases opposed on the two intertwined chains helps to hold the double helix together, and the pairing of the small pyrimidine bases with the larger purine bases results in the equivalent abundance. However, although hydrogen bonding will undoubtedly contribute to stabilisation of the double helix, the strength of binding of the two strands is too great to be accounted for solely in this way.

RNA molecules do not associate in pairs to form a similar double helix. The bulk of the extra hydroxyl group in the sugar units appears to restrict the conformational flexibility of the polynucleotide chain, and this hydroxyl group must be principally responsible for the ability of enzymes to differentiate between DNA and RNA. Nevertheless, within some RNA molecules short, rod-like, double helical structures are formed by the single nucleotide chain folding back on itself.

DNA is the store of hereditary information in the cell, which is coded in

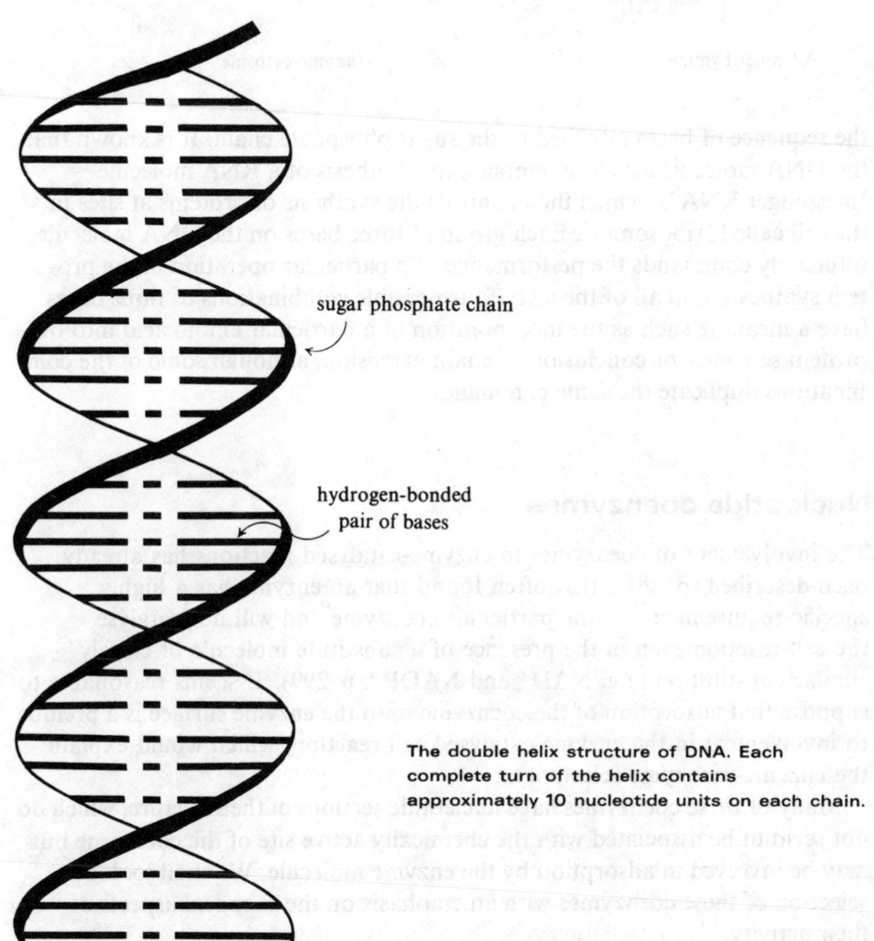

The double helix structure of DNA. Each complete turn of the helix contains approximately 10 nucleotide units on each chain.

Hydrogen-bonded base pairs

the sequence of bases attached to the sugar phosphate chain. It is known that the DNA molecule acts as a template for synthesis of a RNA molecule—'messenger RNA'—which then controls the synthesis of proteins at sites in the cell called 'ribosomes'. Each group of three bases on the DNA molecule ultimately commands the performance of a particular operation in the protein synthesis, and all of the sixty-four possible combinations of three bases have a meaning such as the incorporation of a particular amino acid into the protein sequence or conclusion of chain extension, although some of the combinations duplicate the same command.

Nucleotide coenzymes

The involvement of coenzymes in enzyme-catalysed reactions has already been described (p. 281). It is often found that an enzyme has a highly specific requirement for one particular coenzyme and will not catalyse the cell-reaction even in the presence of a substitute molecule of closely similar constitution (e.g. NAD^+ and $NADP^+$, p. 299). It seems reasonable to suppose that adsorption of the coenzyme onto the enzyme surface is a prelude to involvement in the enzyme-catalysed cell-reaction, which would explain the enzyme-coenzyme selectivity.

Many of these coenzymes have nucleotide sections of the structure, which do not seem to be associated with the chemically active site of the coenzyme but may be involved in adsorption by the enzyme molecule. We shall look at a selection of these coenzymes with an emphasis on the chemical aspects of their activity.

Many of these coenzymes are named after the nucleoside from which they are derived, and the names of the nucleosides are, in their turn, derived from the names of the purine and pyrimidine bases:

NAME OF BASE	NAME OF THE RIBONUCLEOSIDE
Adenine	Adenosine
Guanine	Guanosine
Cytosine	Cytidine
Uracil	Uridine
Thymine	Thymidine

The corresponding deoxyribonucleosides are called 'deoxyadenosine', etc.

Adenosine triphosphate (ATP)

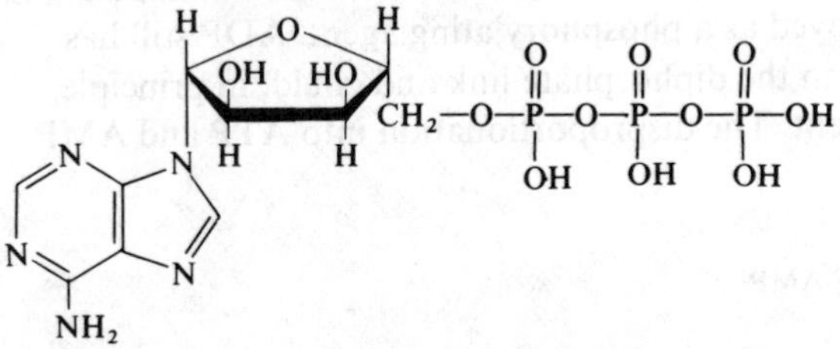

(adenine-D-ribose-phosphate-phosphate-phosphate)*

ATP is universally distributed in living systems and represents a chemical store of energy in the cell. Reactions which, if left to proceed uncontrolled, would evolve large amounts of heat, are linked in living systems to other reactions which utilise the liberated energy in the synthesis of 'energy-rich' compounds (chemists normally refer to these as 'reactive' compounds). ATP is one of the most important of these compounds.

Biological reactions utilising ATP follow one of two paths. Either ATP acts as an acid anhydride and acylates the substrate (phosphorylation) or, more rarely, it can act as an alkylating agent. Numerous examples of phosphorylation are known, ATP being converted into the corresponding diphosphate—adenosine diphosphate (ADP)—by transfer of the terminal phosphate group. For example, D-glucose is phosphorylated to D-glucose-6-phosphate by ATP in the presence of the enzyme hexokinase:

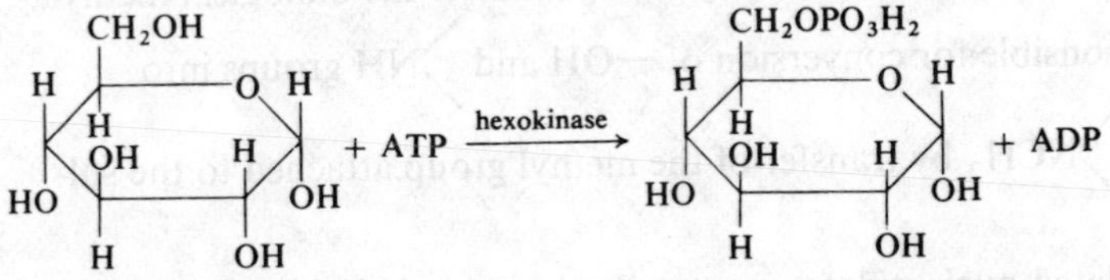

$$\text{D-glucose} + \text{ATP} \xrightarrow{\text{hexokinase}} \text{D-glucose-6-phosphate } (CH_2OPO_3H_2) + \text{ADP}$$

* This and subsequent similar descriptions of nucleotides are not systematic names, but indicate how the complex molecules can be dissected into recognisable fragments.

In some cases ATP can act as a pyrophosphorylating agent transferring the two end phosphate groups and being converted into adenosine monophosphate (AMP):

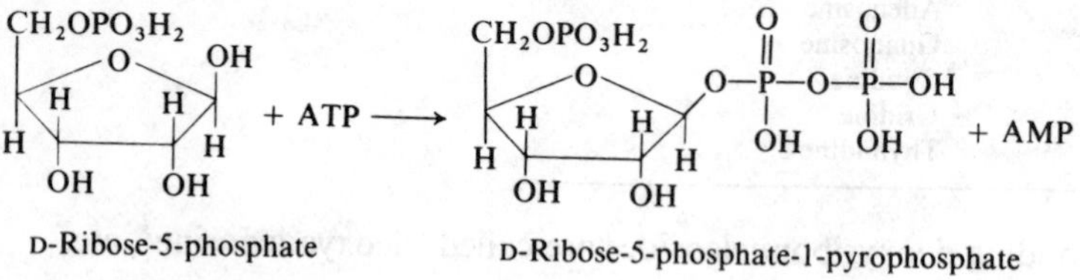

$$\text{D-Ribose-5-phosphate} + \text{ATP} \longrightarrow \text{D-Ribose-5-phosphate-1-pyrophosphate} + \text{AMP}$$

D-Ribose-5-phosphate ($CH_2OPO_3H_2$; OH; H; H; H; H; OH; OH) D-Ribose-5-phosphate-1-pyrophosphate ($CH_2OPO_3H_2$; O—P(=O)(OH)—O—P(=O)(OH)—OH; H; H; H; H; OH; OH)

The monophosphate, being a normal ester of phosphoric acid and not an acid anhydride, is not employed as a phosphorylating agent. ADP still has an acid anhydride structure in the diphosphate link and could, in principle, act as a phosphorylating agent. The disproportionation into ATP and AMP is such a process.

$$\text{ADP} + \text{ADP} \xrightarrow{\text{myokinase}} \text{ATP} + \text{AMP}$$

Other nucleoside triphosphates (e.g. of guanosine or uridine) are known to act as biological phosphorylating agents but are less important than ATP.

An example of ATP acting as an alkylating agent is in its reaction with the amino acid methionine:

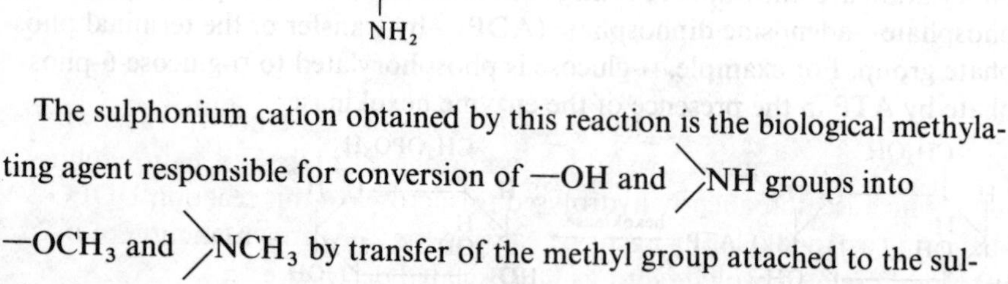

$$CH_3SCH_2CH_2\overset{+}{N}H_3CHCO_2^- + \text{ATP} \longrightarrow \text{adenine–ribose–}CH_2\text{–}\overset{+}{S}(CH_3)\text{–}CH_2CH_2CH(\overset{+}{N}H_3)CO_2^-$$

The sulphonium cation obtained by this reaction is the biological methylating agent responsible for conversion of —OH and >NH groups into —OCH_3 and >NCH_3 by transfer of the methyl group attached to the sulphur atom.

A closely related nucleotide coenzyme is adenosine-3′,5′-monophosphate (‘cyclic AMP’) which is formed from ATP by the enzyme adenylate cyclase.

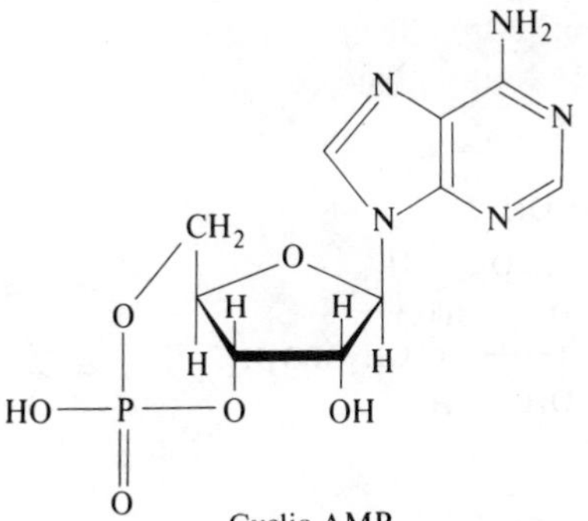

Cyclic AMP

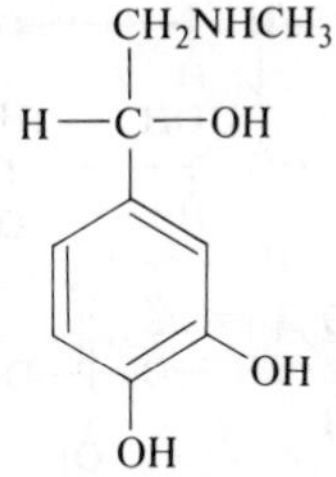

Adrenaline (Epinephrine)
(The natural compound has the *R* configuration)

This is an important compound involved in the regulation of cell reactions. A number of enzymes are known to exist in active and inactive forms, and the overall activity of the cell processes in which they are involved can be controlled by interconversion of the active and inactive forms. Cyclic AMP is required for the activation of a number of such enzymes, and is one of the compounds involved in a complicated sequence of steps linking adrenaline secretion with the stimulation of glycogen degradation and suppression of glycogen synthesis.

Uridinediphosphate–glucose (UDPG)

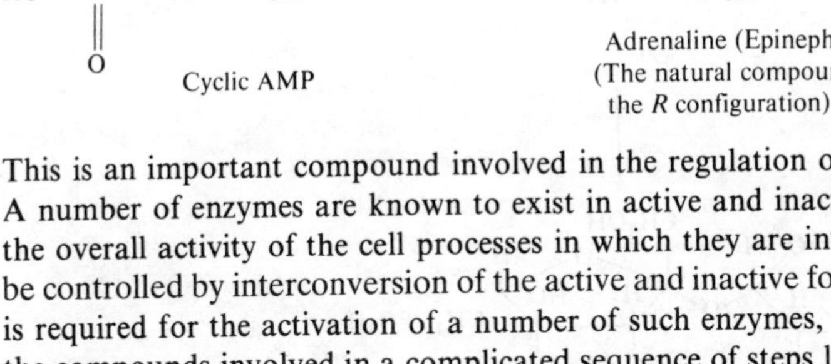

(Uracil-D-ribose-phosphate-phosphate-1α-D-glucose)

UDPG is an intermediate in the synthesis of sucrose in plants. It reacts with D-fructose-6-phosphate to give uridinediphosphate (UDP) and sucrose phosphate, which is subsequently hydrolysed to sucrose. In this reaction UDPG is behaving as an alkylating agent converting the 2-hydroxyl function of D-fructofuranose-6-phosphate into its glucosylated derivative.

Several other nucleotide coenzymes of this type are known and function in a similar manner, e.g. guanosinediphosphate-mannose, uridinediphosphate-galactose, uridinediphosphate-glucosamine.

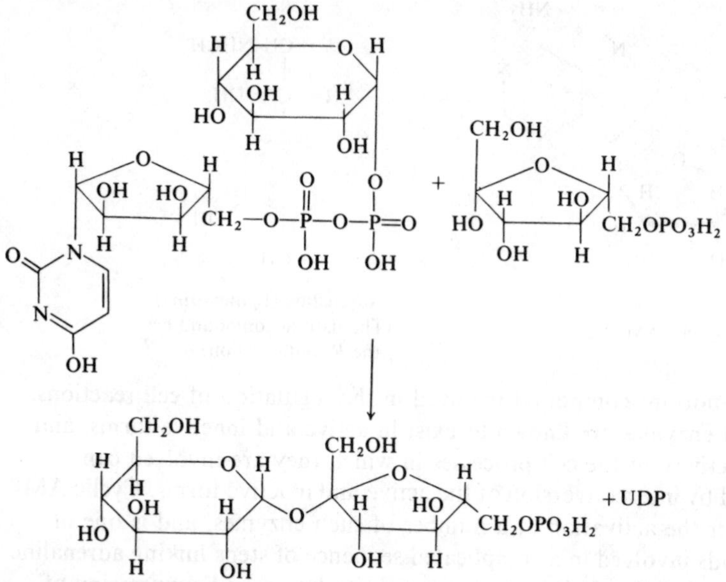

The stereochemistry of the enzyme-catalysed reaction between UDPG and D-fructose-6-phosphate gives an indication of the reaction pathway involved. Both UDPG and sucrose phosphate are α-glucosides so that, overall, the transfer of the glucose moiety occurs with retention of configuration, and any mechanistic rationalisation of the reaction must explain this feature. The S_N1 and S_N2 mechanisms are alternative possibilities for this alkylation, but S_N1 processes are unlikely to occur in living systems, since the high energy carbonium ion intermediates are extremely reactive and their subsequent reactions cannot easily be controlled. The S_N2 mechanism which does not involve such highly reactive intermediates, is more suitable for the close regulation characteristic of cell processes, but is invariably accompanied by invertion of configuration at the site of reaction (p. 203). If UDPG reacted directly with fructose-6-phosphate by an S_N2 process, the result would not be sucrose phosphate but an epimer, glucose 1β——2β fructose-6-phosphate. To explain the production of an α-glucoside it is necessary to propose a two-step process in which both steps proceed with invertion. It is believed that UDPG (an α-glucoside) transfers its glucosyl group to the enzyme (an alkylation reaction) to give a β-glucosyl enzyme, which subsequently alkylates fructose-6-phosphate reforming an α-glucosyl group. The enzyme in this process can be seen to operate not simply as a template for organising the reagents in close proximity, but by having an active chemical role in the transformation.

Nicotinamide-adenine dinucleotide (NAD^+) and nicotinamide-adenine dinucleotide phosphate ($NADP^+$)

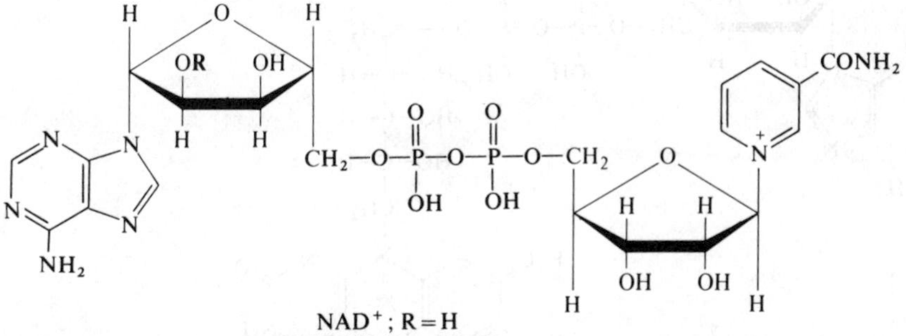

NAD^+; R = H
$NADP^+$; R = PO_3H_2
(NAD = adenine-D-ribose-phosphate-phosphate-D-ribose-nicotinamide)

These two coenzymes (formerly known as diphosphopyridine nucleotide, DPN, and triphosphopyridine nucleotide, TPN, respectively) are frequently involved in enzymic oxidations and reductions. The chemically active group in both cases is the nicotinamide moiety, which can undergo a reversible re-

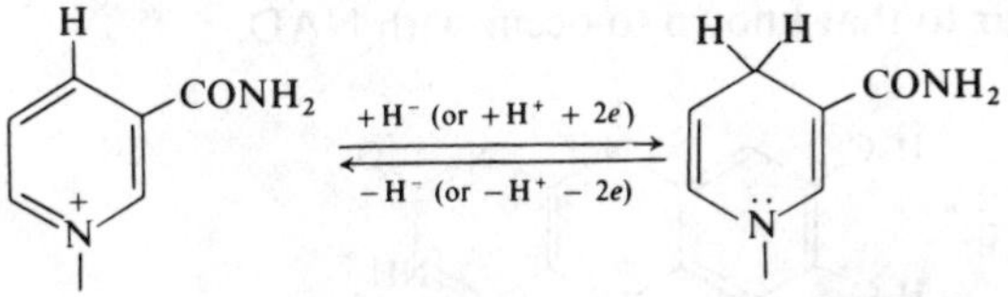

$$\text{Pyridinium cation} \underset{-H^- \ (\text{or } -H^+ - 2e)}{\overset{+H^- \ (\text{or } +H^+ + 2e)}{\rightleftharpoons}} \text{1,4-Dihydropyridine derivative}$$

duction in which the pyridinium ring is reduced to a dihydropyridine by reaction with a hydride ion or the chemically equivalent hydrogen ion and two electrons. These coenzymes and their reduced derivatives can therefore act as acceptors or donors of H^- or electrons. The shorthand representation of these reactions is:

$$H^+ + 2e + NAD^+ \rightleftharpoons NADH$$
$$H^+ + 2e + NADP^+ \rightleftharpoons NADPH$$

Although the coenzymes differ only marginally in their structure, there is absolute specificity for one or the other in enzymic reactions. In general, degradative processes involve NAD^+ and NADH, whilst synthetic processes utilise $NADP^+$ and NADPH. The complete specificity for one or other coenzyme enables these processes to be controlled independently. (For stereochemical aspects of reactions utilising these coenzymes see p. 316.)

Flavine-adenine dinucleotide (FAD)

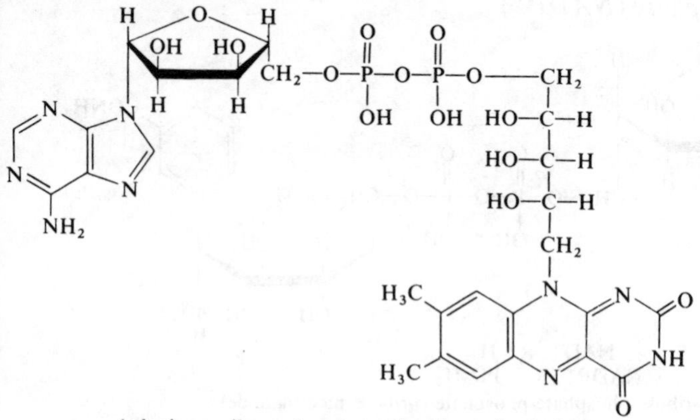

(adenine-D-ribose-phosphate-phosphate-riboflavine)

FAD is composed of a nucleotide unit joined to riboflavine (p. 281). This is another reduction-oxidation coenzyme existing in oxidised and reduced forms like NAD and NADP. If the alloxazine system (p. 280) is represented by the most 'aromatic' canonical structure, the reduction can be reasonably represented by a process similar to that known to occur with NAD.

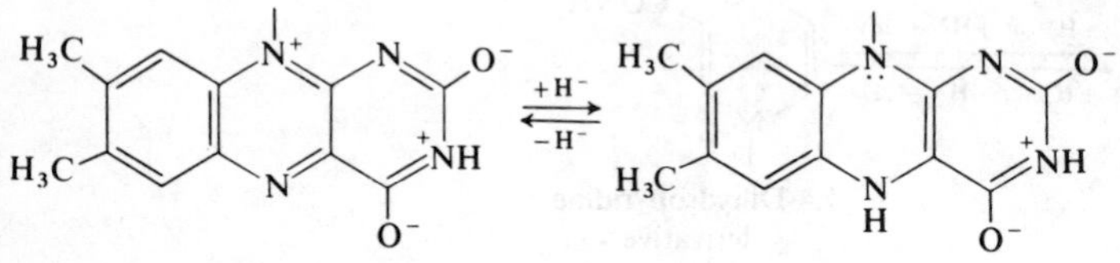

$$\xrightleftharpoons[-H^-]{+H^-}$$

Coenzyme A (CoASH)

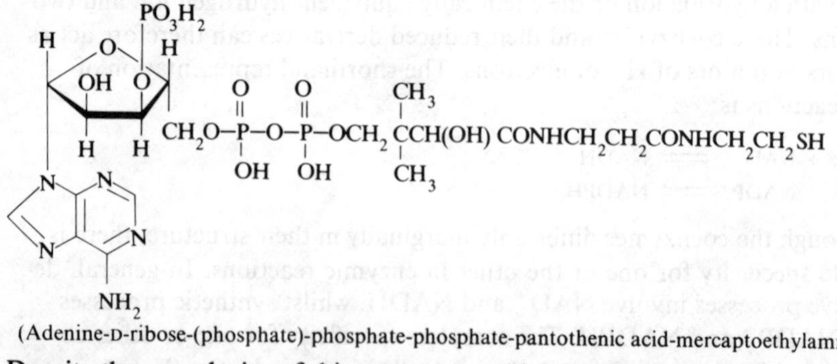

$CH_2O-P-O-P-OCH_2CCH(OH)CONHCH_2CH_2CONHCH_2CH_2SH$

(Adenine-D-ribose-(phosphate)-phosphate-phosphate-pantothenic acid-mercaptoethylamine)

Despite the complexity of this coenzyme, which is widely involved in the metabolism of carboxylic acids (e.g. p. 234), its chemical behaviour in enzymic reactions is that of a simple thiol. It should be noted that this molecule contains a β-amino acid (β-alanine) as part of its structure. Pantothenic acid (Vitamin B_5) has the structure:

$$HOCH_2C(CH_3)_2CH(OH)CONHCH_2CH_2CO_2H$$

and can be seen to occur as a part of the coenzyme structure containing the β-amino acid unit.

20

Lipids

The smaller organic molecules found in living tissues can be divided into two broad groups. On the one hand there are the water-soluble materials such as amino acids and sugars which are insoluble in non-hydroxylic solvents such as chloroform or ether. The other group, comprising the fat-soluble materials which are soluble in chloroform, ether, and other organic solvents, but are usually insoluble in water, are known by the generic name of **lipids**. Clearly, so crude a distinction as solubility in a particular group of solvents implies no specific structural feature in common, but within this broad range of substances are to be found many series of compounds with common functional groups and general constitutional similarities. The low solubility in water implies that, in lipids, highly polar or hydrogen-bonding groups are either absent or constitute only a very small part of the whole molecule whereas non-polar (i.e. hydrocarbon) groups will predominate. Amongst the compounds included in the lipids are many of immense biological importance such as Vitamins A and D (p. 331) and steroidal hormones (p. 328) which occur only in

CH_3 CH_3 CH_3 CH_3 CH_2OH CH_3

Vitamin A_1

minute traces and collectively do not account for more than a tiny fraction of the total lipid content of any living system. Those constituents of the lipids which are most abundant are associated with very few general functions. One group of lipids functions as the protective coatings on the cell walls of bacteria, the leaves of higher plants, the cuticle of insects, and the skin of vertebrates.

The depot fats are another group which form the store of metabolic fuel in living systems, and the third major group is that of the phospholipids, which are important components of biological membranes.

Fatty acids

Although the fatty acids (long-chain aliphatic carboxylic acids) occur in the free state only in trace amounts, they are one of the groups of simple molecules from which many lipids are constructed. The acyl moieties most frequently encountered in the major lipid groups are those derived from straight-chain aliphatic acids with even numbers of carbon atoms, usually in the range C_{14}–C_{22}, with C_{16} and C_{18} being the most abundant. Derivatives of fully saturated and mono- and polyunsaturated acids are found, but derivatives of carboxylic acids with C≡C groups are rare as are those of acids with branched chains or more elaborate structures. Amongst the unsaturated acids concerned, those with *cis (Z)*-stereochemistry about the double bond(s) are more common than the *trans (E)*-stereoisomers, and non-conjugated, polyunsaturated acids are more abundant than their conjugated isomers. Polyunsaturated acyl groups containing sequences of $CH{=}CH{-}CH_2$ groups are fairly common. Some of the commoner fatty acids involved in lipid formation are listed in the table.

Common fatty acids involved in lipid formation

NO. OF C ATOMS	CONSTITUTION	SYSTEMATIC NAME	TRIVIAL NAME
Saturated acids			
10	$CH_3(CH_2)_8CO_2H$	Decanoic acid	Capric
12	$CH_3(CH_2)_{10}CO_2H$	Dodecanoic acid	Lauric
14	$CH_3(CH_2)_{12}CO_2H$	Tetradecanoic acid	Myristic
16	$CH_3(CH_2)_{14}CO_2H$	Hexadecanoic acid	Palmitic
18	$CH_3(CH_2)_{16}CO_2H$	Octadecanoic acid	Stearic
20	$CH_3(CH_2)_{18}CO_2H$	Eicosanoic acid	Arachidic
22	$CH_3(CH_2)_{20}CO_2H$	Docosanoic acid	Behenic
24	$CH_3(CH_2)_{22}CO_2H$	Tetracosanoic acid	Lignoceric
26	$CH_3(CH_2)_{24}CO_2H$	Hexacosanoic acid	Cerotic
Unsaturated acids			
16	$CH_3(CH_2)_5CH{=}CH(CH_2)_7CO_2H$	*Z*-Hexadec-9-enoic acid	Palmitoleic
18	$CH_3(CH_2)_7CH{=}CH(CH_2)_7CO_2H$	*Z*-Octadec-9-enoic acid	Oleic

18	$CH_3(CH_2)_5CH{=}CH(CH_2)_9CO_2H$ *E*-Octadec-11-enoic acid	Vaccenic
18	$CH_3(CH_2)_4(CH{=}CHCH_2)_2(CH_2)_6CO_2H$ Octadeca-9(*Z*),12(*Z*)-dienoic acid	Linoleic
18	$CH_3CH_2(CH{=}CHCH_2)_3(CH_2)_6CO_2H$ Octadeca-9(*Z*),12(*Z*),15(*Z*)-trienoic acid	Linolenic
18	$CH_3(CH_2)_3(CH{=}CH)_3(CH_2)_7CO_2H$ Octadeca-9(*Z*),11(*E*),13(*E*)-trienoic acid	α-Eleostearic
20	$CH_3(CH_2)_4(CH{=}CHCH_2)_4CH_2CH_2CO_2H$ Eicosa-5(*Z*),8(*Z*),11(*Z*),14(*Z*)-tetraenoic acid	Arachidonic
24	$CH_3(CH_2)_7CH{=}CH(CH_2)_{13}CO_2H$ *Z*-Tetracosa-15-enoic acid	Nervonic

Plant and animal waxes. These compounds, which form the protective, waterproof covering on many plants and animals, are usually the esters of long-chain fatty acids and alcohols with a large hydrocarbon group. Beeswax contains the palmitate esters of long-chain alcohols, and leaf waxes consist of esters of fatty acids and alcohols having up to thirty-four carbon atoms. Wool-wax, lanolin, contains the esters of lanosterol (p. 327).

$CH_3(CH_2)_{14}CO_2(CH_2)_{25}CH_3$ A component of beeswax

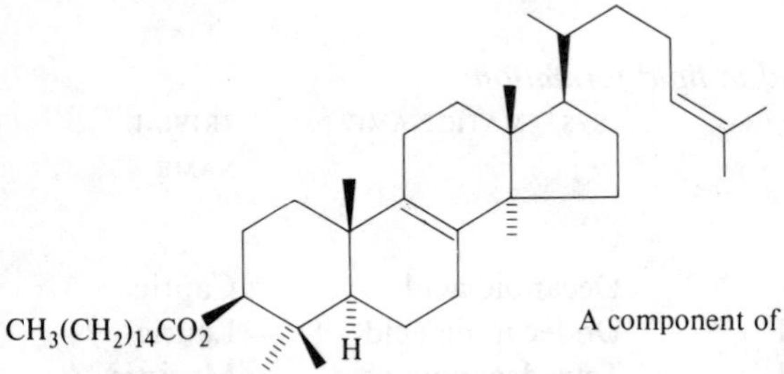

$CH_3(CH_2)_{14}CO_2$ ... H

A component of wool-wax

In all these compounds, save for the two oxygen atoms of the ester function, the whole of the molecule consists of hydrocarbon groups. It is not very surprising that their physical properties resemble those of the larger hydrocarbons e.g. paraffin waxes.

Depot fats. The depot fats, which form one of the metabolic fuel reserves of living systems, are predominantly triacyl derivatives of glycerol (p. 67). In general the triacylglycerols from animal sources differ from those of many plant oils in the higher proportion of saturated acyl groups in the animal fat. There is a clear correlation between the extent of unsaturation and the melting point in triacylglycerols, the highly unsaturated seed-oils having very low melting points, whereas animal fats are usually solid at ambient temperature.

The commercial catalytic hydrogenation of plant oils to form margarine results in the production of a commodity with physical properties resembling those of a typical animal fat. This difference in physical properties can be related to the different shapes of the saturated and unsaturated fatty acid molecules, which is most clearly shown by considering the shapes of the molecules when the carbon chain is in the fully extended conformation.

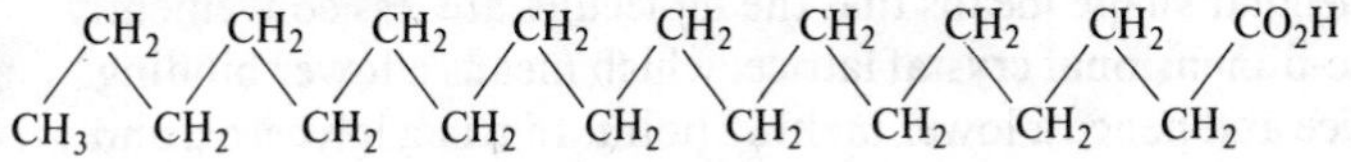

$CH_3(CH_2)_{16}CO_2H$ Stearic acid

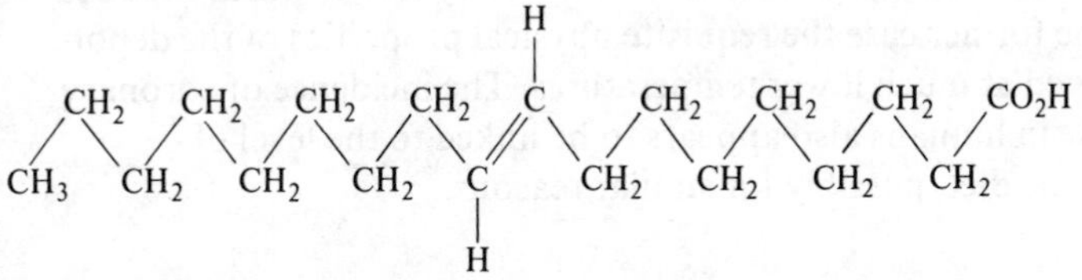

$CH_3(CH_2)_7CH{=}CH(CH_2)_7CO_2H$ Elaidic acid

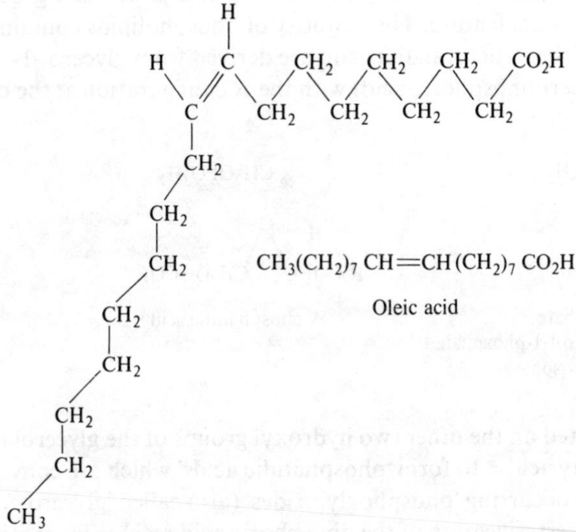

$CH_3(CH_2)_7CH{=}CH(CH_2)_7CO_2H$

Oleic acid

Whereas the *E* configuration about the double bond in elaidic acid produces scarcely any significant change in the molecular shape compared with the saturated analogue (stearic acid), the *Z* configuration of the unsaturated group in oleic acid produces a very pronounced kink in the molecule. Multiple unsaturation, such as in linolenic and arachidonic acids, will enhance the irregularity in the shape of an acyl chain so that more extensive unsaturation accompanied by *Z* stereochemistry at the double bonds will lead to more irregularity in the molecular shape of derivatives such as the triacylglycerols. Irregularity in molecular shape means that the molecules are less conveniently packed into a three-dimensional crystal lattice, which means a lower binding energy for the lattice and hence a lower melting point. In triacylglycerols and other lipids containing more than one long-chain acyl group a wide variation in molecular shape and associated properties can be achieved by introduction of *Z*-unsaturation into the acyl hydrocarbon chain. It is not very surprising that fish-oils are more highly unsaturated than the body fats of warm-blooded animals since in the former case the requisite physical properties of the depot fats must be achieved at much lower temperatures. The incidence of coronary and arterial disease in humans also appears to be linked to the level of unsaturated fat in the diet, possibly for similar reasons.

Phospholipids

The phospholipids are a large group of fairly complex molecules occurring extensively in biological membranes and having a phosphate ester group as the common constitutional feature. The majority of phospholipids contain glycerol as one of the structural units and are derived from glycerol-1-phosphate (α-glycerophosphoric acid) with the *R* configuration at the chiral centre.

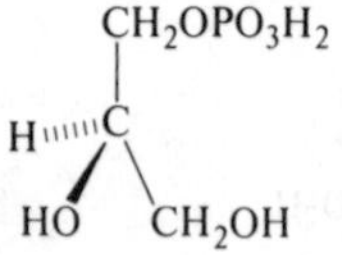

$CH_2OPO_3H_2$ / H····C / HO, CH_2OH

R-Glycerol-1-phosphate
(Also called D-Glycerol-1-phosphate
and L-Glycerol-3-phosphate)

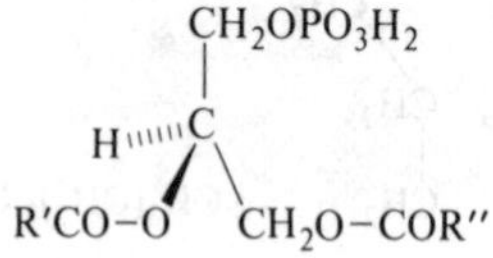

$CH_2OPO_3H_2$ / H····C / R′CO–O, CH_2O–COR″

A phosphatidic acid

This is then acylated on the other two hydroxyl groups of the glycerol residue by long-chain fatty acids* to form ‘phosphatidic acids’ which are converted into the naturally occurring ‘phosphoglycerides’ (also called ‘glycero-phosphatides’) by esterification of the phosphoric acid residue by another

* It is often found that the 2-hydroxyl group of the glycerol residue is acylated by an unsaturated acid.

Types of phosphoglycerides

$$\begin{array}{l} CH_2-CH-CH_2-O-P(=O)(OH)-X \\ \quad | \qquad | \\ \quad O \qquad O \\ \quad CO \quad\ CO \\ \quad R^1 \quad\ \ R^2 \end{array}$$

	$-X$
Phosphatidyl ethanolamine (Cephalin)	$-OCH_2CH_2NH_2$
Phosphatidyl choline (Lecithin)	$-OCH_2CH_2\overset{+}{N}(CH_3)_3$
Phosphatidyl glycerol	$-OCH_2CHOHCH_2OH$
Phosphatidyl 3-*O*-aminoacylglycerol	$-OCH_2CHOHCH_2O-CO-CH(NH_2)-R$
Phosphatidyl serine	$-OCH_2CH(\overset{+}{N}H_3)CO_2^-$
Phosphatidyl inositol	$-O-$inositol

Inositol = [cyclohexane ring bearing six OH groups]

Cardiolipin (1,3-bisphosphatidyl glycerol)

$$CH_2(OCOR^1)-CH(OCOR^2)-CH_2-O-P(=O)(OH)-O-CH_2CH(OH)CH_2-O-P(=O)(OH)-O-CH_2-CH(OCOR^3)-CH_2(OCOR^4)$$

molecule containing a hydroxyl group. Several types of phosphoglyceride are known, distinguished by the nature of this final section of the molecule, which is always a highly polar or hydrogen-bonding group. The table summarises the constitutional feature of the more important types of phosphoglyceride.

A related group of phospholipids are the plasmalogens, differing from the phosphatidic acid derivatives by having an unsaturated ether group at position 3 of the glycerol residue in place of the normal acyl group.

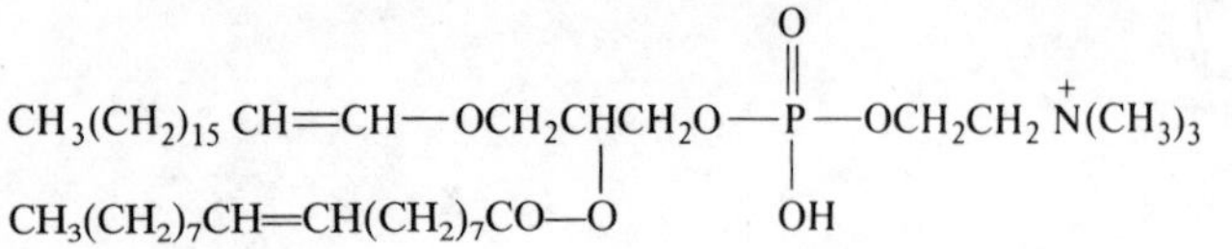

$$\begin{array}{l} CH_3(CH_2)_{15}CH{=}CH{-}OCH_2\underset{\displaystyle CH_3(CH_2)_7CH{=}CH(CH_2)_7CO{-}O}{\underset{|}{C}H}CH_2O{-}\overset{\displaystyle O}{\overset{\|}{\underset{\underset{\displaystyle OH}{|}}{P}}}{-}OCH_2CH_2\overset{+}{N}(CH_3)_3 \end{array}$$

A plasmalogen

Glycolipids are constituents of plant membranes and although not phosphate esters are conveniently mentioned here. In glycolipids a 1,2-diacylglycerol is linked through the 3-hydroxyl group to a sugar, often D-galactose, which provides a highly hydrogen-bonding terminal group.

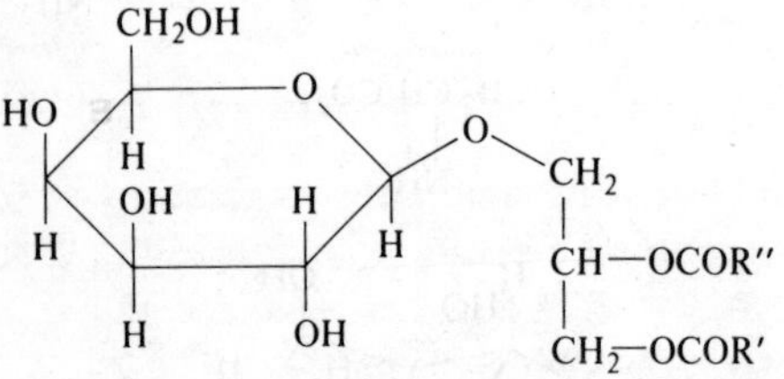

A glycolipid

Sphingolipids are derived from sphingosine or its dihydro-derivative sphinganine. Sphingosine is a long-chain molecule whose terminal three carbon atoms have functionality resembling glycerol whilst the rest of the molecule has a long hydrocarbon chain similar to that of a fatty acid.

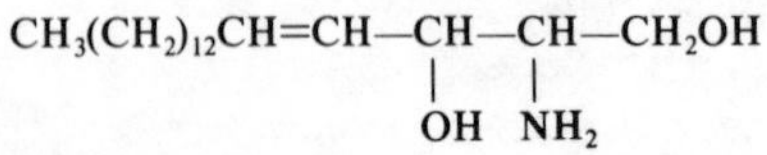

$$CH_3(CH_2)_{12}CH{=}CH{-}\underset{\displaystyle OH}{\underset{|}{CH}}{-}\underset{\displaystyle NH_2}{\underset{|}{CH}}{-}CH_2OH$$

Sphingosine
2(*S*)-Amino-1,3(*R*)-dihydroxyoctadec-4(*E*)-ene

In sphingolipids the 3-hydroxyl function of sphingosine or dihydrosphingosine is usually unchanged and the amino group is acylated by a long-chain fatty acid. This ‘ceramide’ is then linked to a variety of groups through the terminal hydroxyl group.

$$
\begin{array}{l}
\qquad\qquad\qquad\qquad\qquad\quad OH \\
\qquad\qquad\qquad\qquad\qquad\quad | \\
CH_3(CH_2)_{12}CH{=}CH{-}CH \\
\qquad\qquad\qquad\qquad\qquad\quad | \\
CH_3(CH_2)_{22}CO{-}NH{-}CH{-}CH_2OH
\end{array}
$$

A ceramide

Spingomyelins have a choline phosphate unit joined onto this terminal hydroxyl group (cf. lecithin), cerebrosides have a D-galactose residue attached here (cf. glycolipids), whilst gangliosides have complex oligosaccharide residues joined onto the 1 position of the ceramide.

$$
\begin{array}{l}
\qquad\qquad\qquad\qquad\qquad\quad OH \\
\qquad\qquad\qquad\qquad\qquad\quad | \\
CH_3(CH_2)_{12}CH{=}CH{-}CH \\
\qquad\qquad\qquad\qquad\qquad\quad | \\
\qquad\quad R\,CO{-}NH{-}CH \qquad\quad O \\
\qquad\qquad\qquad\qquad\qquad\quad | \qquad\qquad \| \qquad + \\
\qquad\qquad\qquad\qquad\qquad\quad CH_2{-}O{-}P{-}OCH_2CH_2N(CH_3)_3 \\
\qquad\qquad\qquad\qquad\qquad\qquad\qquad\quad | \\
\qquad\qquad\qquad\qquad\qquad\qquad\qquad\quad OH
\end{array}
$$

A sphingomyelin

Lipids and the structure of biological membranes

As previously stated, phospholipids, glycolipids, and sphingolipids occur extensively in the membranes of living systems and are almost completely absent from depot fats. Whilst the precise function of phospholipids etc. in membranes is still unclear, it is well understood why these types of compound are found in association with that type of cellular structure. All of the types of lipid described above under the heading of phospholipids have molecular constitutions in which a large non-polar, hydrocarbon section of the molecule has a relatively small, polar or hydrogen-bonding portion. The general pattern is like a tuning fork with long non-polar prongs and a short highly polar leg. All the phospholipids have an acidic hydroxyl group on the phosphorus atom which will be dissociated at *p*H7 to give a charged centre, and additionally cephalins, lecithins, phosphatidyl serine, and sphingomyelins will be zwitterionic at ambient *p*H.

To understand why these compounds are involved in the formation of membranes it is necessary to look at some of the factors relating to the phenomenon of solubility. The extent to which a solute will dissolve in a solvent is determined by the relative strengths of solute–solute interaction in the solid state and solvent–solvent and solute–solvent interaction in the liquid phase. Polar solutes usually have strong forces binding the crystal lattice (e.g. electrostatic interaction in ionic or zwitterionic solids, or multiple hydrogen bonding as in the sugars) and they are unlikely to be easily dispersed in non-polar solvents where solute–solvent interactions will be weak and there will be little energy provided from this source to compensate for the energy required to remove molecules from the crystal lattice. Conversely, highly polar solvents are unlikely to dissolve non-polar solutes since insertion of non-polar solute molecules between polar solvent molecules would disrupt the relatively strong interactions between solvent molecules without any significant compensating solute–solvent interaction. So for those substances which dissolve as dispersed, isolated molecules there is a well recognised qualitative relationship between the solubility and the relative polarities of solute and solvent.

A more complex situation exists where one considers the case of a molecule which contains both highly polar and non-polar regions. In a polar solvent, such as water, the solvent will interact with (solvate) the polar end of the solute tending to cause this part of the molecule to dissolve, whilst the low solute–solvent interaction will tend to exclude the non-polar part of the molecule from the solvent medium. Such solutes, under these conditions, tend to form clusters known as **micelles** in which the non-polar ends of the solute molecule crowd together whilst the polar ends form an outer layer interacting with the polar medium. The net effect is to form a non-polar globule with a polar surface.

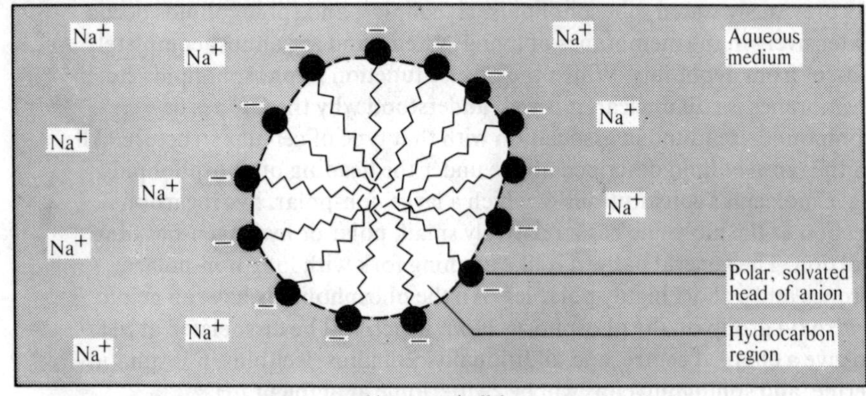

'A soap micelle'

The alkali metal salts of long-chain fatty acids (soaps), e.g. $Na^{+}\ {}^{-}O_2C(CH_2)_{16}CH_3$, are the most familiar example of this type of molecule, the polar end of the molecule being the carboxylate anion and the long hydrocarbon chain being the non-polar section. The detergent properties of these salts are due to the ability of the micelle to include grease and other non-polar material inside the hydrocarbon region, where it is, in effect, dissolved in a hydrocarbon solvent. The droplets are stabilised against coagulation by the surface electrical charge which repels any other approaching micelle with similar surface charge. In this way grease can be dispersed as an emulsion of droplets, stabilised in water by the soap. Long-chain quaternary ammonium salts (e.g. cetylpyridinium chloride, $CH_3(CH_2)_{15}\overset{+}{N}C_5H_5\ Cl^-$) form cationic detergents which act in precisely the same way.

The formation of micelles does not require an electrical charge on the solute. Whereas the triacylglycerols are insoluble in water, the mono- and diacylglycerols dissolve forming micelles. In these cases the water soluble (hydrophilic) section is the free hydroxyl group which can hydrogen bond to water and is responsible for the miscibility of glycerol and water. The water-insoluble (hydrophobic) sections in these glycerides are the long hydrocarbon chains of the fatty acid residues, as in the case of the soaps.

Globular clusters are not the only type of structure formed by molecules with both hydrophilic and hydrophobic sections. Phospholipids and related compounds readily and spontaneously form monolayers on the surface of aqueous media, and bilayers in the aqueous media, the effect being like an extended micelle. The polar ends of the phospholipids are solvated by the water whilst the hydrocarbon tails of the fatty acid residues etc. form a non-polar, electrically insulating layer, impervious to the passage of charged species such as Na^+ and K^+.

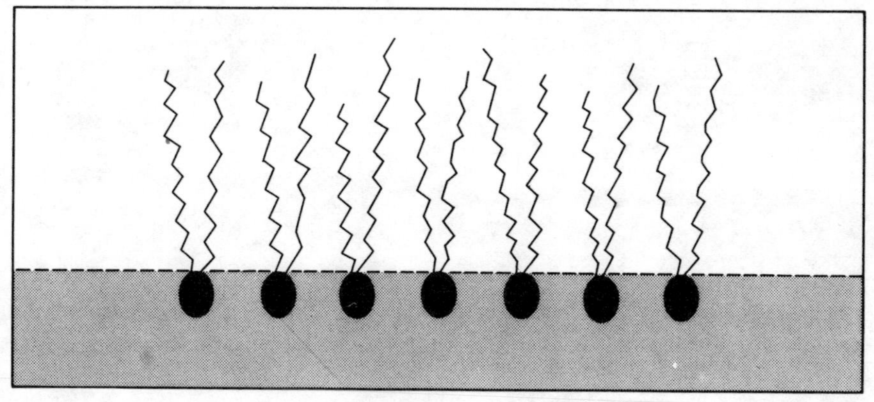

Lipid monolayer on a liquid surface

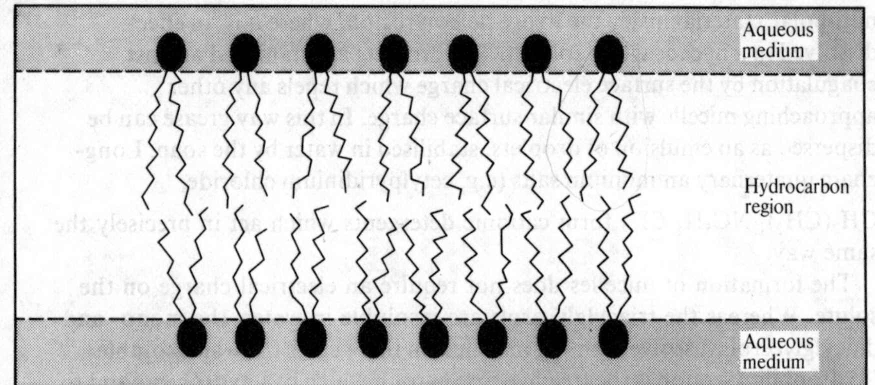

Lipid bilayer

Biological membranes are not composed solely of phospholipid but contain on average 60 per cent protein and 40 per cent lipid, the lipid content including varying amounts of steroids notably cholesterol (p. 327). Despite the complexity of composition, the simple picture above is still valid since proteins also have non-polar and polar regions which can interact with the bilayer and aqueous medium respectively. There are apparently no covalent links between the protein and lipid constituents of the membranes and individual molecules may have some freedom of movement within the membrane. However, these lipoprotein complexes have considerable stability despite the absence of covalent bonding. How these membranes function in the chemistry of the cell is still far from clear and is well beyond the scope of this text.

21

Stereochemical aspects of enzymic reactions

The stereoselectivity of enzymes has already been mentioned on numerous occasions, e.g. in connection with the biological resolution of racemic mixtures (p. 177), the specificity of maltase and emulsin (p. 254), and the structural and stereochemical requirements of proteolytic enzymes (p. 268). It is generally accepted that enzymic catalysis proceeds via adsorption of the substrate onto the surface of the large protein molecule, and the stereospecificity can be understood if the enzyme is considered to have receptor sites capable of binding or accommodating particular types of group. Taking the case of a single asymmetrically substituted carbon atom as an example, an enzyme with receptors for three of the four groups could distinguish between the two enantiomers, since the correct enantiomer could be adsorbed with three groups attached to the receptor sites, whilst the other enantiomer could, at best, bind to only two of the sites. The 'attachment' to these sites might be either by covalent bond formation, hydrogen bonding, interaction of ionic or polar groups, or even just by the filling of cavities which will accommodate groups of a particular shape or less than a certain size.

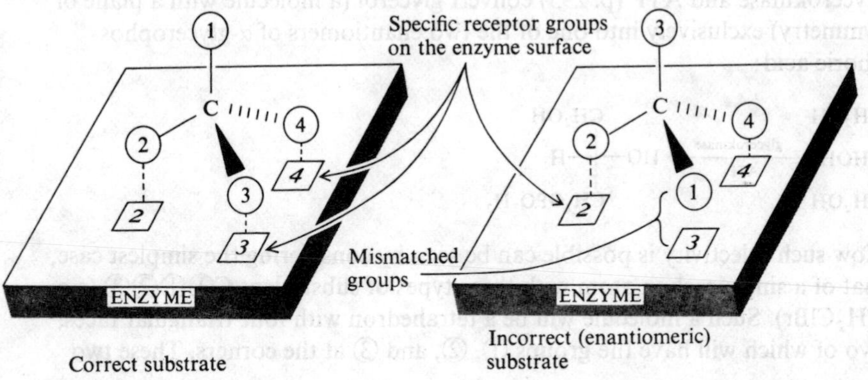

The steric course of enzyme-catalysed reactions, leading stereospecifically from symmetrical starting materials to optically pure chiral products, can also be understood in similar terms, since the three-dimensional structure of the enzyme-substrate complex may determine the direction from which reagents approach the adsorbed molecule, and thereby determine the absolute stereochemistry of the product. For example the stereospecific reduction of pyruvic acid to L-lactic acid, catalysed by lactic dehydrogenase, may be schematically described by the following diagram:

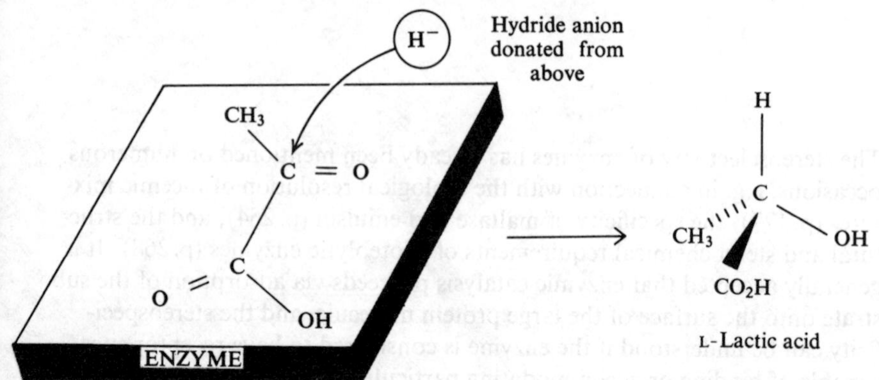

Pyruvic acid adsorbed on the enzyme surface

Prochirality in enzymic reactions

Besides the ability to distinguish between enantiomers, enzymes are frequently able to distinguish between apparently identical groups in a molecule. Thus glycerokinase and ATP (p. 295) convert glycerol (a molecule with a plane of symmetry) exclusively into one of the two enantiomers of α-glycerophosphoric acid:

$$\begin{array}{l} CH_2OH \\ | \\ CHOH \\ | \\ CH_2OH \end{array} \xrightarrow[\text{ATP}]{\text{glycerokinase}} \begin{array}{c} CH_2OH \\ | \\ HO-\!\!\!-\!\!\!+\!\!\!-\!\!\!-H \\ | \\ CH_2OPO_3H_2 \end{array}$$

How such selectivity is possible can be seen by considering the simplest case, that of a single carbon atom with three types of substituent C①①②③ (e.g. CH_2ClBr). Such a molecule will be a tetrahedron with four triangular faces, two of which will have the groups ①, ②, and ③ at the corners. These two faces will differ from each other, since the sequence ①–②–③ is clockwise on one face and anticlockwise on the other. In fact the two faces have an object–mirror-image relationship. If such a molecule is adsorbed on an enzyme which

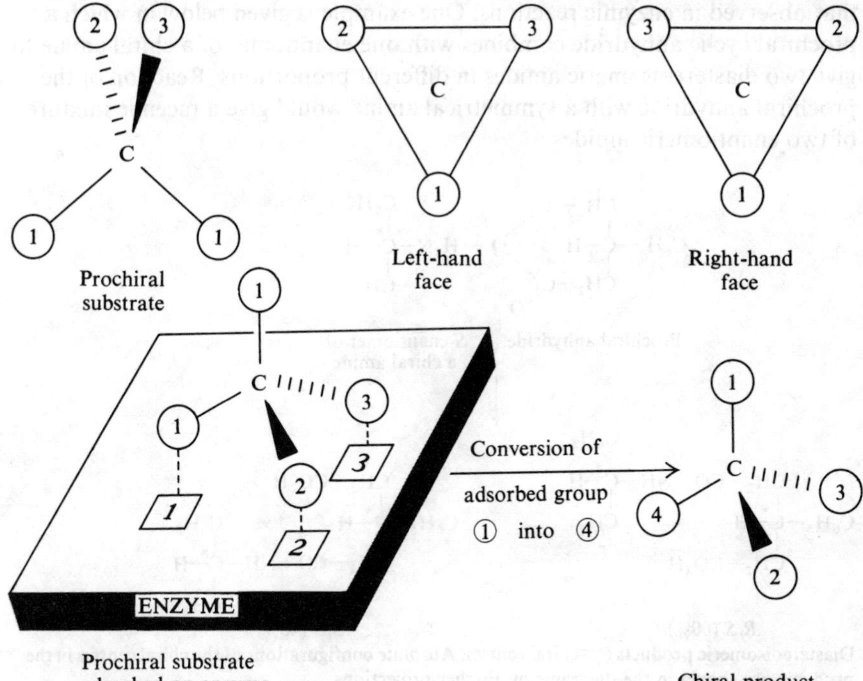

has receptors for groups ①, ②, and ③ arranged in an anticlockwise sequence (see the diagram), then the face of the substrate with the clockwise sequence will be adsorbed, since of the two faces in question only this one will permit interaction between three matched sets of receptor and group. When thus adsorbed, the two groups ① are in totally different environments, one being held onto the enzyme surface and the other projecting away from it. In these circumstances it is not very surprising that one of the groups ① may undergo reaction exclusively, leading to a chiral product.

Molecules of the type C①①②③ are known as **prochiral**.* It should be noted that in an isolated prochiral molecule the two groups ① are identical and indistinguishable by normal chemical or physical means. Distinction between the two groups is possible only when the molecule is brought into an unsymmetrical or chiral environment by adsorption on an enzyme surface or encounter with a chiral reagent.

* The term '*meso*' was previously employed, but is now obsolete in this context.

In principle, prochiral molecules in non-enzymic reactions should react preferentially at one of the two similar groups when combining with a chiral reagent, but the selectivity in such cases is usually very low compared with that observed in enzymic reactions. One example is given below in which a prochiral cyclic anhydride combines with one enantiomer of a chiral amine to give two diastereoisomeric amides in different proportions. Reaction of the prochiral anhydride with a symmetrical amine would give a racemic mixture of two enantiomeric amides.

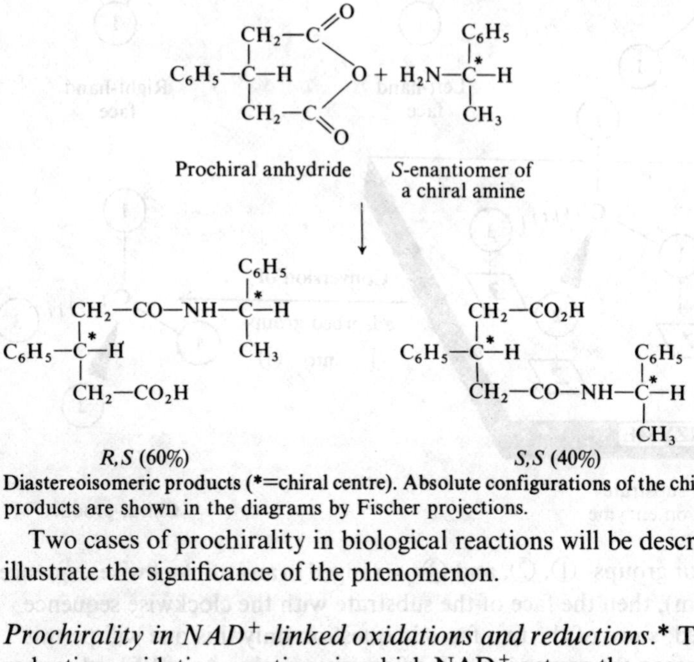

Prochiral anhydride — S-enantiomer of a chiral amine

$$C_6H_5-CH(CH_2-CO)_2O + H_2N-\overset{*}{C}H(C_6H_5)-CH_3 \longrightarrow$$

R,S (60%) — S,S (40%)

Diastereoisomeric products (*=chiral centre). Absolute configurations of the chiral centres in the products are shown in the diagrams by Fischer projections.

Two cases of prochirality in biological reactions will be described briefly to illustrate the significance of the phenomenon.

*Prochirality in NAD^+-linked oxidations and reductions.** The biological reduction-oxidation reactions in which NAD^+ acts as the acceptor and NADH as the donor of hydride ion can be expressed by:

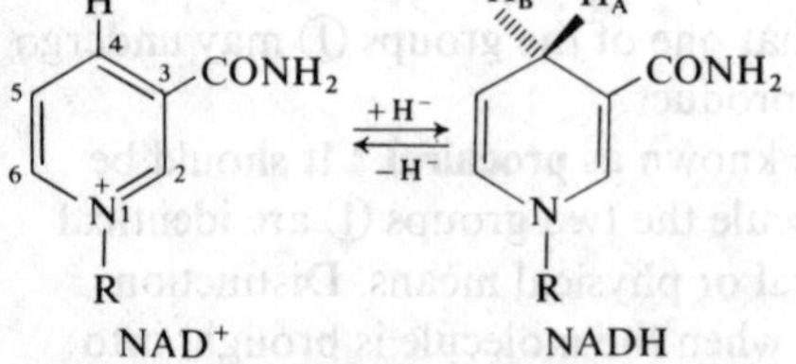

$$NAD^+ \underset{-H^-}{\overset{+H^-}{\rightleftharpoons}} NADH$$

NAD^+ — NADH

R = remainder of the molecule (p. 299)

* The subsequent discussion will be concerned with NAD^+ and NADH. Exactly similar considerations apply to $NADP^+$ and NADPH (p. 299).

The aromatic (pyridinium) ring of NAD^+ has two faces designated A and B related as object and mirror-image:

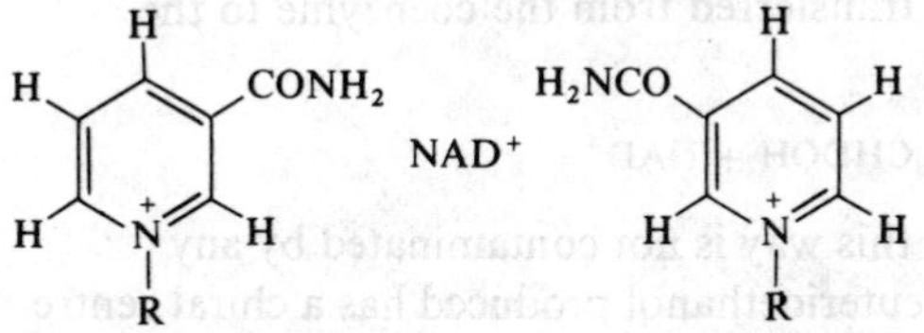

NAD^+

A face B face

Donation of a hydride ion can take place onto either face of the ring, and the resultant CH_2 group at position 4 is a prochiral centre, NADH likewise having two faces to the reduced pyridine ring with an object–mirror-image relationship.

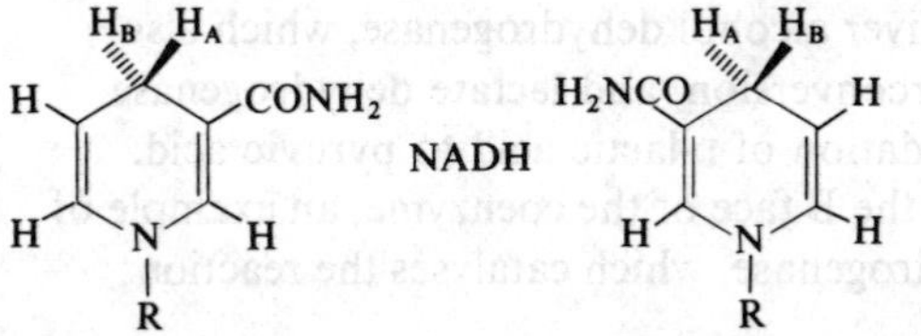

NADH

A face B face

It is found that NAD^+-linked enzymes are usually specific for one or other of the two faces of the pyridine ring. Thus yeast alcohol dehydrogenase reacts with deuterium-labelled ethanol in the presence of NAD^+ to give deuterio-ethanal and stereospecifically labelled NADH in which the hydrogen atom on the A face of NADH has been replaced by deuterium. The abbreviated nomenclature of these deuterated coenzymes and their absolute structures are:

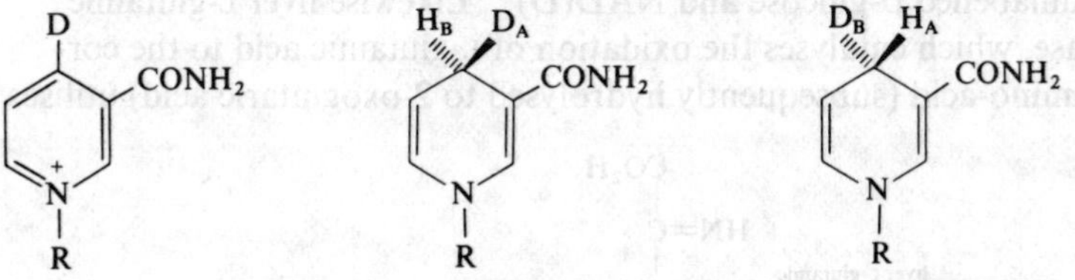

$NAD(D)^+$ $NADD_A$ $NADD_B$

(In the structural formulae D = deuterium.)

The reaction of yeast alcohol dehydrogenase (YAD) may be represented by:

$$CH_3CD_2OH + NAD^+ \xrightarrow{YAD} CH_3CDO + NADD_A + H^+$$

(1,1-Dideuterioethanol) (1-Deuterioethanal)

Yeast alcohol dehydrogenase, like all catalysts, accelerates the reverse reaction, and if the $NADD_A$ from this reaction is then utilised to reduce unlabelled ethanal, the deuterium is transferred from the coenzyme to the substrate:

$$CH_3CHO + NADD_A + H^+ \xrightarrow{YAD} CH_3CHDOH + NAD^+$$

Note that the NAD^+ produced in this way is not contaminated by any $NAD(D)^+$. Note also that the 1-deuterioethanol produced has a chiral centre (since hydrogen and deuterium are not identical) and this reduction gives exclusively the *S*(-) enantiomer. Reduction of 1-deuterioethanal, CH_3CDO, by NADH and yeast alcohol dehydrogenase would give the mirror-image product. Addition of deuteriobenzaldehyde, C_6H_5CDO, to actively fermenting yeast results in the stereospecific formation of *R*-C_6H_5CHDOH.

Other enzymes which transfer and abstract hydride ion utilising the A face of NAD^+ and NADH are horse-liver alcohol dehydrogenase, which also catalyses the ethanol-ethanal interconversion, and lactate dehydrogenase which catalyses the reversible oxidation of L-lactic acid to pyruvic acid. Another group of enzymes utilise the B face of the coenzyme, an example of these is beef-liver D-glucose dehydrogenase, which catalyses the reaction:

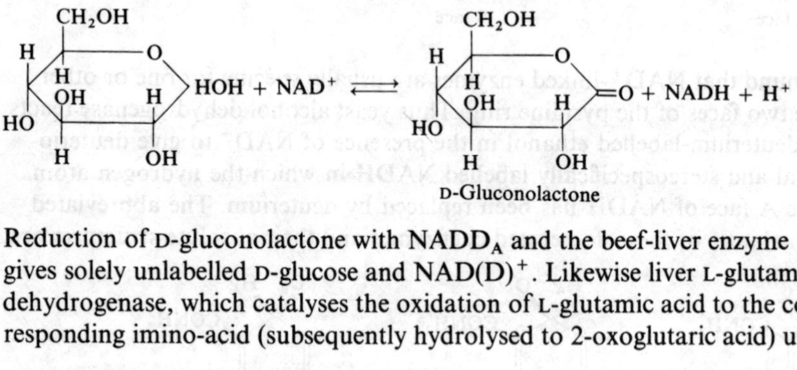

D-glucose (ring: CH_2OH, H, O, H, OH, H, HOH, HO, H, OH) $+ NAD^+ \rightleftharpoons$ D-gluconolactone (ring: CH_2OH, H, O, H, OH, H, =O, HO, H, OH) $+ NADH + H^+$

D-Gluconolactone

Reduction of D-gluconolactone with $NADD_A$ and the beef-liver enzyme gives solely unlabelled D-glucose and $NAD(D)^+$. Likewise liver L-glutamic dehydrogenase, which catalyses the oxidation of L-glutamic acid to the corresponding imino-acid (subsequently hydrolysed to 2-oxoglutaric acid) utilises

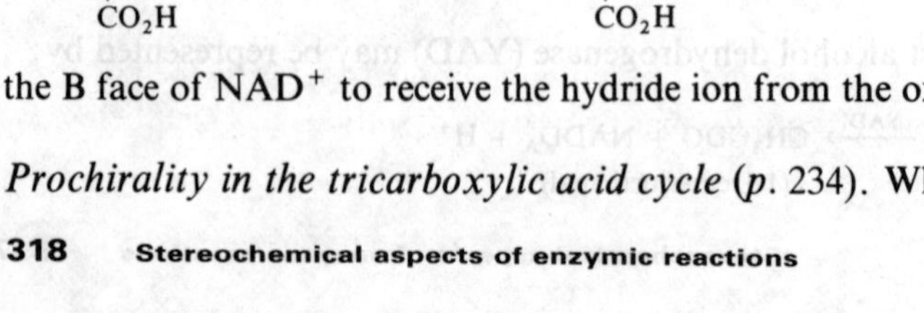

$$HO_2C{-}CH(NH_2){-}CH_2{-}CH_2{-}CO_2H + NAD^+ \underset{\text{dehydrogenase}}{\overset{\text{liver L-glutamic}}{\rightleftharpoons}} HO_2C{-}C({=}NH){-}CH_2{-}CH_2{-}CO_2H + NADH + H^+$$

the B face of NAD^+ to receive the hydride ion from the oxidation.

Prochirality in the tricarboxylic acid cycle (*p*. 234). Whilst studying the pathway of the tricarboxylic acid cycle by use of radioactive isotopes, a puzzling labelling pattern was observed. Thus if oxaloacetic acid, labelled as shown in the scheme below, was converted into labelled citric acid by the appropriate enzymes, and the fate of this labelled citric acid followed through the subsequent steps of the cycle, it was found that at the stage of oxidation of 2-oxoglutaric acid all the ^{14}C label was lost as carbon dioxide. At the time the observations were made, it was thought that, since citric acid is a molecule with a plane of symmetry, formation of the double bond during production of *cis*-aconitic acid should involve either CH_2CO_2H group with equal

$O{=}C(CO_2H)-CH_2-\overset{*}{C}O_2H$ —[$CH_3COSCoA$ + condensing enzyme]→ $CH_2CO_2H-C(OH)CO_2H-CH_2\overset{*}{C}O_2H$ —[enzymic dehydration (aconitase)]→ $(HO_2C-CH_2)(HO_2C)C{=}CH-\overset{*}{C}O_2H$

thermal dehydration → $CH_2CO_2H-CCO_2H{=}CH\overset{*}{C}O_2H$ + $CHCO_2H{=}CCO_2H-CH_2\overset{*}{C}O_2H$

enzymes of the tricarboxylic acid cycle → $CH_2CO_2H-CH_2-C(=O)-\overset{*}{C}O_2H$ → $CH_2CO_2H-CH_2CO_2H$ + $\overset{*}{C}O_2$

probability (as would undoubtedly be the case if dehydration were achieved by heating), leading to a mixture of aconitic acid molecules labelled on either of the terminal carboxyl groups. If such a mixture of labelled molecules were allowed to continue through the steps of the tricarboxylic acid cycle, one type of labelled molecule would give $^{14}CO_2$ and unlabelled succinic acid, whilst the other would give $^{12}CO_2$ and radioactive succinic acid, i.e. only half the ^{14}C radioactivity introduced as label in oxaloacetic acid should be lost.

It can, however, be seen that citric acid is a prochiral molecule, representation as $C(CH_2CO_2H)_2(OH)CO_2H$ showing the characteristic substitution pattern on the central carbon atom. The adsorption of citric acid on the enzyme (aconitase) will differentiate between the two chemically similar CH_2CO_2H groups resulting in the dehydration occurring exclusively in one direction. It should also be noted that the $—CH_2$ groups are also prochiral centres, and the formation of *cis*-aconitic acid from citric acid by aconitase occurs with the specific loss of one of the two hydrogen atoms of the appropriate methylene group. In the Fischer projection of citric acid given below, the labels show the origin and fate of the various atoms.

This section of the molecule comes from oxaloacetic acid

$$
\begin{array}{c}
CO_2H \\
| \\
H—C—H \\
| \\
HO_2C—C—OH \\
| \\
H—C—H \\
| \\
CO_2H
\end{array}
$$

← This hydrogen atom is lost in the enzymic dehydration to *cis*-aconitic acid.

This section of the molecule comes from acetyl CoA

Numerous other cases of prochirality are known in biological reactions. All are explicable in terms of the simple picture described earlier.

Designation of prochirality

Recognition of the phenomenon of enantiomerism leads to the necessity for a systematic designation of the absolute configurations of the enantiomers of chiral molecules (the *R*/*S* system). Similarly, the ability of enzymes and other chiral reagents to differentiate between alternative sites of reaction in prochiral molecules leads to a need for systematic distinction between those features of prochiral molecules which in an achiral environment would be identical. Prochirality arises in two distinct ways, each requiring its own nomenclature.

Prochirality at sp^3 *hybridised centres* involves carbon atoms with a substitution pattern CA_2XY, in which conversion of one of the groups A into Z will generate a new chiral centre (e.g. glycerol → *R*-glycerol-1-phosphate, p. 314). We can distinguish between two identical groups A by arbitrarily assigning one of these a higher priority than the other and then using the Cahn-Ingold-Prelog rules (p. 183) to designate the 'configuration' of the prochiral centre. That group, which when given higher priority, results in notional *R*-chirality is designated '*pro-R*'. The other group, '*pro-S*', would, if given higher priority, result in notional *S*-chirality. The diagrams below show the designations applied to the methyl groups of propan-2-ol and the prochiral hydrogen atoms of ethanol.

CH_3 (*pro-R*), CH_3 (*pro-S*), C, HO, H

CH_3, OH, C, H *pro-R*, H *pro-S*

It should be noted that if the *pro-R* group of such a prochiral molecule undergoes a chemical change generating a new chiral centre, it does not automatically follow that this will have *R*-chirality.

Prochirality at sp^2 *hybridised centres* involves unsaturated molecules which can give rise to chiral centres in reactions which convert the sp^2 hybridised atoms to sp^3 hybridisation. The enzyme catalysed reduction of pyruvic acid

O
||
C
/ \
CH_3 CO_2H ⟶ HO H C CH_3 CO_2H *R* or H OH C CH_3 CO_2H *S*

is such a process. Reduction of pyruvic acid with $FADH_2$ (p. 300) and D-lactate dehydrogenase from *E. coli* gives *R*-lactic acid, whilst reduction with muscle lactate dehydrogenase and NADH gives *S*-lactic acid. In these and similar processes the new chiral centre arises by bonding of an incoming species ($H:^-$ in the example above) to one of the two faces of the sp^2 hybridised centre, and it is these alternative faces which must be distinguished. Using the Cahn-Ingold-Prelog rules, a priority sequence is assigned to the atoms or groups bonded to the sp^2 hybridised centres, exactly as for chiral centres. One face will have a clockwise sequence of decreasing priority and is designated the '*re*' face, the other with an anticlockwise sequence is known as the '*si*' face. Thus looking at the reduction of pyruvic acid we can see that donation of hydride onto the *re*-face of the carbonyl group gives *S*-lactic acid.

O^1 C CH_3 (2) H (3)
anticlockwise sequence
si-face

O^1 C H (3) CH_3 (2)
clockwise sequence
re-face

Where C=C groups are involved in enzymic reactions the prochirality at both ends of the double bond must be considered. Thus fumaric acid has one face with *re, re*-prochirality and the other with *si, si*. Maleic acid has *re, si*-prochirality on both faces. Hydration of fumaric acid in the tricarboxylic acid cycle (p. 234) gives *S*-malic acid, and enzyme-catalysed addition of D_2O gives a product with the *R*-configuration at the CHD group. From the Newman projection shown, it can be seen that the stereochemistry of this hydration of fumaric acid involves *anti*-addition of D_2O (and therefore also

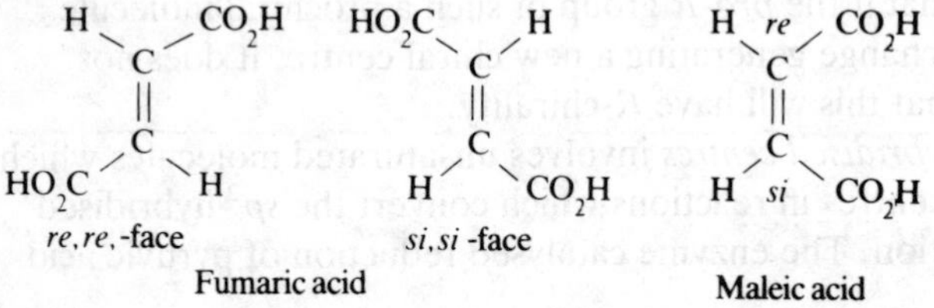

re,re,-face — si,si -face

Fumaric acid

Maleic acid

of H_2O) with the hydroxyl group becoming attached to the *si*-face and the hydrogen being linked to the *re*-face of the prochiral centres involved.

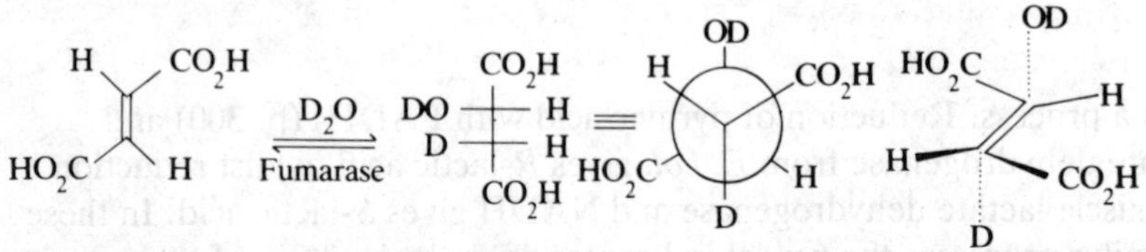

D_2O Fumarase

Although this *anti*-stereochemistry for the total reaction coincides with that observed for most non-enzymic electrophilic additions to alkenes (p. 207), no secure conclusions about the detailed mechanism of the reaction can be made on this basis alone since the possibility of a stepwise process involving an intermediate species covalently linked to the enzyme cannot be excluded.

Problems

1. Which of the following molecules have prochiral centres? In each case mark the prochiral centre(s) with an asterisk.

   CH_3NH_2 $CH_3CH_2CH_2OH$ $CH_3CO_2CH_2CH_3$

   $(CH_3)_2CHCH_2CH_2CH_2CN$ CH_2O

   $CH_3CH_2CCH_3$ $CH_3CH{=}C(CH_3)_2$

   $CH_3CH{=}CHCH_3$ (*E* and *Z* stereoisomers)

2. Designate the prochirality of the H_A and H_B atoms in NADH (p. 317) and the A and B faces of NAD^+ at position 4 of the pyridine ring.
3. In the tricarboxylic cycle (p. 234) *cis*-aconitic acid is hydrated to form isocitric acid. Is the addition of water a *syn* or *anti*-process overall?

22

Some physiologically active compounds

Living organisms produce an immense range of organic compounds which have provided stimuli to organic chemists for over a century. Some of these are small molecules like sugars or hydroxy-acids whilst others are large polymers such as proteins, polysaccharides and nucleic acids, which are characteristic components of all living systems. Between these extremes lie many molecules of medium size and complexity, some of which have very potent physiological effects, such as the vitamins. Compounds of this type have often provided the basis for research aimed at the production of drugs in which the desired physiological effects of naturally occurring compounds are shown with greater potency and specificity by synthetic substances of related structure. It is an implicit premise of this type of work that physiological activity is intimately related to molecular structure, and comparisons of structure-activity relationships within large groups of compounds are slowly leading to the understanding of the molecular topography of some receptor sites in living tissues which interact with both the natural compound and its synthetic analogues.

In this chapter four groups of physiologically active compounds of moderately complex structure will be described which have provided a powerful stimulus to pharmaceutical research, even though the mode of action at a molecular level has not always been clear. The natural synthetic pathways by which these substances are formed will also be briefly described. Although in some cases the structures involved are quite complex, nevertheless a good deal of the chemistry can be predicted on the basis of knowledge about simple functional groups. Thus oxidation of primary or secondary alcohol functions or esterification of carboxylic acid groups is likely to proceed normally however complex the molecular environment in which these functional groups occur.

Eicosanoids

The eicosanoids are a group of widely distributed lipids which are schematically derived from the straight chain C_{20} carboxylic acid,

$CH_3(CH_2)_{18}CO_2H$, eicosanoic acid. The true biological precursors of these compounds are the multiply unsaturated C_{20} acids.

Prostaglandins are derivatives of prostanoic acid and were first isolated from semen. They have a wide range of action in the body including the

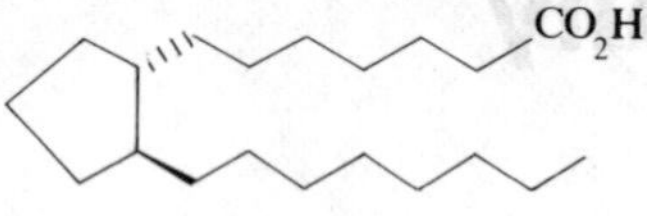

CO_2H

Prostanoic acid

$OCOCH_3$

CO_2H

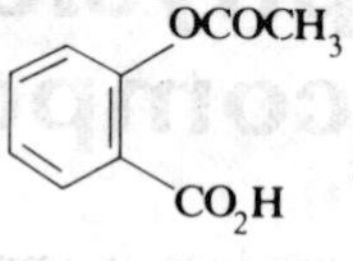

Aspirin

Acetylsalicylic acid

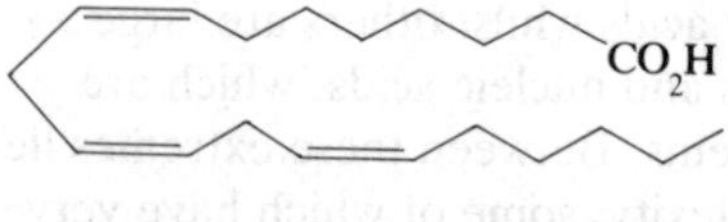

8(*Z*),11(*Z*),14(*Z*) – Eicosatrienoic acid 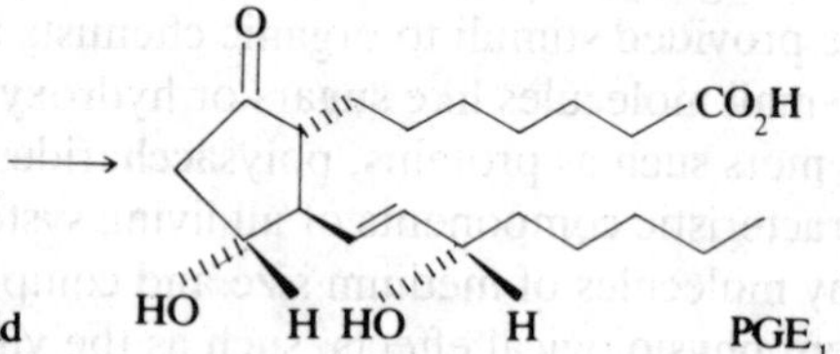 → PGE$_1$

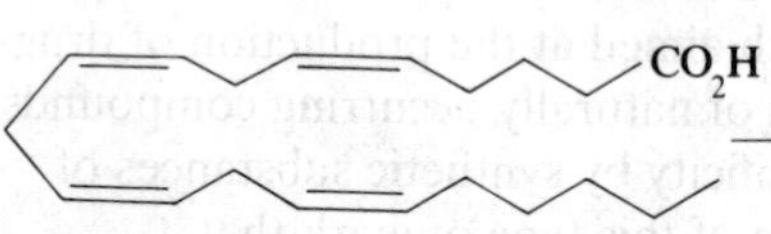

5(*Z*), 8(*Z*), 11(*Z*), 14(*Z*) – Eicosatetraenoic acid

Arachidonic acid → PGE$_2$, 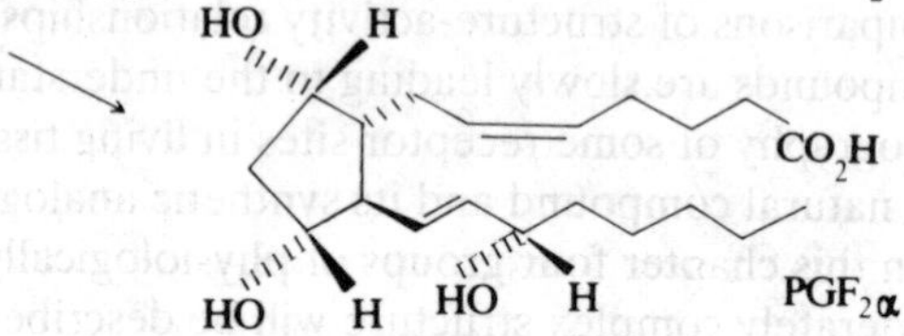 PGF$_{2\alpha}$

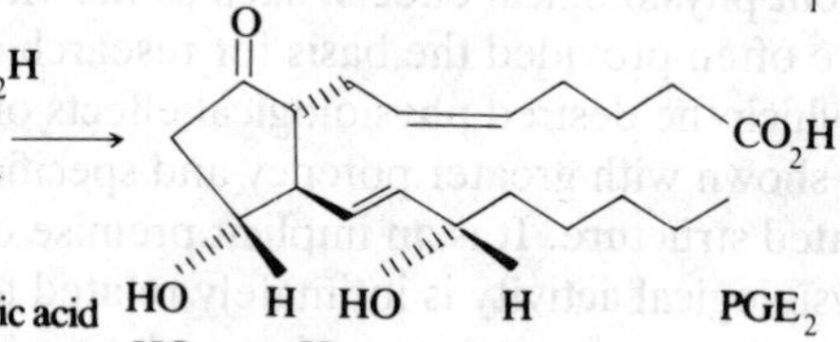

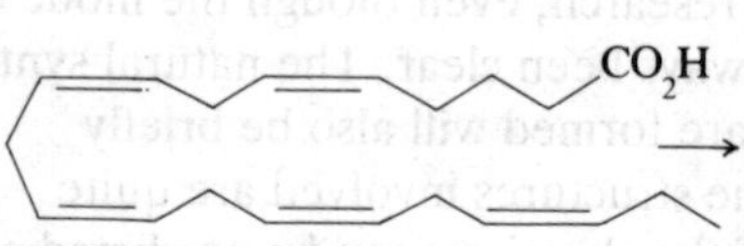

5(*Z*), 8(*Z*), 11(*Z*), 14(*Z*), 17(*Z*) – Eicosapentaenoic acid → PGE$_3$

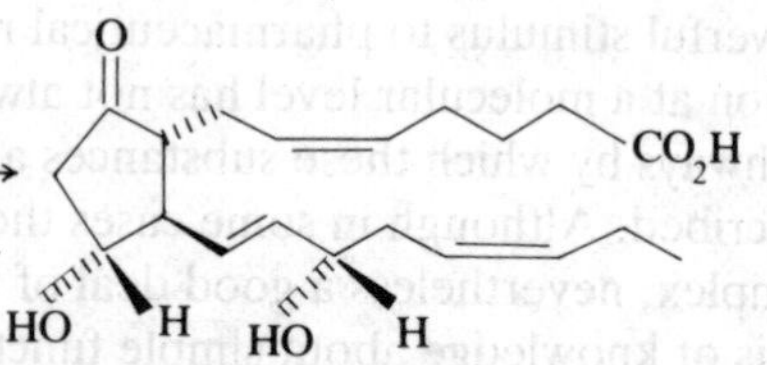

stimulation of smooth muscle, dilation and contraction of arteries with consequent effects on blood pressure, inhibition of gastric secretion, and induction of labour. They are also implicated in inflammatory reactions and it is believed that aspirin owes its analgesic effect to inhibition of prostaglandin synthesis. The individual compounds in these groups of eicosanoids are known by a code of letters which indicates the structure (PG = prostaglandin), functionality, and unsaturation in the various molecules. Four of the prostaglandins are shown in the diagram with the C_{20} acids which are their biological precursors. Note that these precursors are all multiply unsaturated with *Z*-stereochemistry at all the double bonds.

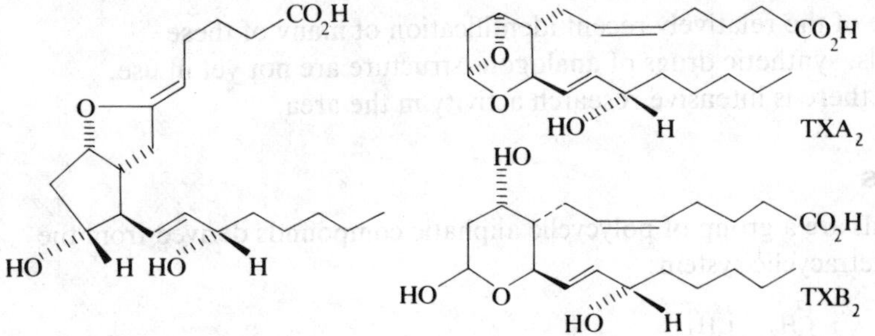

Prostacyclin

TXA$_2$

TXB$_2$

Thromboxanes

Prostacyclin and Thromboxanes are structurally related to the prostaglandins, although in thromboxanes the continuous C_{20} chain of the prostaglandin skeleton has been broken by insertion of an oxygen atom. The two types of compound have complementary physiological action, prostacyclin inhibiting platelet aggregation and causing dilation of blood vessels whilst thromboxanes have precisely the opposite effects.

Leucotrienes are another group of eicosanoids with pronounced physiological activity, also derived from arachidonic acid. A mixture of related compounds (LTC_4, LTD_4, and LTE_4) constitute the 'slow reacting substance of anaphylaxis', a very potent factor for contraction of smooth muscle and enhancement of vascular permeability associated with severe allergic reaction. All of these compounds have a small peptide unit linked to the C_{20} chain through the sulphur atom of cysteine. In the case of LTC_4 this peptide is glutathione. A closely related, simpler eicosanoid (LTB_4) is responsible for attracting white blood cells to the sites of inflammation.

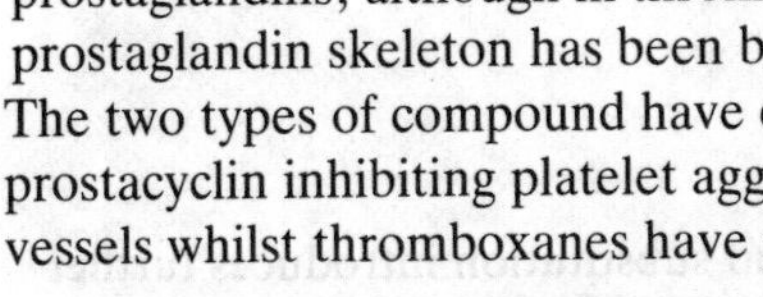

Leucotrienes

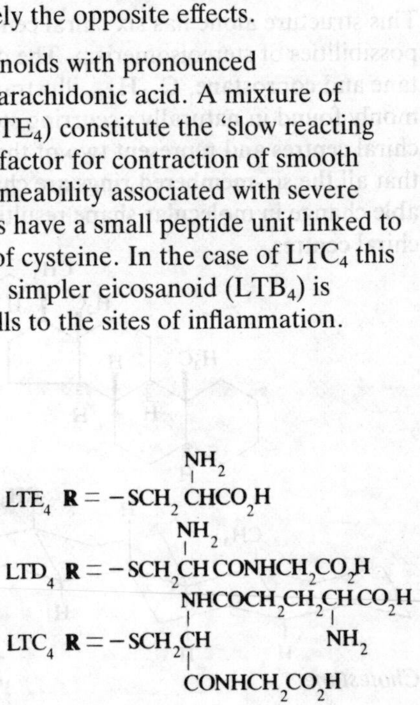

LTE_4 **R** = $-SCH_2CH(NH_2)CO_2H$

LTD_4 **R** = $-SCH_2CH(NH_2)CONHCH_2CO_2H$

LTC_4 **R** = $-SCH_2CH(NHCOCH_2CH_2CH(NH_2)CO_2H)CONHCH_2CO_2H$

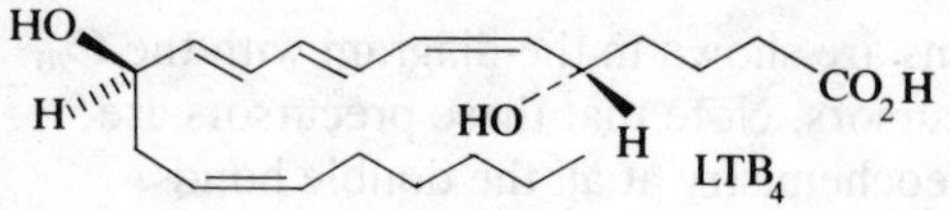

LTB_4

Because of the relatively recent identification of many of these eicosanoids, synthetic drugs of analogous structure are not yet in use. However, there is intensive research activity in the area.

Steroids

The steroids are a group of polycyclic aliphatic compounds derived from the saturated tetracyclic system:

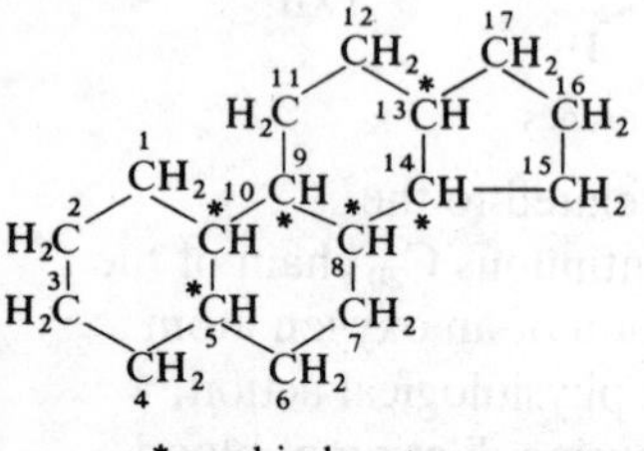

* = chiral centre

This structure alone has six chiral centres, and substitution introduces further possibilities of stereoisomerism. The diastereoisomeric hydrocarbons cholestane and coprostane, $C_{27}H_{48}$, illustrate the absolute configurations commonly found in naturally occurring steroids. These hydrocarbons have eight chiral centres and represent two of the 256 (2^8) possible stereoisomers. Note that all the six-membered rings are chair shaped, and note also the considerable change in molecular shape resulting from epimerisation at one of the chiral centres.

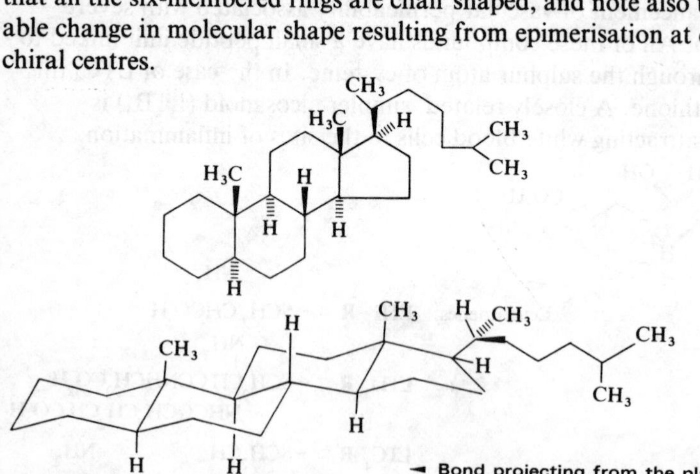

Cholestane

◄ Bond projecting from the plane of the paper

IIIIII Bond projecting backwards into the paper

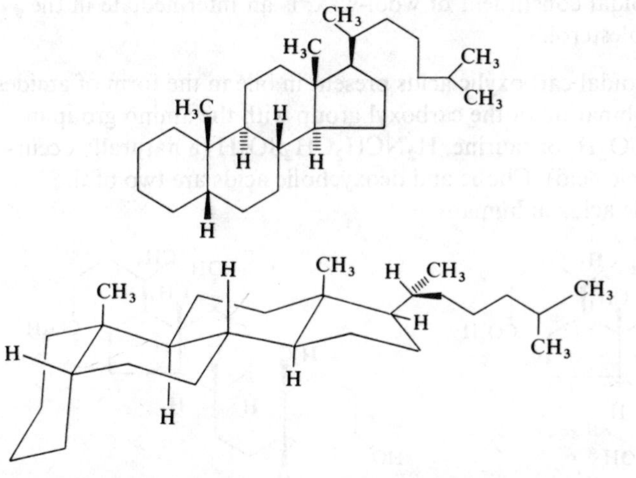

Coprostane (5β-cholestane)

Coprostane (5β-cholestane)

Steroids have two distinct faces to the polycyclic skeleton designated 'α' and 'β'. The face from which the methyl substituents project in cholestane is the β face.

Because the six-membered rings in these structures are fused together, the inversion of one chair conformation into another, as found in simple cyclohexane derivatives, is no longer possible in steroids, in which the four fused rings form a rigid central framework to the molecules.

A brief survey will now be made of the structures of some of the more important naturally occurring steroids.

Cholesterol is one of the most important of the physiologically active steroids, being synthesised in the liver from ethanoic (acetic) acid, via mevalonic acid (p. 177) and squalene (p. 34). It is present in all tissues, particularly

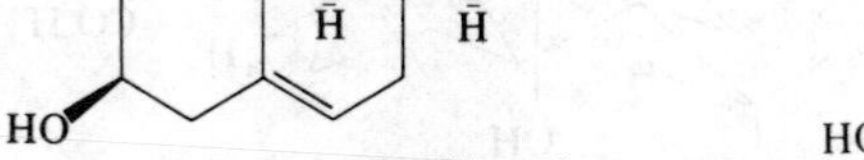

Cholesterol

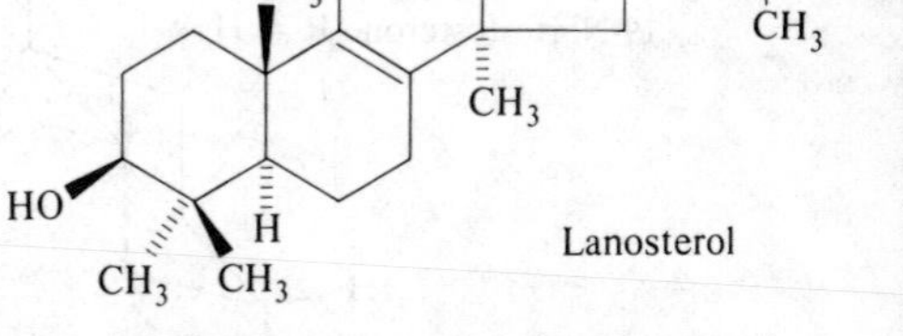

Lanosterol

in the brain, and though its precise role is somewhat obscure it is the precursor from which all other physiologically active steroids are formed in the body.

Lanosterol, a steroidal constituent of wool-wax, is an intermediate in the biosynthesis of cholesterol.

Bile acids are steroidal carboxylic acids present in bile in the form of amides resulting from combination of the carboxyl group with the amino group in glycine, $H_2NCH_2CO_2H$, or taurine, $H_2NCH_2CH_2SO_3H$ (a naturally occurring amino-sulphonic acid). Cholic and deoxycholic acids are two of the more important bile acids in humans.

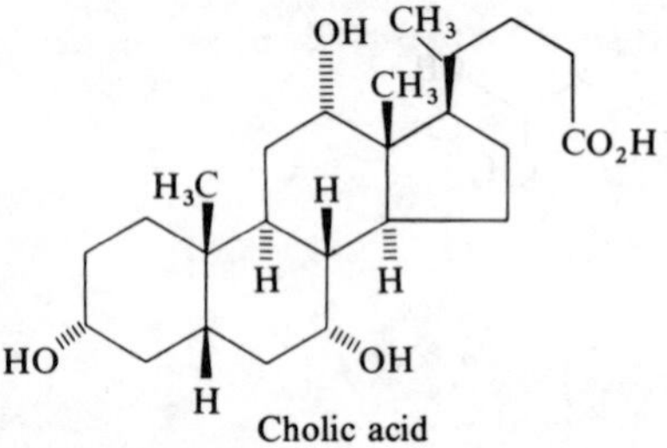

Cholic acid

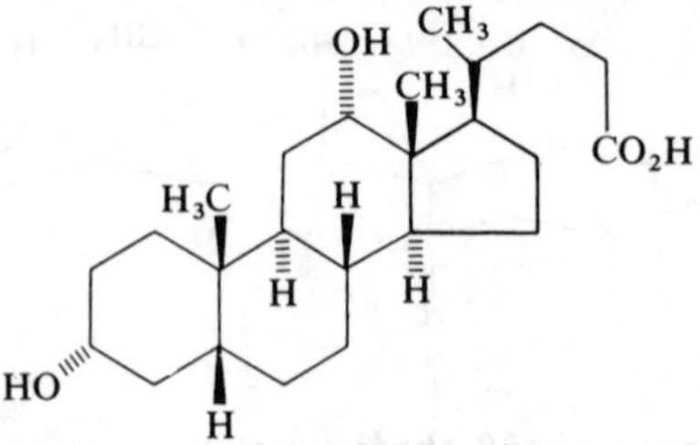

Deoxycholic acid

These compounds assist the digestive process by emulsification of fats in the intestine, i.e. they are intestinal detergents. In the perspective diagram of cholic acid below, it can be seen that all the hydroxyl groups on the steroidal skeleton project from the α face making it hydrophilic, whilst the β face contains exclusively hydrocarbon groups and is hydrophobic.

Sex hormones are produced in the testes and ovaries and are responsible for the development and functioning of the reproductive system. The principal androgen (human male sex hormone) is testosterone, which, in addition to its effect on the reproductive system, has a marked anabolic (tissue-building)

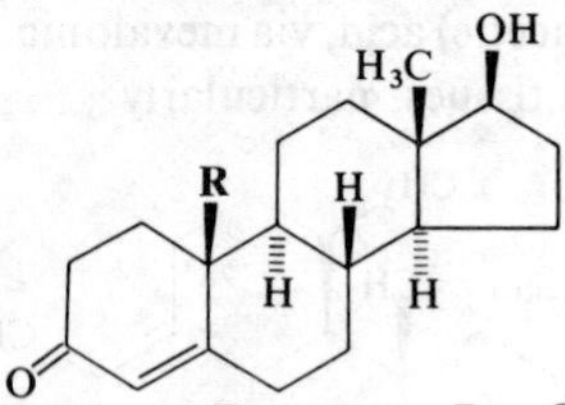

Testosterone R = CH_3
19-Nortestosterone R = H

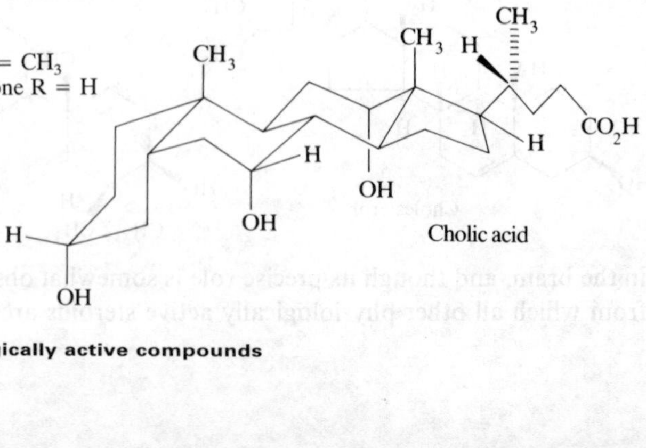

Cholic acid

effect, being responsible for the characteristic muscularity of men. Drugs with structures similar to testosterone find therapeutic use and have been used by some groups of athletes (shot putters, weight lifters, and body builders) to promote muscular growth. 19-Nortestosterone, which has been used for this purpose, also suppresses sperm production in men and is potentially a male contraceptive drug. The higher level of free testosterone in the blood is also responsible for the forceful masculine personality characteristics of ambition and leadership, which often develop in post-menopausal women on account of the change in hormone levels as the ovaries atrophy.

The reproductive system in women is controlled by two types of hormone.

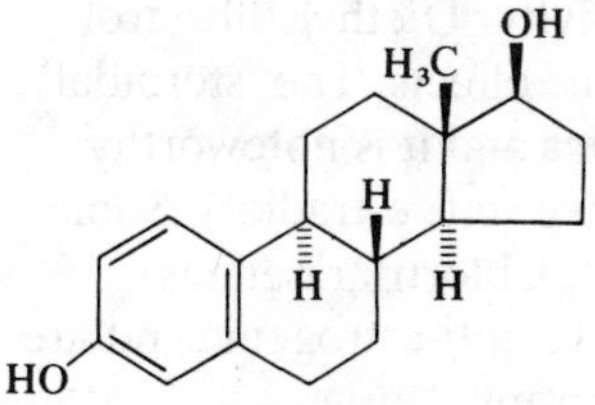

Estradiol

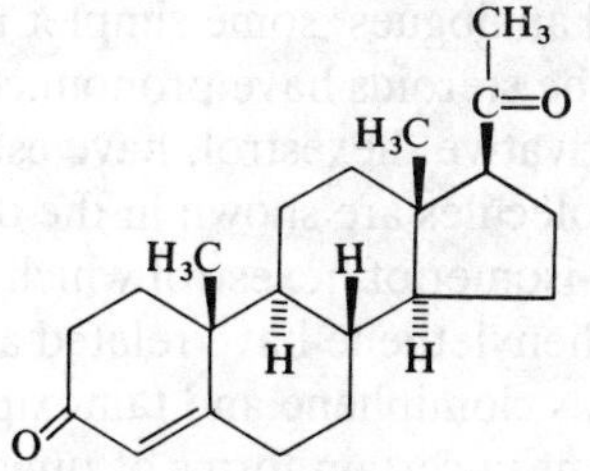

Progesterone

Estrogens (or œstrogens) control the cyclic changes in the uterus, whilst progestogens assist and maintain pregnancy after conception. Estradiol, the principal estrogen, is secreted by the ovaries during development and release of the

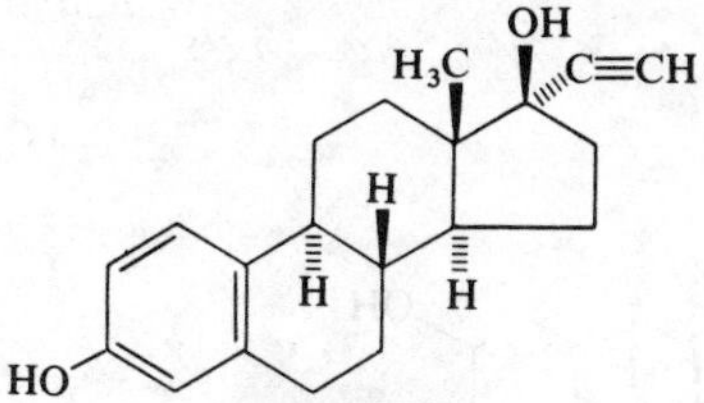

Ethynylestradiol

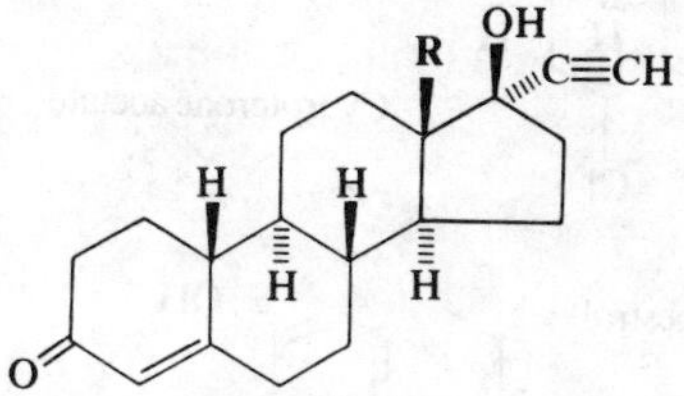

Norethisterone (**R**=CH_3)
Norgestrel (**R**=C_2H_5)

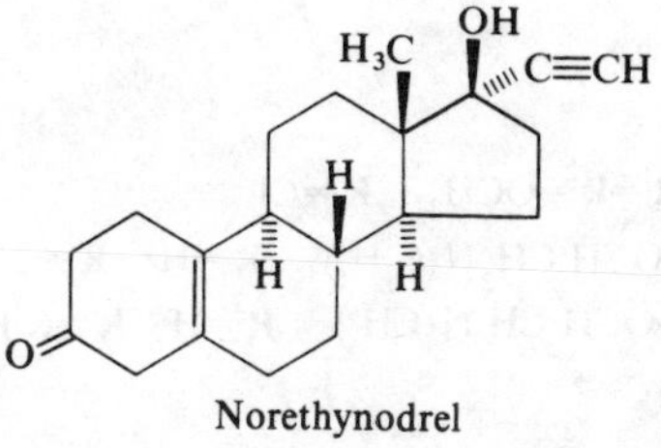

Norethynodrel

Four steroidal oral contraceptives. Note the close similarity between these structures and those of the sex hormones.

ovum, and prepares the lining of the uterus for implantation of a fertilised ovum. If conception does not occur, the estradiol secretion falls off and the uterine lining is sloughed off (menstruation). Development and release of another ovum is then accompanied by further secretion of estradiol. If fertilisation occurs, then progesterone, the principal progestogen, is produced in the corpus luteum and later in the placenta and maintains the pregnancy. Many of the compounds tested and used as oral contraceptives have structures allied to estradiol and progesterone.

Cyproterone acetate is a steroidal drug closely resembling progesterone, which is used for its anti-androgenic properties.

Beside steroidal analogues, some simpler molecules which can adopt shapes similar to the steroids have pronounced activity. Diethylstilbestrol and its dihydroderivative, hexestrol, have estrogenic effects. The 'steroidal' shapes of these molecules are shown in the diagrams and it is noteworthy that it is the *meso*-isomer of hexestrol which is active (c.f. estradiol). Some derivatives of triphenylethene have related activity. Chlortrianisene is estrogenic, whereas clomiphene and tamoxiphen are anti-estrogenic and are used on this account in certain forms of tumour therapy.

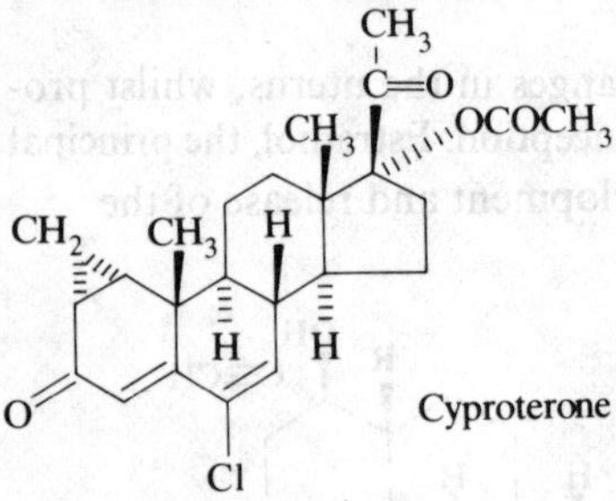

Cyproterone acetate

Diethylstilbestrol

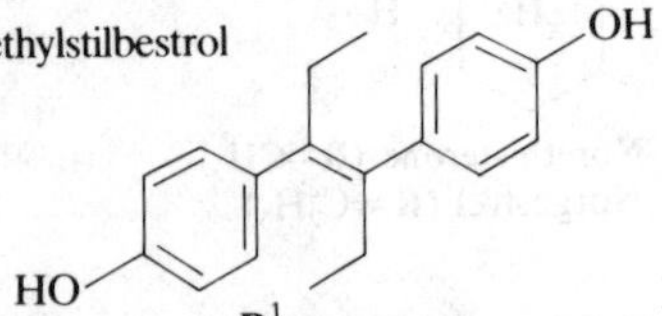

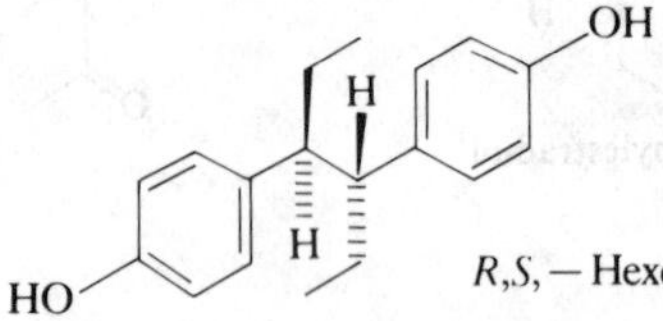

R,S,— Hexestrol

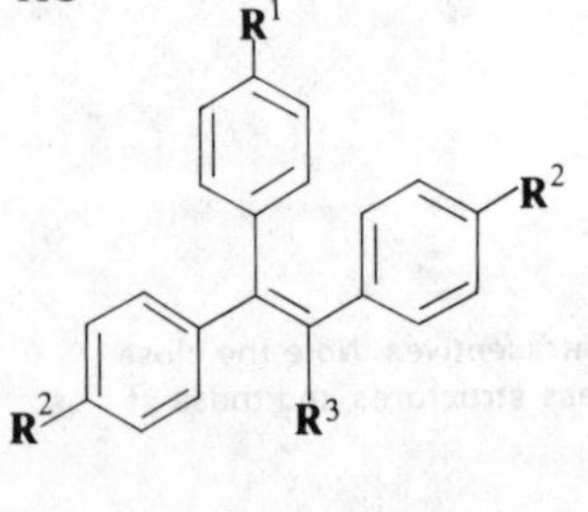

Chlortrianisene $\mathbf{R}^1=\mathbf{R}^2=OCH_3$ $\mathbf{R}^3=Cl$

Clomiphene $\mathbf{R}^1=OCH_2CH_2N(C_2H_5)_2$ $\mathbf{R}^2=H$ $\mathbf{R}^3=Cl$

Tamoxiphen $\mathbf{R}^1=OCH_2CH_2N(CH_3)_2$ $\mathbf{R}^2=H$ $\mathbf{R}^3=C_2H_5$

Hormones of the adrenal cortex. A range of steroidal hormones is produced by the adrenal cortex, which have very profound physiological effects. A distinction is often made between the mineralocorticoids such as deoxycorticosterone and aldosterone which control the concentrations of sodium

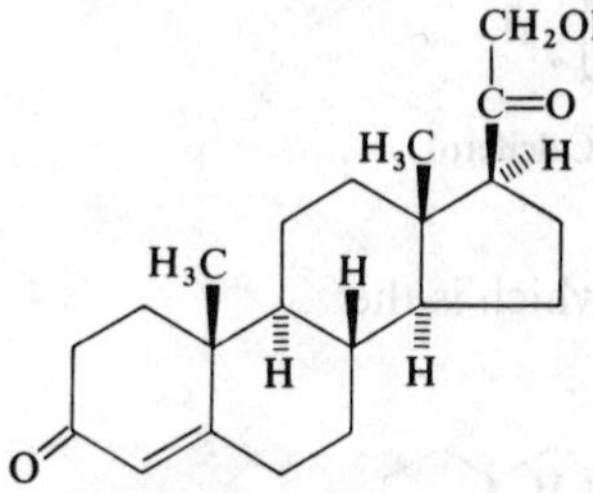

Deoxycorticosterone

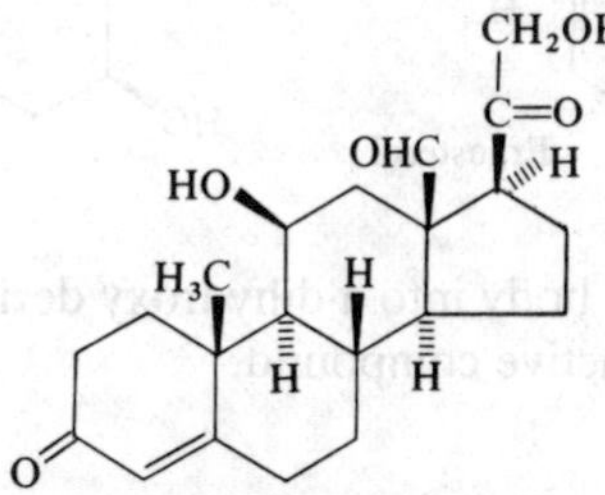

Aldosterone

and potassium ions in the tissues, and the glucocorticoids such as cortisol, which stimulate the storage of glycogen in the liver and have anti-inflammatory properties.

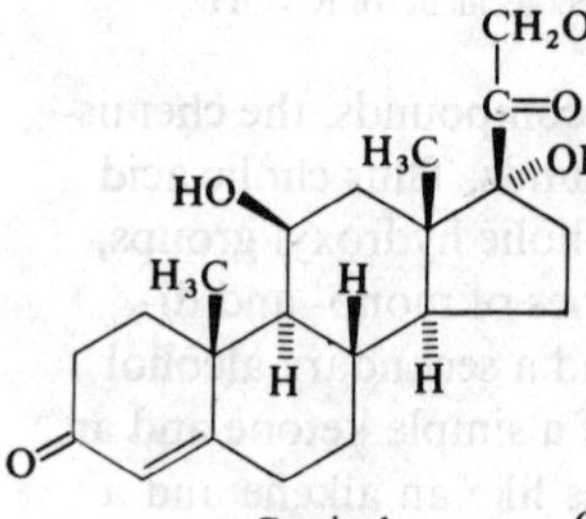

Cortisol

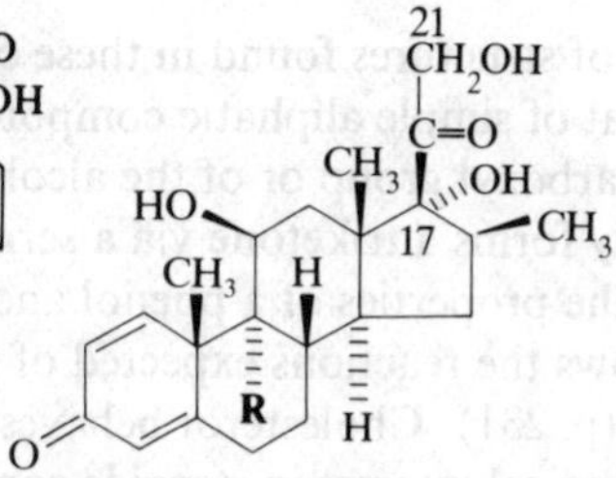

Beclomethasone R=Cl

Betamethasone R=F

Beclomethasone and betamethasone are synthetic glucocorticoid steroids used in the form of acyl derivatives (e.g. propionates of the hydroxyl groups at positions 17 and 21) as topical anti-inflammatory drugs.

Calciferol (*vitamin* D_2) is an important dietary factor necessary for the correct development of bones. Rickets is the disease caused by a deficiency of this vitamin. Vitamin D_2 is not itself a steroid, but arises by the action of sunlight or ultraviolet light on ergosterol, the principal steroid of yeast. Dehydrocholesterol, on irradiation with ultraviolet light, similarly forms cholecalciferol (vitamin D_3) which is also antirachitic. It is known that this is

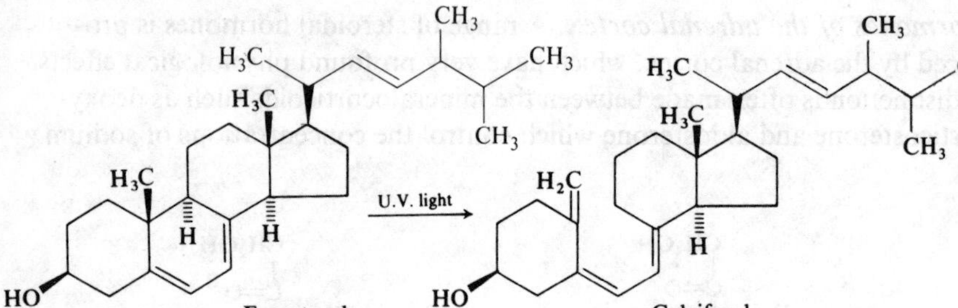

Ergosterol $\xrightarrow{\text{U.V. light}}$ Calciferol

converted in the body into a dihydroxy derivative which is the physiologically active compound.

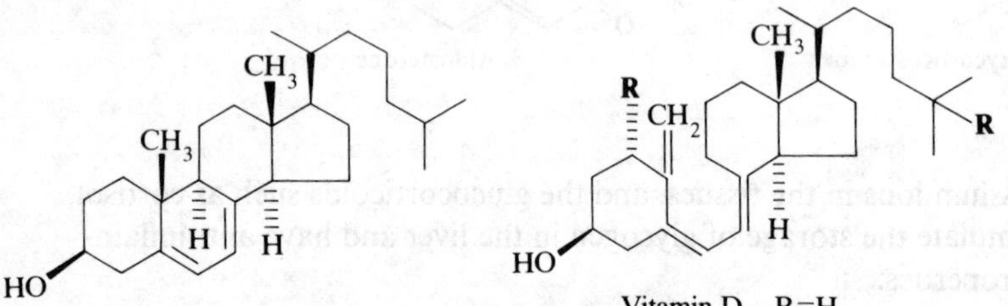

Dehydrocholesterol

Vitamin D_3 R=H

Dihydroxycholecalciferol R=OH

Despite the complexity of structures found in these compounds, the chemistry is predominantly that of simple aliphatic compounds. Thus cholic acid can form esters of the carboxyl group or of the alcoholic hydroxyl groups, and oxidation ultimately forms a triketone via a series of mono- and diketones. Estradiol has the properties of a phenol and a secondary alcohol whilst progesterone shows the reactions expected of a simple ketone and an $\alpha\beta$-unsaturated ketone (p. 231). Cholesterol behaves like an alkene and a secondary alcohol. Biological interest in steroids concerns the relationship between structure and physiological activity and the methods of synthesis in the body. Chemically they are of interest both for their own sake and also because the rigidity of the fused-ring skeleton is of value in the investigation of stereochemical requirements of reactions.

Opioids

The narcotic and analgesic effects of opium, the dried juice of the opium poppy, have been known for over 6000 years, and from this source the principal alkaloid* morphine was first isolated in 1803. Morphine constitutes up to 10% of the dry weight of opium, and lesser quantities of other alkaloids codeine, thebaine, and papaverine are also present. After the invention of the hypodermic needle, morphine was very extensively used for the relief of pathological and surgical pain, since it gives analgesia without loss of consciousness. However, in addition to its analgesic effects, arising from the blocking of nerve impulses in the spinal cord, morphine interacts with a variety of sites in the brain leading to a complex pattern of activity including drowsiness, mental clouding, and changes of mood which may involve euphoria (a sense of blissful well-being) or sometimes dysphoria (anxiety and fearfulness). Respiration is depressed, a common cause of death from an overdose, as also is the cough reflex, but nausea and vomiting are not infrequent. Tolerance of the drug develops easily leading to the need for increasing doses, and addiction comes from physical and psychological dependence caused by acute withdrawal symptoms (cramps, vomiting, muscular spasms, and dysphoria). Heroin, the diacetyl derivative of morphine, is a more potent analgesic and more addictive than morphine since it penetrates to the brain more readily where it is hydrolysed to morphine. Codeine, a monomethyl derivative of morphine, is about one tenth as potent as an analgesic but much less addictive and is used in a synergic combination with aspirin.

Morphine $R^1=R^2=H$ $R^3=CH_3$

Codeine $R^1=R^3=CH_3$ $R^2=H$

Heroin $R^1=R^2=CH_3CO$ $R^3=CH_3$

Nalorphine $R^1=R^2=H$ $R^3=CH_2CH=CH$

Pholcodeine $R^1=$ O NCH$_2$CH $R^2=H$ $R^3=CH_3$

* Alkaloid—a name applied to a broad class of complex, basic (i.e. nitrogen containing), organic compounds of plant origin.

Biosynthesis of morphine. Studies with isotopically labelled compounds have shown that morphine and the other alkaloids of opium are produced from tyrosine by the route outlined below:

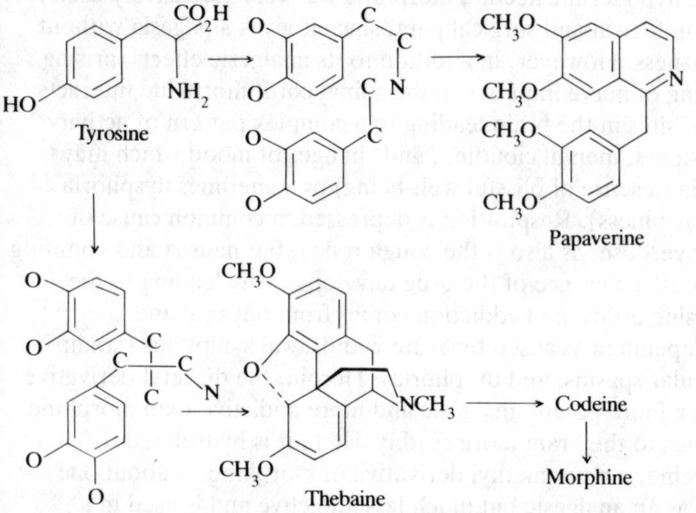

CO_2H NH_2 HO
Tyrosine
CH_3O CH_3O N CH_3O CH_3O
Papaverine
CH_3O NCH_3 CH_3O
Thebaine
Codeine
Morphine

Oxidation of the aromatic ring of tyrosine gives a dihydroxy compound, which undergoes further modification. Combination of two of these tyrosine-derived units leads to the formation of papaverine—an alkaloid with muscle-relaxant but no analgesic properties—and morphine. Thebaine and codeine are intermediates on the biosynthetic path to morphine.

Opioid drugs. Many attempts have been made to synthesise compounds with structural resemblance to morphine in the search for analgesics free of addictive and other unwanted effects. Among the simpler structures methadone, pethidine (meperidine), and dextropropoxyphene have analgesic properties but the first two are addictive and the last has unpleasant side-effects. These compounds can adopt conformations which give an overall morphine-like shape to the molecule, shown below for pethidine. Other synthetic approaches are based on modification or simplification of the morphine structure. Replacement of the *N*-methyl group of morphine by an allyl group gives nalorphine which is a morphine antagonist reversing many of the effects of morphine and giving dysphoria and sometimes hallucinations. Given to an addict, antagonists will produce withdrawal symptoms. Naloxone is a non-analgesic morphine antagonist used to treat a morphine overdose. Naltrexone, with a closely similar structure, has analgesic properties but is otherwise an antagonist.

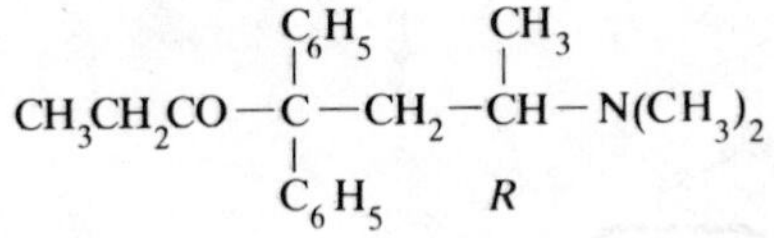

$CH_3CH_2CO-C(C_6H_5)_2-CH_2-CH(CH_3)-N(CH_3)_2$ Methadone

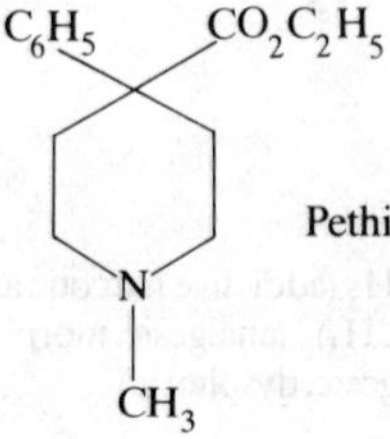

R

Pethidine

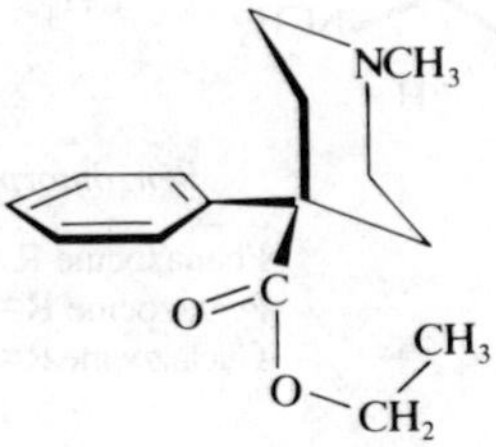

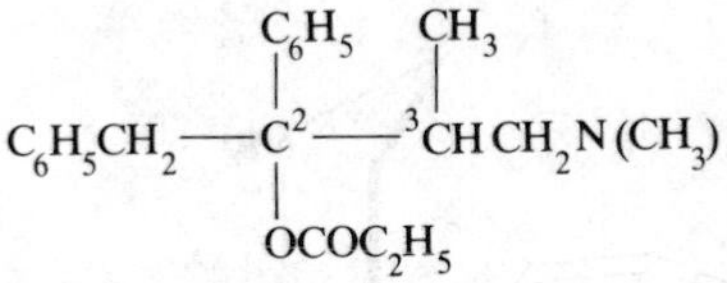

Dextropropoxyphene (2*S*, 3*R*)
Levopropoxyphene (2*R*, 3*S*)

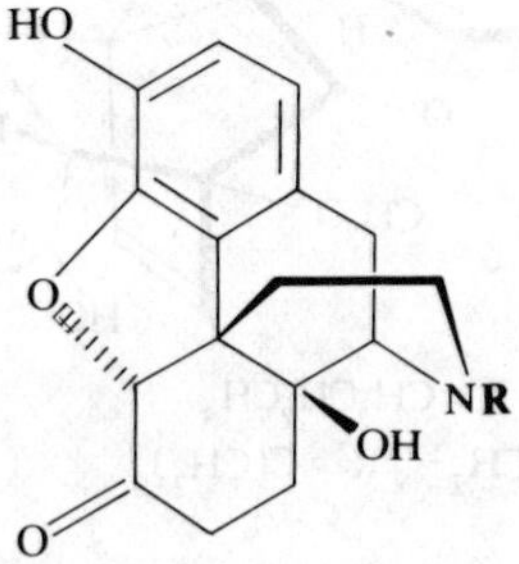

$C_6H_5CH_2-C^2(C_6H_5)(OCOC_2H_5)-{}^3CH(CH_3)CH_2N(CH_3)$

Naloxone $R=CH_2CH=CH_2$
Naltrexone $R=CH_2$-cyclopropyl

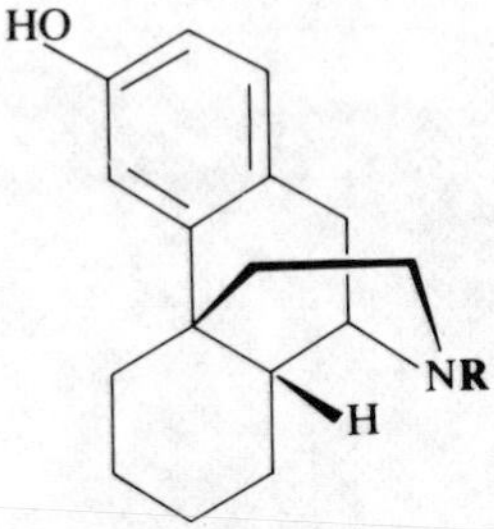

Morphinans

Levorphanol $R=CH_3$ (narcotic)

Levallorphan $R=CH_2CH=CH_2$ (morphine antagonist)

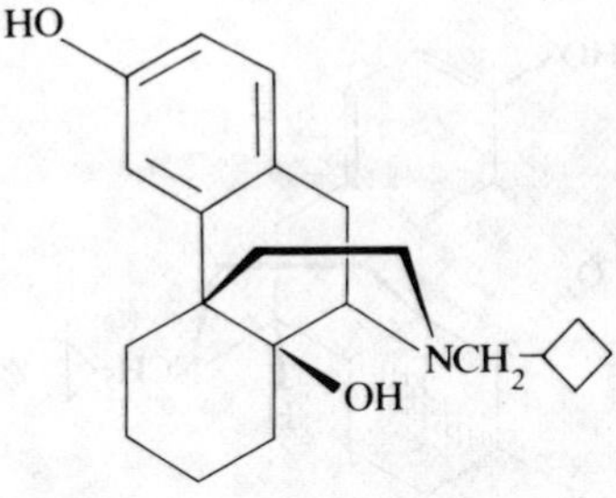

Butorphanol (potent non-addictive analgesic)

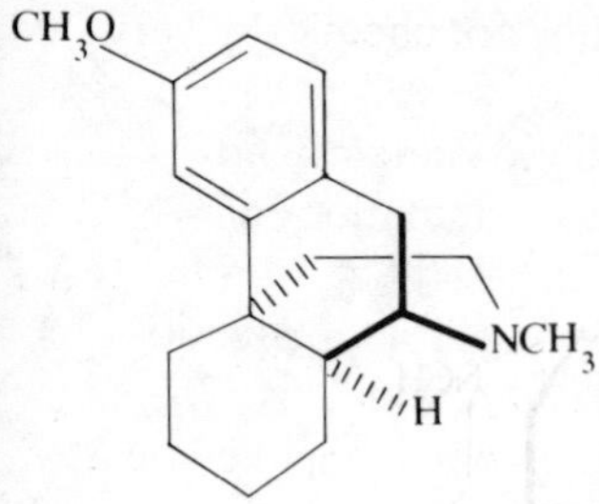

Dextromethorphan

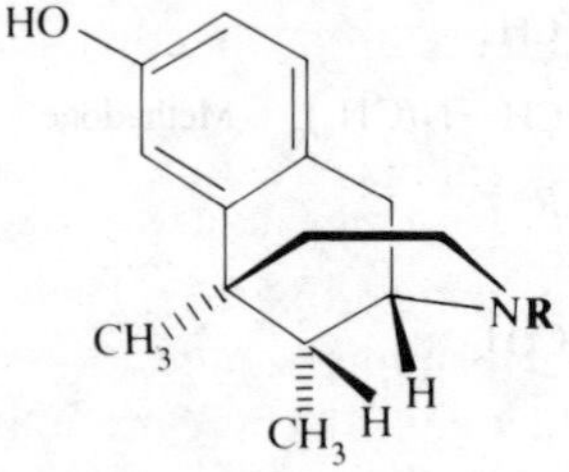

Benzomorphans

Phenazocine $R=CH_2CH_2C_6H_5$ (addictive narcotic analgesic)
Pentazocine $R=CH_2CH=C(CH_3)_2$ (analgesic, morphine antagonist)
Cyclazocine $R=CH_2$–◁ (analgesic, dysphoric)

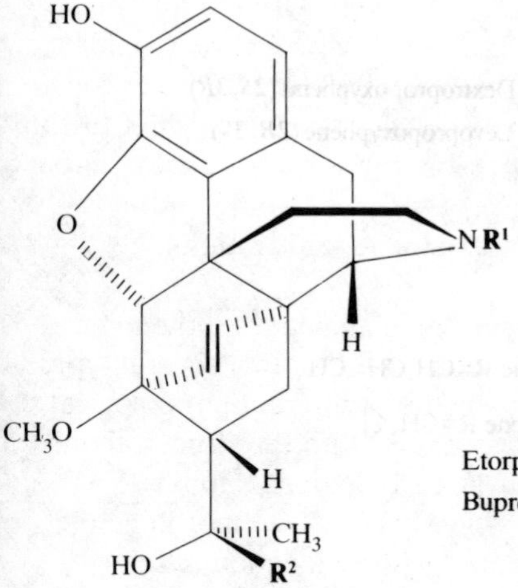

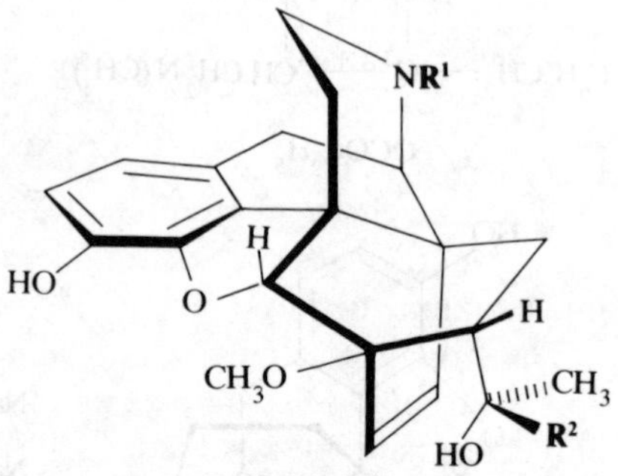

Etorphine $R^1=CH_3$ $R^2=CH_2CH_2CH_3$
Buprenorphine $R^1=CH_2$–◁ $R^2=C(CH_3)_3$

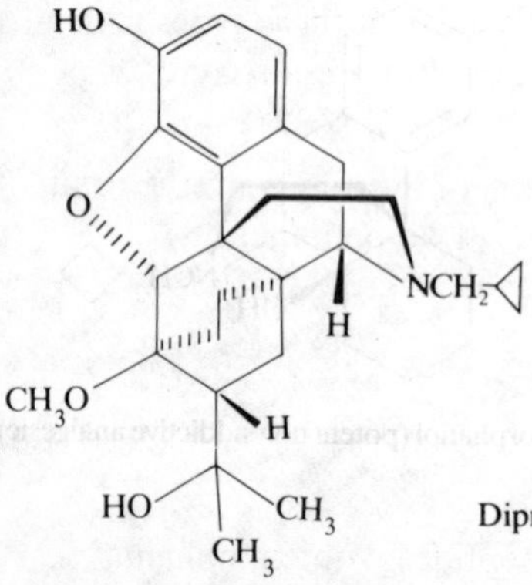

Diprenorphine

Interestingly, the allyl and cyclopropylmethyl groups are chemically very similar.

Simplification of the morphine structure leads to morphinans and benzomorphans, derivatives of which have varying combinations of morphine-like and antagonistic properties. Dextromethorphan is of interest in this group being a cough-suppressant (antitussive) without analgesic properties and having the opposite configuration to morphine. Levopropoxyphene, the enantiomer of dextropropoxyphene, is likewise a non-analgesic antitussive.

Some of the most powerful opioids currently in use are derived from thebaine, a minor alkaloid of opium devoid of morphine-like properties. Etorphine has up to 80,000 times the potency of morphine and is used only in veterinary practice, e.g. for anaesthetising big game with tranquilising darts. An equally potent antagonist, diprenorphine, with a closely similar structure is used to counteract the effects of etorphine. Buprenorphine is a related drug used in humans with a complex dose-effect relationship. It is up to 500 times as potent an analgesic as morphine, but is non-addictive and at high dose levels acts as an antagonist. It can be administered orally and finds use in emergencies such as traffic accidents where administration by injection may be impossible.

The structure-activity relationships in morphine analogues show some obvious trends. The complex and variable effects suggest that several types of receptor site are involved, each responsible for a different physiological response. Analgesic activity seems to be associated with the structural unit C_6H_5—C—C—C—NR_2 in which the benzylic carbon atom is fully substituted. In compounds with structures closely related to morphine the free phenolic hydroxyl group is an important factor, alkylation as in codeine or pholcodeine reducing analgesic activity markedly. Allyl or related groups on the nitrogen atom are strongly associated with morphine antagonism. It is probably the combination of factors of these types which, in part, leads to the dose-response characteristics of buprenorphine.

Enkephalins. The complex and profound effects of morphine on the brain led to the search for naturally occurring opioids. Two pentapeptides, 'enkephalins', of closely similar structure were discovered which interact strongly with the morphine receptor sites:

H-Tyr-Gly-Gly-Phe- { -Met-OH Met-enkephalin / or / -Leu-OH Leu-enkephalin }

and a number of other more complex opioid peptides with enkephalin sequences in larger polypeptide chains also have morphine-like effects. These polypeptides, whose production is believed to be stimulated by mild irritation such as scratching or acupuncture, are very short lived in the body, being rapidly hydrolysed. Attempts have been made to synthesise analogous molecules which will be resistant to hydrolysis and therefore have more prolonged effect. One such derivative is:

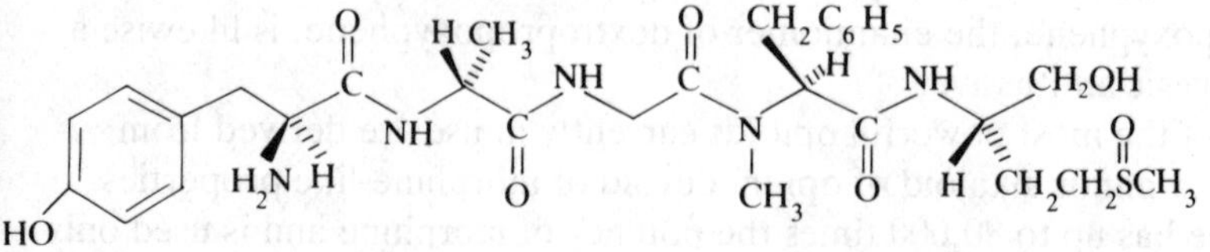

L – Tyrosine – D – Alanine — Glycine – *N* – Methyl – L – phenylalanine – Methioninol – *S* – oxide

in which met-enkephalin is modified by replacement of one of the glycine units by D-alanine, methylation of the phenylalanine amine group, and replacement of methionine by the sulphoxide of the corresponding alcohol. This compound is over 20,000 times as active as morphine in tests. In the enkephalins and other opioids peptides it is tempting to suppose that the phenolic group of the tyrosine moiety interacts with the same receptor site in the brain as the phenolic ring in morphine and its analogues.

β-Lactam Antibiotics

The accidental discovery of the antibacterial secretions of a mould (*Penicillium notatum*) is part of the folk lore of science. The isolation of the active material and identification of its structure initiated an immense programme of research into microbial antibiotics. Nowadays deep fermentation using *P. chrysogenum* annually produces hundreds of tons of penicillin G and related compounds. The originally isolated material, which is no longer of significance, differed in the nature of the side chain R group.

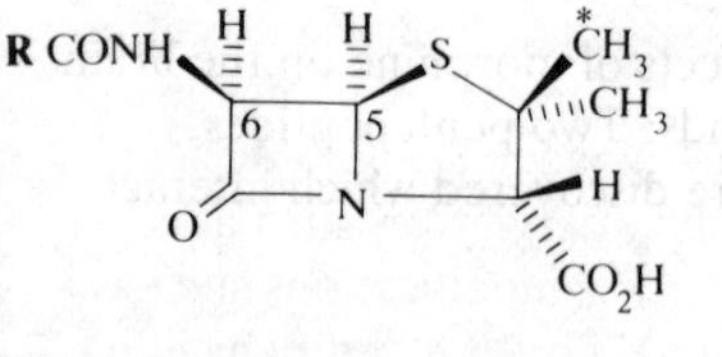

pro-S CH_3 group

Penicillin G	$R=C_6H_5CH_2$
Penicillin V	$R=C_6H_5OCH_2$
Penicillin H	$R=H_2NCH(CH_2)_3$ (CO_2H)
Ampicillin	$R=C_6H_5CH(NH_2)$
Methicillin	R= (2,6-dimethoxyphenyl, OCH_3, OCH_3)

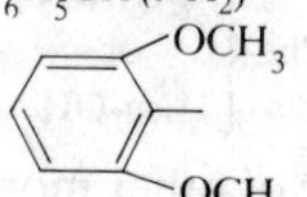

Commercial production of penicillin G is achieved by addition of phenylacetic acid ($C_6H_5CH_2CO_2H$) to the fermentation broth.

Tests on an immense number of penicillin derivatives have shown that antibiotic activity is dependent upon the four membered β-lactam ring in conjunction with the adjacent carboxylic acid group. However, the β-lactam ring is rather easily hydrolysed by chemical means forming penicilloic acids

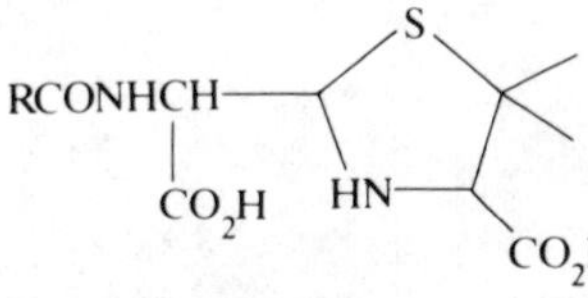

Penicilloic acid

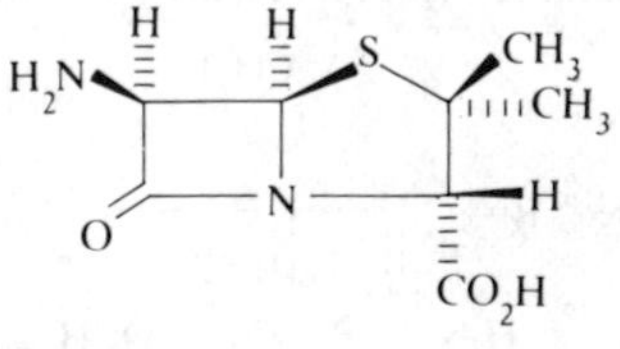

6 — Aminopenicillanic acid

devoid of antibacterial activity. Enzymic hydrolysis of penicillins gives two types of product depending upon the enzymes employed. β-Lactamases give penicilloic acids, but other enzymes will cleave the side chain forming 6-aminopenicillanic acid. This compound, obtainable directly from fermentation, is the precursor for many 'semi-synthetic' β-lactam antibiotics as the free amino group can now be converted into a wide range of side chains. Very few penicillins can be made like penicillin G by addition of carboxylic acids to the fermentation broth, penicillin V is another such example. Most of the penicillin derivatives in current use are semi-synthetic and have a wider spectrum of activity and resistance to hydrolysis of the β-lactam ring by both chemical and enzymic means. Of the simple derivatives ampicillin, although rather easily hydrolysed by β-lactamases, is stable under the acidic stomach conditions and can be administered orally rather than by injection. Methicillin is relatively resistant to enzymic hydrolysis.

Cephalosporins are a group of antibiotics, also isolated from mould cultures, having the structure shown which is closely related to that of penicillins. As with penicillins, the deacylated derivative, 7-aminocephalosporanic acid is the starting point for the synthesis of many useful derivatives such as cefotaxime, and additionally compounds with other modifications such as a methoxyl group at position 7 of cephalosporin (or position 6 of penicillin) have been investigated. Modification of the side chain at position 3 in cephalosporin has also given other series of therapeutically valuable compounds.

Cephalosporin C R=$H_2NCH(CH_2)_3$ with CO_2H on the CH

Cefotaxime R= (2-amino-thiazol-4-yl)–C(=N–OCH_3)–

Biosynthesis of penicillins and cephalosporins. The biological precursor of the naturally occurring penicillins and cephalosporins is a relatively simple tripeptide formed from L-cysteine, D-valine, and a less common amino acid L-2-aminoadipic acid. This undergoes cyclisation to form penicillin N, which can be rearranged to form cephalosporin C. Note that the two methyl

→ Penicillin N → Cephalosporin C

L – 2 – Aminoadipic acid – L – Cysteine – D – Valine

groups in penicillins are part of a prochiral group. It is known from isotopic labelling studies that it is the *pro-S* methyl group which is incorporated into the enlarged ring in cephalosporins. Subsequent to the formation of penicillin N or cephalosporin C further enzyme-catalysed exchange of the acyl side chain may occur, as in the formation of penicillin G.

Method of action of β-lactam antibiotics. β-Lactam antibiotics are effective against only a limited range of microbial pathogens and their activity depends on interference with cell-wall construction. All single celled organisms require a mechanically robust cell wall to resist the substantial osmotic pressure arising from the high concentration of dissolved species inside the cell. Cells which are not normally subject to this osmotic pressure and do not have strengthened walls, such as red blood cells, normally burst when put into water.

Gram positive and negative bacteria have cell walls in which the mechanical strength comes from a network composed of sugars and amino acids (a peptidoglycan). A polysaccharide chain constructed from alternate 1β-4 linked *N*-acetylglucosamine (NAG) and *N*-acetylmuramic acid (NAM) (p. 260) units is cross linked by a branched polypeptide chain linked to the carboxylic acid groups of the NAM residues. The whole structure is like a string bag which reinforce the lipid membrane inside. If the cell is to grow and divide this peptidoglycan must be extended or modified and it is the enzymes controlling the synthesis of new cell wall peptides which are the targets for β-lactam antibiotics. These compounds are absorbed by the enzyme, presumably because of their peptide-like structure, and then acylate the active site by a process which opens the β-lactam ring, hence the ineffectiveness of the penicilloic acids. Faults in the cell wall arising from inhibition of these enzymes eventually cause the cell to burst from the osmotic pressure.

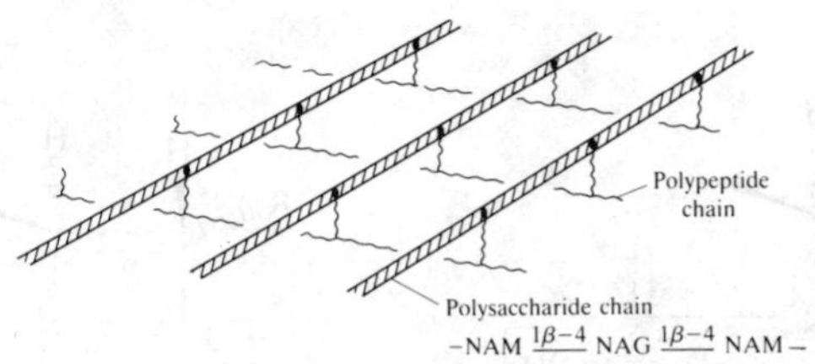

Peptidoglycan of bacterial cell wall

β-Lactamase inhibitors. In order to have their desired effect, β-lactam antibiotics must pass through the cell wall and bind to the appropriate enzymes. Gram negative bacteria are, on the whole, less susceptible than Gram positive bacteria partly because the structure of the cell wall makes penetration more difficult. Additionally, many bacteria produce β-lactamases, enzymes which very efficiently hydrolyse the β-lactam ring forming harmless products. This latter difficulty has been overcome by the discovery of compounds such as the naturally occurring clavulanic acid and the synthetic sulbactam, which have no antibacterial action but are very effective inhibitors of β-lactamases. Instead of giving a β-lactamase sensitive drug, such as ampicillin, in huge doses in the hope that some will escape

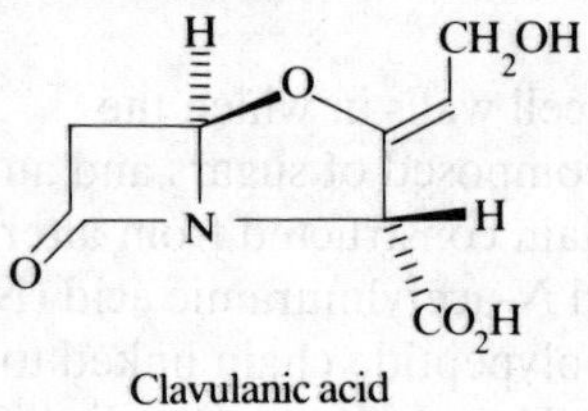

Clavulanic acid

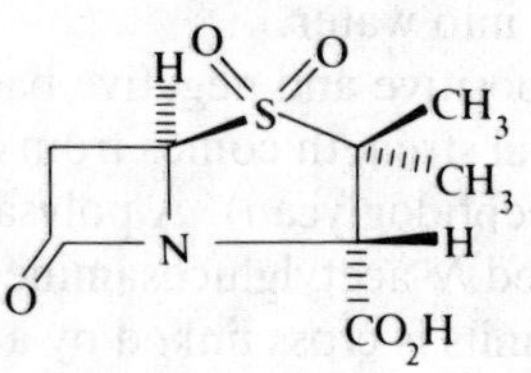

Sulbactam

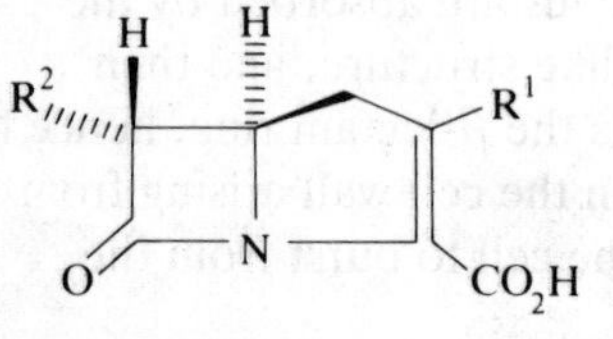

Thienamycins

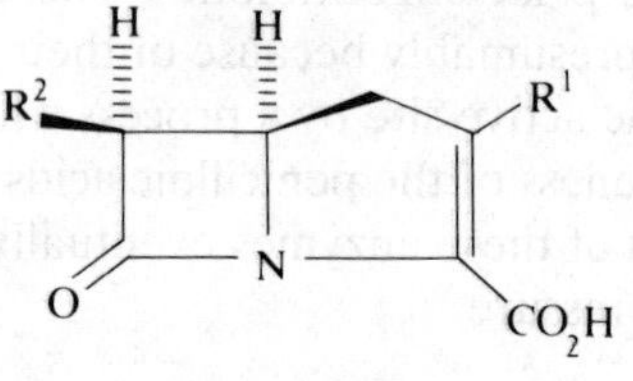

Olivanic acids

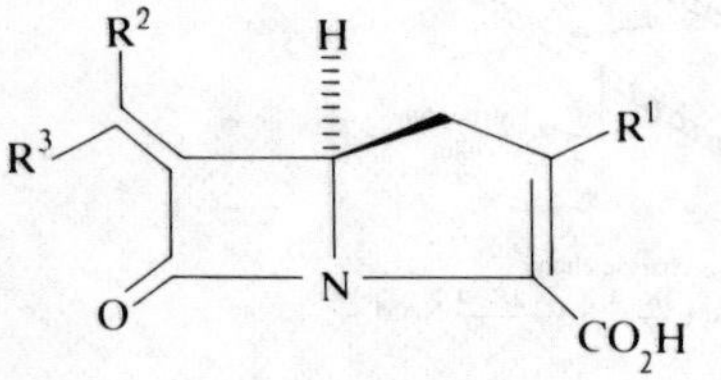

Asparenomycins

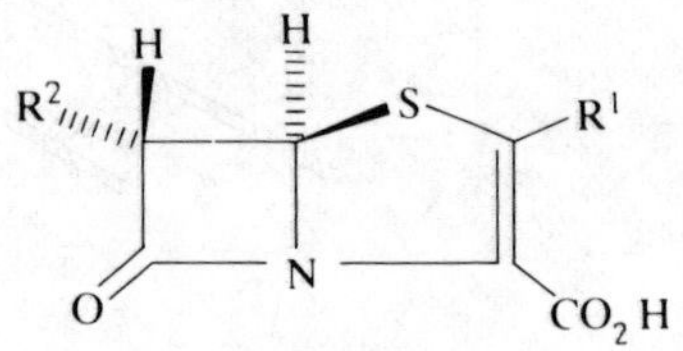

Penems

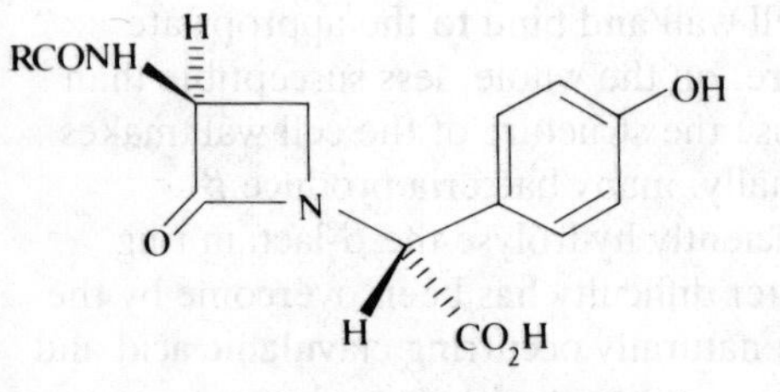

Nocardicins

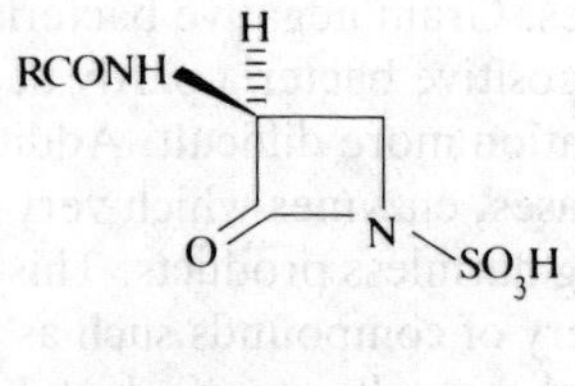

Monobactams

hydrolysis and reach the target enzymes, a normal dose in conjunction with a β-lactamase inhibitor will have the desired effect.

Other β-Lactam antibiotics. Following the success of penicillins and cephalosporins a wide range of other β-lactam derivatives with antibiotic activity has been discovered from both natural and synthetic sources. Thienamycins, olivanic acids and asparenomycins are series of penicillin-like compounds collectively known as carbapenems, originating from naturally occurring substances after which the groups are named. Penems and cephalosporin analogues with O or CH_2 replacing the sulphur atom in the ring are wholly synthetic series of β-lactam antibiotics. The nocardicins and monobactams are based upon naturally occurring monocyclic antibiotic β-lactams, and the latter compounds are the only ones so far known which lack the carboxylic acid group adjacent to the β-lactam ring and still possess antibacterial activity.

Index